9

MARXISM

MARXISM

ESSENTIAL WRITINGS

EDITED BY

DAVID McLELLAN

OXFORD UNIVERSITY PRESS

1988

Oxford University Press, Walton Street, Oxford OX2 6DP
Oxford New York Toronto
Delhi Bombay Calcutta Madras Karachi
Petaling Jaya Singapore Hong Kong Tokyo
Nairobi Dar es Salaam Cape Town
Melbourne Auckland
and associated companies in
Beirut Berlin Ibadan Nicosia

Oxford is a trade mark of Oxford University Press

Published in the United States
by Oxford University Press, New York

British Library Cataloguing in Publication Data
Marxism: essential writings.
1. Communism
I. McLellan, David
335.4   HX73
ISBN 0–19–827518–8
ISBN 0–19–827517–X Pbk

Library of Congress Cataloging in Publication Data
Data available

Set by Butler & Tanner Ltd
Printed in the United States of America

PREFACE

The following selections are intended to present the reader with key passages from the most important Marxist writings from Marx himself up to the present day. Since this is a book of primary source materials, I have tried to keep interpretation to a minimum. Of course, the very activity of selection involves certain priorities: there is, for example, very little on economics or the finer points of Western 'critical' Marxism in the following pages. And all the subheadings are mine. In the latter half of the book I have particularly tried to give the reader a representative sample of the various Marxisms of the contemporary world: Russian, Yugoslav, Chinese, Latin American, and African. I have added short lists of further reading and a full index which gives brief biographical details of persons mentioned in the text and enables the reader to follow a given theme through the various authors.

D.M.

ACKNOWLEDGEMENTS

THE editor is grateful for permission from the following publishers to reprint:

Unwin Hyman Ltd. for K. Kautsky, *The Dictatorship of the Proletariat*; University of Michigan Press for L. Trótsky, *Literature and Revolution*, and R. Luxemburg, *The Russian Revolution*; Lawrence and Wishart Ltd. for A. Gramsci, *Selections from Prison Notebooks*, V. Lenin, *Selected Works*, and G. Plekhanov, *Fundamental Problems of Marxism*; Pathfinder Press for L. Trotsky, *The Permanent Revolution*, copyright Merit Publishers, 1969; Merlin Press for G. Lukacs, *History and Class Consciousness*; Penguin Books Ltd. for *Mao Tse-tung Unrehearsed*; Summerfield Press for E. Kardelj, *Democracy and Socialism*; University of Nebraska Press for E. Guevara, *Guerrilla Warfare*; MIT Press for E. Guevara, *Selected Works*; Routledge and Kegan Paul PLC for H. Marcuse, *One-dimensional Man*.

The editor has made every effort to contact all relevant publishers. Should any have been omitted, he would be glad to hear from them.

CONTENTS

PART I: CLASSICAL MARXISM

1. **Karl Marx**   3
   1. The Materialist Concept of History   3
   2. The *Communist Manifesto*   20
   3. The Destiny of Capitalism   49

2. **Friedrich Engels**   62
   1. Socialism: Utopian and Scientific   62
   2. The Materialist Conception of History   69
   3. Revolution: Peaceful or Violent?   71

3. **Eduard Bernstein**   76
   Evolutionary Socialism   76

4. **Karl Kautsky**   87
   Socialist Democracy   87

5. **Rosa Luxemburg**   108
   1. Social Reform or Revolution?   108
   2. The Mass Strike   120
   3. Lenin's Centralism   124
   4. The Russian Revolution   127

6. **Georgy Plekhanov**   134
   1. The Role of Personality in History   134
   2. The Marxist Conception of Historical Determinism   138

7. **Vladimir Lenin**   145
   1. The Proletarian Party   146
   2. Revolution: Bourgeois or Proletarian?   150
   3. Imperialism   153
   4. The State and Revolution   163
   5. Against 'Left-wing' Communism   177
   6. On Soviet Bureaucracy   184

8. **Leon Trotsky**   188
   1. The Prospect of Revolution   188
   2. Permanent Revolution   195
   3. The Revolution Betrayed   202
   4. The Socialist Society of the Future   223

 9. **Nikolai Bukharin**   226
    1. The Communist Revolutionary State   227
    2. Dialectics and Revolution   232

10. **Georg Lukacs**   246
    History and Class-consciousness   246

11. **Antonio Gramsci**   264
    1. Intellectuals and Hegemony   264
    2. Revolution in the West   268
    3. The Culture of Marxism   272

PART II: CONTEMPORARY MARXISM

12. **Josef Stalin: Soviet Marxism**   285
    1. Lenin's Theory of Revolution   285
    2. The Dictatorship of the Proletariat   297
    3. Socialism in One Country   307
    4. Revolution from Above   308
    5. Wars Between Capitalist Nations   309

13. **Mao Zedong: Chinese Marxism**   314
    1. The Peasantry as a Revolutionary Force   315
    2. On Guerrilla Warfare   317
    3. On Contradiction   318
    4. On the Correct Handling of Contradictions Among the
       People   325
    5. Difficulties of the Cultural Revolution   335

14. **Herbert Marcuse: Western Marxism**   340
    One-dimensional Society   340

15. **Edvard Kardelj: Yugoslav Marxism**   357
    Democracy and Socialism   357

16. **Ernesto Guevara: Latin American Marxism**   373
    1. On Guerrilla Strategy   373
    2. Socialist Humanism and Revolution   377

17. **Amilcar Cabral: African Marxism**   392
    Class and Revolution in Africa   392

    NAME INDEX   411
    SUBJECT INDEX   418

PART I
CLASSICAL MARXISM

KARL MARX

Now so-called 'Marxism' in France is in any case a very singular product—to such an extent that Marx said to Lafargue: 'What is certain is that *I* am not a Marxist'.

Letter of Engels to Bernstein, in K. Marx and F. Engels, *Werke* (Berlin, 1967), xxxv. 388.

KARL MARX (1818–83) *was a German social scientist and revolutionary who founded the movement bearing his name. On being debarred in his native Germany from academic and journalistic careers, Marx moved in 1843 to Paris where he became a convert to Communism. After the failure of the 1848 revolutions, Marx emigrated definitively to London, where he composed his works on political economy. In the first extract below, from* The German Ideology *of 1845, Marx explains his materialist conception of history. The second extract contains Marx's famous summary of what he called 'a guiding thread for my studies': the historical primacy of the economic basis over the ideological superstructure. The third selection is* the Communist Manifesto, *a propaganda sheet hastily composed on the eve of the 1848 revolutions and the most widely read of Marxist writings. The next two extracts are from subsequent prefaces to the* Manifesto *and indicate the ways in which Marx would have wanted to modify it: in the Preface of 1872 he suggests that the State must be dismantled by the working class, not just taken over. And in the Preface of 1882, he discusses the possibility of the revolution starting in Russia. The final extract is from Marx's major work* Capital *and presents a vivid picture of the genesis and destiny of capitalism.*

1. The Materialist Concept of History

(a) FROM *The German Ideology*

THE premises from which we begin are not arbitrary ones, not dogmas, but real premises from which abstraction can only be made in the imagination. They are the real individuals, their activity, and the material conditions under which they live, both those which they find already existing and those produced by their activity. These premises can thus be verified in a purely empirical way.

The first premiss of all human history is, of course, the existence of living human individuals. Thus the first fact to be established is the physical organization of these individuals and their consequent relation to the rest of nature. Of course, we cannot here go either into the actual physical nature of man, or into the natural conditions in which man finds himself—geological, oro-hydrographical, climatic, and so. The writing of history must always set out from these natural bases and their modification in the course of history through the action of men.

Men can be distinguished from animals by consciousness, by religion, or anything else you like. They themselves begin to distinguish themselves from animals as soon as they begin to produce their means of subsistence, a step which is conditioned by their physical organization. By producing their means of subsistence men are indirectly producing their actual material life.

The way in which men produce their means of subsistence depends first of all on the nature of the actual means of subsistence they find in existence and have to reproduce. This mode of production must not be considered simply as being the production of the physical existence of the individuals. Rather it is a definite form of activity of these individuals, a definite form of expressing their life, a definite mode of life on their part. As individuals express their life, so they are. What they are, therefore, coincides with their production, both with *what* they produce and with *how* they produce. The nature of individuals thus depends on the material conditions determining their production.

This production only makes its appearance with the increase of population. In its turn this presupposes the intercourse of individuals with one another. The form of this intercourse is again determined by production.

The relations of different nations among themselves depend upon the extent to which each has developed its productive forces, the division of labour, and internal intercourse. This statement is generally recognized. But not only the relation of one nation to others, but also the whole internal structure of the nation itself depends on the stage of development reached by its production and its internal and external intercourse. How far the productive forces of a nation are developed is shown most manifestly by the degree to which the division of labour has been carried. Each new productive force, in so far as it is not merely a quantitative extension of productive forces already known (for instance the bringing into cultivation of fresh land), causes a further development of the division of labour.

The division of labour inside a nation leads at first to the separation of industrial and commercial from agricultural labour, and hence to the separation of town and country and to the conflict of their interests. Its further development leads to the separation of commercial from industrial labour. At the same time, through the division of labour inside these various branches there develop various divisions among the individuals co-operating in definite kinds of labour. The relative position of these individual groups is determined by the methods employed in agriculture, industry, and commerce (patriarchalism, slavery, estates, classes). These same conditions are to be seen (given a more developed intercourse) in the relations of different nations to one another.

The various stages of development in the division of labour are just so many different forms of ownership, i.e. the existing stage in the division of labour determines also the relations of individuals to one another with reference to the material, instrument, and product of labour.

The first form of ownership is tribal ownership. It corresponds to the undeveloped stage of production, at which a people lives by hunting and fishing, by the rearing of beasts, or, in the highest stage, agriculture. In the latter case it presupposes a great mass of uncultivated stretches of land. The division of labour is at this stage still very elementary and is confined to a further extension of the natural division of labour existing in the family. The social structure is, therefore, limited to an extension of the family; patriarchal family chieftains, below them the members of the tribe, finally slaves. The slavery latent in the family only develops gradually with the increase of population, the growth of wants, and with the extension of external relations, both of war and of barter.

The second form is the ancient communal and State ownership which proceeds especially from the union of several tribes into a city by agreement or by conquest, and which is still accompanied by slavery. Beside communal ownership we already find movable, and later also immovable, private property developing, but as an abnormal form subordinate to communal ownership. The citizens hold power over their labouring slaves only in their community, and on this account alone, therefore, they are bound to the form of communal ownership. It is the communal private property which compels the active citizens to remain in this spontaneously derived form of association over against their slaves. For this reason the whole structure of society based on this communal ownership, and with it the power of the people, decays in the same measure as, in particular, immovable private property evolves. The division of labour is already more developed. We already find the antagonism of town and country; later the antagonism between those states which represent town interests and those which represent country interests, and inside the towns themselves the antagonism between industry and maritime commerce. The class relation between citizens and slaves is now completely developed.

With the development of private property, we find here for the first time the same conditions which we shall find again, only on a more extensive scale, with modern private property. On the one hand, the concentration of private property, which began very early in Rome (as the Licinian agrarian law proves) and proceeded very rapidly from the time of the civil wars and especially under the emperors; on the other hand, coupled with this, the transformation of the plebeian small peasantry into a proletariat, which, however, owing to its intermediate position between propertied citizens and slaves, never achieved an independent development.

The third form of ownership is feudal or estate property. If antiquity started out from the town and its little territory, the Middle Ages started out from the country. This differing starting-point was determined by the sparseness of the population at that time, which was scattered over a large area and which received no large increase from the conquerors. In contrast to Greece and Rome, feudal development at the outset, therefore, extends over a much wider territory, prepared by the Roman conquests and the spread of agriculture at first associated with it. The last centuries of the declining Roman Empire and its conquest by the barbarians destroyed a number of productive forces; agriculture had declined, industry had decayed for want of a market, trade had died out or been violently suspended, the rural and urban population had decreased. From these conditions and the mode of organization of the conquest determined by them, feudal property developed under the influence of the Germanic military constitution. Like tribal and communal ownership, it is based again on a community; but the directly producing class standing over against it is not, as in the case of the ancient community, the slaves, but the enserfed small peasantry. As soon as feudalism is fully developed, there also arises antagonism towards the towns. The hierarchical structure of landownership, and the armed bodies of retainers associated with it, gave the nobility power over the serfs. This feudal organization was, just as much as the ancient communal ownership, an association against a subjected producing class; but the form of association and the relation to the direct producers were different because of the different conditions of production.

This feudal system of landownership had its counterpart in the towns in the shape of corporative property, the feudal organization of trades. Here property consisted chiefly in the labour of each individual person. The necessity for association against the organized robber barons, the need for communal covered markets in an age when the industrialist was at the same time a merchant, the growing competition of the escaped serfs swarming into the rising towns, the feudal structure of the whole country: these combined to bring about the guilds. The gradually accumulated small capital of individual craftsmen and their stable numbers, as against the growing population, evolved the relation of journeyman and apprentice, which brought into being in the towns a hierarchy similar to that in the country.

Thus the chief form of property during the feudal epoch consisted on the one hand of landed property with serf labour chained to it, and on the other of the labour of the individual with small capital commanding the labour of journeymen. The organization of both was determined by the restricted conditions of production—the small-scale and primitive cultivation of the land and the craft type of industry. There was little division of labour in the heyday of feudalism. Each country bore in itself the antithesis of town and country; the division into estates was certainly strongly marked; but apart from the differentiation of princes, nobility, clergy, and peasants in the country, and masters, journeymen, apprentices, and soon also the rabble of casual labourers in the towns, no division of importance took place. In agriculture it was rendered difficult by the strip-system, beside which the cottage industry of the peasants themselves emerged. In industry there was no division of labour at all in the individual trades themselves, and very little between them. The separation of industry and commerce was found already in existence in older towns; in the newer it only developed later, when the towns entered into mutual relations.

The grouping of larger territories into feudal kingdoms was a necessity for the landed nobility as for the towns. The organization of the ruling class, the nobility, had, therefore, everywhere a monarch at its head.

The fact is, therefore, that definite individuals who are productively active in a definite way enter into these definite social and political relations. Empirical observation must in each separate instance bring out empirically, and without any mystification and speculation, the connection of the social and political structure with production. The social structure and the State are continually evolving out of the life-process of definite individuals, but of individuals, not as they may appear in their own or other people's imagination, but as they really are, i.e. as they operate, produce materially, and hence as they work under definite material limits, presuppositions, and conditions independent of their will.

The production of ideas, of conceptions, of consciousness, is at first directly interwoven with the material activity and the material intercourse of men, the language of real life. Conceiving, thinking, the mental intercourse of men, appear at this stage as the direct efflux of their material behaviour. The same applies to mental production as expressed in the language of politics, laws, morality, religion, metaphysics, etc. of a people. Men are the producers of their conceptions, ideas, etc.—real, active men, as they are conditioned by a definite development of their productive forces and of the intercourse corresponding to these, up to its furthest forms. Consciousness can never be anything else than conscious existence, and the existence of men is their actual life-process. If in all ideology men and their circumstances appear upside-down as in a camera obscura, this phenomenon arises just as much from their historical life-process as the inversion of objects on the retina does from their physical life-process.

In direct contrast to German philosophy which descends from heaven to earth, here we ascend from earth to heaven. That is to say, we do not set out from what men say, imagine, conceive, nor from men as narrated, thought of, imagined, conceived, in order to arrive at men in the flesh. We set out from real, active men, and on the basis of their real life-process we demonstrate the development of the ideological reflexes and echoes of this life-process. The phantoms formed in the human brain are also, necessarily, sublimates of their material life-process, which is empirically verifiable and bound to material premisses. Morality, religion, metaphysics, all the rest of ideology and their corresponding forms of consciousness, thus no longer retain the semblance of independence. They have no history, no development; but men, developing their material production and their material intercourse, alter, along with this their real existence, their thinking and the products of their thinking. Life is not determined by consciousness, but consciousness by life. In the first method of approach the starting-point is consciousness taken as the living individual; in the second method, which conforms to real life, it is the real living individuals themselves, and consciousness is considered solely as their consciousness.

This method of approach is not devoid of premisses. It starts out from the real premisses and does not abandon them for a moment. Its premisses are men, not in any fantastic isolation and rigidity, but in their actual, empirically perceptible process of development under definite conditions. As soon as this active life-process is described, history ceases to be a collection of dead facts as it is with the empiricists (themselves still abstract), or an imagined activity of imagined subjects, as with the idealists.

Where speculation ends—in real life—there real, positive science begins: the representation of the practical activity, of the practical process of development of men. Empty talk about consciousness ceases, and real knowledge has to take its place. When reality is depicted, philosophy as an independent branch of knowledge loses its medium of existence. At the best its place can only be taken by a summing-up of the most general results, abstractions which arise from the observation of the historical development of men. Viewed apart from real history, these abstractions have in themselves no value whatsoever. They can only serve to facilitate the arrangement of historical material, to indicate the sequence of its separate strata. But they by no means afford a recipe or schema, as does philosophy, for neatly trimming the epochs of history. On the contrary, our difficulties begin only when we set about the observation and the arrangement—the real depiction—of our historical material, whether of a past epoch or of the present. The removal of these difficulties is governed by premises which it is quite impossible to state here, but which only the study of the actual life-process and the activity of the individuals of each epoch will make evident....

The social power, i.e. the multiplied productive force, which arises through the co-operation of different individuals as it is determined by the division of labour, appears to these individuals, since their co-operation is not voluntary but has come about naturally, not as their own united power, but as an alien force existing outside them, of the origin and goal of which they are ignorant, which they thus cannot control, which on the contrary passes through a peculiar series of phases and stages independent of the will and the action of man, nay even being the prime governor of these.

How otherwise could, for instance, property have had a history at all, have taken on different forms, and landed property, for example, according to the different premises given, have proceeded in France from parcellation to centralization in the hands of a few, in England from centralization in the hands of a few to parcellation, as is actually the case today? Or how does it happen that trade, which after all is nothing more than the exchange of products of various individuals and countries, rules the whole world through the relation of supply and demand—a relation which, as an English economist says, hovers over the earth like the Fates of the ancients, and with invisible hand allots fortune and misfortune to men, sets up empires and overthrows empires, causes nations to rise and to disappear—while with the abolition of the basis of private property, with the communistic regulation of production (and, implicit in this, the destruction of the alien relation between men and what they themselves produce), the power of the relation of supply and demand is dissolved into nothing, and men get exchange, production, the mode of their mutual relation, under their own control again?

This 'alienation' (to use a term which will be comprehensible to the philosophers) can, of course, only be abolished given two practical premises. For it to become an 'intolerable' power, i.e. a power against which men make a revolution, it must necessarily have rendered the great mass of humanity 'propertyless', and produced, at the same time, the contradiction of an existing world of wealth and culture, both of which conditions presuppose a great increase in productive power, a high degree of its development. And, on the other hand, this development of productive forces (which itself implies the actual empirical existence of men in their world-historical, instead of local, being) is an absolutely necessary practical premiss because without it want is merely made general, and with destitution the struggle for necessities and all the old filthy business would necessarily be reproduced; and furthermore, because only with this universal development of productive forces is a universal intercourse between men established, which produces in all nations simultaneously the phenomenon of the 'propertyless' mass (universal competition), makes each nation dependent on the revolutions of the others, and finally has put world-historical, empirically universal individuals in place of local ones. Without this, (1) Communism could only exist as a local event; (2) the forces of intercourse themselves could not have developed as universal, hence intolerable powers: they would have remained home-bred conditions surrounded by superstition; and (3) each extension of intercourse would abolish local Communism. Empirically, Communism is only possible as the act of the dominant peoples 'all at once' and simultaneously, which presupposes the universal development of productive forces and the world intercourse bound up with Communism. Moreover, the mass of propertyless workers—the utterly precarious position of labour-power on a mass scale cut off from capital or from even a limited satisfaction and, therefore, no longer merely temporarily deprived of work itself as a secure source of life—presupposes the world market through competition. The proletariat can thus only exist world-historically, just as Communism, its activity, can only have a 'world-historical' existence. World-historical existence of individuals means existence of individuals which is directly linked up with world history.

Communism is for us not a state of affairs which is to be established, an ideal to which reality will have to adjust itself. We call Communism the real movement which abolishes the present state of things. The conditions of this movement result from the premisses now in existence.

... In history up to the present it is certainly an empirical fact that separate individuals have, with the broadening of their activity into world-historical activity, become more and more enslaved under a power alien to them (a pressure which they have conceived of as a dirty trick on the part of the so-called universal spirit, etc.), a power which has become more and more enormous and, in the last instance, turns out to be the world market. But it is just as empirically established that, by the overthrow of the existing state of society by the Communist revolution (of which more below) and the abolition of private property which is identical with it, this power, which so baffles the German theoreticians, will be dissolved; and that then the liberation of each single individual will be accomplished in the measure in which history becomes transformed into world history. From the above it is clear that the real intellectual wealth of the individual depends entirely on the wealth of his real connections. Only then will the separate individuals be liberated from the various national and local barriers, be brought into practical connection with the material and intellectual production of the whole world, and be put in a position to acquire the capacity to enjoy this all-sided production of the whole earth (the creations of man). All-round dependence, this natural form of the world-historical co-operation of individuals, will be transformed by this Communist revolution into the control and conscious mastery of these powers, which, born of the action of men on one another, have till now overawed and governed men as powers completely alien to them. Now this view can be expressed again in speculative-idealistic, i.e. fantastic, terms as 'self-generation of the species' ('society as the subject'), and thereby the consecutive series of interrelated individuals connected with each other can be conceived as a single individual, which accomplishes the mystery of generating itself. It is clear here that individuals certainly make one another, physically and mentally, but do not make themselves either in the nonsense of Saint Bruno, or in the sense of the 'Unique', of the 'made' man.

This conception of history depends on our ability to expound the real process of production, starting out from the material production of life itself, and to comprehend the form of intercourse connected with this and created by this mode of production (i.e. civil society in its various stages) as the basis of all history; and to show it in its action as State, to explain all the different theoretical products and forms of consciousness, religion, philosophy, ethics, etc., etc., and trace their origins and growth from that basis; by which means, of course, the whole thing can be depicted in its totality (and therefore, too, the reciprocal action of these various sides on one another). It has not, like the idealistic view of history, in every period to look for a category, but remains constantly on the real ground of history; it does not explain practice from the idea but explains the formation of ideas from material practice; and accordingly it comes to the conclusion that all forms and products of consciousness cannot be dissolved by mental criticism, by resolution into 'self-consciousness' or transformation into 'apparitions', 'spectres', 'fancies', etc., but only by the practical overthrow of the actual social relations which gave rise to this idealistic humbug; that not criticism but revolution is the driving force of history, also of religion, of philosophy, and all other types of theory. It shows that history does not end by being resolved into 'self-consciousness' as 'spirit of the spirit', but that in it at each stage there is found a material result: a sum of productive forces, a historically created relation of individuals to nature and to one another, which is handed down to each generation from its predecessor; a mass of productive forces, capital funds and conditions, which, on the one hand, is indeed modified by the new generation, but also, on the other, prescribes for it its conditions of life and gives it a definite development, a special character. It shows that circumstances make men just as much as men make circumstances.

This sum of productive forces, capital funds, and social forms of inter-course, which every individual and generation finds in existence as some-thing given, is the real basis of what the philosophers have conceived as 'substance' and 'essence of man', and what they have deified and attacked; a real basis which is not in the least disturbed, in its effect and influence on the development of men, by the fact that these philosophers revolt against it as 'self-consciousness' and the 'Unique'. These conditions of life, which different generations find in existence, decide also whether or not the periodically recurring revolutionary convulsion will be strong enough to overthrow the basis of the entire existing system. And if these material elements of a complete revolution are not present (namely, on the one hand the existing productive forces, on the other the formation of a revolutionary mass, which revolts not only against separate conditions of society up till then, but against the very 'production of life' till then, the 'total activity' on which it was based), then, as far as practical development is concerned, it is absolutely immaterial whether the idea of this revolution has been expressed a hundred times already, as the history of Communism proves.

In the whole conception of history up to the present this real basis of history has either been totally neglected or else considered as a minor matter quite irrelevant to the course of history. History must, therefore, always be written according to an extraneous standard; the real production of life seems to be primeval history, while the truly historical appears to be separated from ordinary life, something superterrestrial. With this the relation of man to nature is excluded from history and hence the antithesis of nature and history is created. The exponents of this conception of history have consequently only been able to see in history the political actions of princes and States, religious and all sorts of theoretical struggles, and in particular in each historical epoch have had to share the illusion of that epoch. For instance, if an epoch imagines itself to be actuated by purely 'political' or 'religious' motives, although 'religion' and 'politics' are only forms of its true motives, the historian accepts this opinion. The 'idea', the 'conception' of the people in question about their real practice, is trans-formed into the sole determining, active force, which controls and deter-mines their practice. When the crude form in which the division of labour appears with the Indians and Egyptians calls forth the caste system in their State and religion, the historian believes that the caste system is the power which has produced this crude social form. While the French and the English at least hold by the political illusion, which is moderately close to reality, the Germans move in the realm of the 'pure spirit', and make religious illusion the driving force of history. The Hegelian philosophy of history is the last consequence, reduced to its 'finest expression', of all this German historiography, for which it is not a question of real, nor even of political, interests, but of pure thoughts, which consequently must appear to Saint Bruno as a series of 'thoughts' that devour one another and are finally swallowed up in 'self-consciousness'. . . .

In reality and for the practical materialist, i.e. the Communist, it is a question of revolutionizing the existing world, of practically attacking and changing existing things. When occasionally we find such views with Feuerbach, they are never more than isolated surmises and have much too little influence on his general outlook to be considered here as anything else than embryos capable of development. Feuerbach's 'conception' of the sensuous world is confined on the one hand to mere contemplation of it, and on the other to mere feeling; he says 'Man' instead of 'real historical man'. 'Man' is really 'the German'. In the first case, the contemplation of the sensuous world, he necessarily lights on things which contradict his consciousness and feeling, which disturb the harmony he presupposes, the harmony of all parts of the sensuous world and especially of man and nature. To remove this disturbance, he must take refuge in a double perception, a profane one which only perceives the 'flatly obvious' and a higher, philosophical, one which perceives the 'true essence' of things. He does not see how the sensuous world around him is not a thing given direct from all eternity, remaining ever the same, but the product of industry and of the state of society; and, indeed, in the sense that it is an historical product, the result of the activity of a whole succession of generations, each standing on the shoulders of the preceding one, developing its industry and its intercourse, modifying its social system according to the changed needs. Even the objects of the simplest 'sensuous certainty' are only given him through social development, industry, and commercial intercourse. The cherry-tree, like almost all fruit-trees, was, as is well known, only a few centuries ago transplanted by commerce into our zone, and therefore only by this action of a definite society in a definite age it has become 'sensuous certainty' for Feuerbach.

Incidentally, when we conceive things thus, as they really are and happened, every profound philosophical problem is resolved, as will be seen even more clearly later, quite simply into an empirical fact. For instance, the important question of the relation of man to nature (Bruno [Bauer] goes so far as to speak of 'the antitheses in nature and history' (p. 110), as though these were two separate 'things' and man did not always have before him an historical nature and a natural history), out of which all the 'unfathomably lofty works' on 'substance' and 'self-consciousness' were born, crumbles of itself when we understand that the celebrated 'unity of man with nature' has always existed in industry and has existed in varying forms in every epoch according to the lesser or greater development of industry, just like the 'struggle' of man with nature, right up to the development of his productive powers on a corresponding basis. Industry and commerce, production and the exchange of the necessities of life, themselves determine distribution, the structure of the different social classes, and are, in turn, determined by it as to the mode in which they are carried on; and so it happens that in Manchester, for instance, Feuerbach sees only factories and machines, where a hundred years ago only spinning-wheels and weaving-looms were to be seen, or in the Campagna of Rome he finds only pasture lands and swamps, where in the time of Augustus he would have found nothing but the vineyards and villas of Roman capitalists. Feuerbach speaks in particular of the perception of natural science; he mentions secrets which are disclosed only to the eye of the physicist and chemist; but where would natural science be without industry and commerce? Even this 'pure' natural science is provided with an aim, as with its material, only through trade and industry, through the sensuous activity of men. So much is this activity, this unceasing sensuous labour and creation, this production, the basis of the whole sensuous world as it now exists, that, were it interrupted only for a year, Feuerbach would not only find an enormous change in the natural world, but would very soon find that the whole world of men and his own perceptive faculty, nay his own existence, were missing. Of course, in all this the priority of external nature remains unassailed, and all this has no application to the original men produced by *generatio aequivoca* [spontaneous generation]; but this differentiation has meaning only in so far as man is considered to be distinct from nature. For that matter, nature, the nature that preceded human history, is not by any means the nature in which Feuerbach lives, it is nature which today no longer exists anywhere (except perhaps on a few Australian coral islands of recent origin) and which, therefore, does not exist for Feuerbach.

Certainly Feuerbach has a great advantage over the 'pure' materialists in that he realizes how man too is an 'object of the senses'. But apart from the fact that he only conceives him as an 'object of the senses', not as 'sensuous activity', because he still remains in the realm of theory and conceives of men not in their given social connection, not under their existing conditions of life, which have made them what they are, he never arrives at the really existing active men, but stops at the abstraction 'man', and gets no further than recognizing 'the true, individual, corporeal man' emotionally, i.e. he knows no other 'human relationships' 'of man to man' than love and friendship, and even then idealized. He gives no criticism of the present conditions of life. Thus he never manages to conceive the sensuous world as the total living sensuous activity of the individuals composing it; and therefore when, for example, he sees instead of healthy men a crowd of scrofulous, overworked, and consumptive starvelings, he is compelled to take refuge in the 'higher peception' and in the ideal 'compensation in the species', and thus to relapse into idealism at the very point where the Communist materialist sees the necessity, and at the same time the condition, of a transformation both of industry and of the social structure.

As far as Feuerbach is a materialist he does not deal with history, and as far as he considers history he is not a materialist. With him materialism and history diverge completely, a fact which incidentally is already obvious from what has been said. . . .

The ideas of the ruling class are in every epoch the ruling ideas, i.e. the class which is the ruling material force of society is at the same time its ruling intellectual force. The class which has the means of material production at its disposal has control at the same time over the means of mental production, so that thereby, generally speaking, the ideas of those who lack the means of mental production are subject to it. The ruling ideas are nothing more than the ideal expression of the dominant material relationships, the dominant material relationships grasped as ideas; hence of the relationships which make the one class the ruling one, therefore, the ideas of its dominance. The individuals composing the ruling class possess among other things consciousness, and therefore think. In so far, therefore, as they rule as a class and determine the extent and compass of an epoch, it is self-evident that they do this in its whole range, hence among other things rule also as thinkers, as producers of ideas, and regulate the production and distribution of the ideas of their age; thus their ideas are the ruling ideas of the epoch. For instance, in an age and in a country where royal power, aristocracy, and bourgeoisie are contending for mastery and where, therefore, mastery is shared, the doctrine of the separation of powers proves to be the dominant idea and is expressed as an 'eternal law' . . .

Our investigation hitherto started from the instruments of production, and it has already shown that private property was a necessity for certain industrial stages. In *industrie extractive* [raw materials industry] private property still coincides with labour; in small industry and all agriculture up till now property is the necessary consequence of the existing instruments of production; in big industry the contradiction between the instrument of production and private property appears for the first time and is the product of big industry; moreover, big industry must be highly developed to produce this contradiction. And thus only with big industry does the abolition of private property become possible.

In big industry and competition the whole mass of conditions of existence, limitations, biases of individuals, are fused together into the two simplest forms: private property and labour. With money every form of intercourse, and intercourse itself, is considered fortuitous for the individuals. Thus money implies that all previous intercourse was only intercourse of individuals under particular conditions, not of individuals as individuals. These conditions are reduced to two: accumulated labour or private property, and actual labour. If both or one of these ceases, then intercourse comes to a standstill. The modern economists themselves, e.g. Sismondi, Cherbuliez, etc., oppose 'association of individuals' to 'association of capital'. On the other hand, the individuals themselves are entirely subordinated to the division of labour and hence are brought into the most complete dependence on one another. Private property, in so far as within labour itself it is opposed to labour, evolves out of the necessity of accumulation, and has still to begin with, rather the form of the communality; but in its further development it approaches more and more the modern form of private property. The division of labour implies from the outset the division of the conditions of labour, of tools, and materials, and thus the splitting-up of accumulated capital among different owners, and thus, also, the division between capital and labour, and the different forms of property itself. The more the division of labour develops and accumulation grows, the sharper are the forms that this process of differentiation assumes. Labour itself can only exist on the premiss of this fragmentation.

Thus two facts are here revealed. First the productive forces appear as a world for themselves, quite independent of and divorced from the individuals, alongside the individuals: the reason for this is that the individuals, whose forces they are, exist split up and in opposition to one another, while, on the other hand, these forces are only real forces in the intercourse and association of these individuals. Thus, on the one hand, we have a totality of productive forces, which have, as it were, taken on a material form and are for the individuals no longer the forces of the individuals but of private property, and hence of the individuals only in so far as they are owners of private property themselves. Never, in any earlier period, have the productive forces taken on a form so indifferent to the intercourse of individuals as individuals, because their intercourse itself was formerly a restricted one. On the other hand, standing over against these productive forces, we have the majority of the individuals from whom these forces have been wrested away, and who, robbed thus of all real life-content, have become abstract individuals, but who are, however, only by this fact put into a position to enter into relation with one another as individuals.

The only connection which still links them with the productive forces and with their own existence—labour—has lost all semblance of self-activity and only sustains their life by stunting it. While in the earlier periods self-activity and the production of material life were separated, in that they devolved on different persons, and while, on account of the narrowness of the individuals themselves, the production of material life was considered as a subordinate mode of self-activity, they now diverge to such an extent that altogether material life appears as the end, and what produces this material life, labour (which is now the only possible but, as we see, negative form of self-activity), as the means.

Thus things have now come to such a pass that the individuals must appropriate the existing totality of productive forces, not only to achieve self-activity, but also merely to safeguard their very existence. This appropriation is first determined by the object to be appropriated, the productive forces, which have been developed to a totality and which only exist within a universal intercourse. From this aspect alone, therefore, this appropriation must have a universal character corresponding to the productive forces and the intercourse.

The appropriation of these forces is itself nothing more than the development of the individual capacities corresponding to the material instruments of production. The appropriation of a totality of instruments of production is, for this very reason, the development of a totality of capacities in the individuals themselves.

This appropriation is further determined by the persons appropriating. Only the proletarians of the present day, who are completely shut off from all self-activity, are in a position to achieve a complete and no longer restricted self activity, which consists in the appropriation of a totality of productive forces and in the thus postulated development of a totality of capacities. All earlier revolutionary appropriations were restricted; individuals, whose self-activity was restricted by a crude instrument of production and a limited intercourse, appropriated this crude instrument of production, and hence merely achieved a new state of limitation. Their instrument of production became their property, but they themselves remained subordinate to the division of labour and their own instrument of production. In all expropriations up to now, a mass of individuals remained subservient to a single instrument of production; in the appropriation by the proletarians, a mass of instruments of production must be made subject to each individual, and property to all. Modern universal intercourse can be controlled by individuals, therefore, only when controlled by all.

This appropriation is further determined by the manner in which it must be effected. It can only be effected through a union, which by the character of the proletariat itself can again only be a universal one, and through a revolution, in which, on the one hand, the power of the earlier mode of production and intercourse and social organization is overthrown, and, on the other hand, there develops the universal character and the energy of the proletariat, without which the revolution cannot be accomplished; and in which, further, the proletariat rids itself of everything that still clings to it from its previous position in society.

Only at this stage does self-activity coincide with material life, which corresponds to the development of individuals into complete individuals and the casting-off of all natural limitations. The transformation of labour into self-activity corresponds to the transformation of the earlier limited intercourse into the intercourse of individuals as such. With the appropriation of the total productive forces through united individuals, private property comes to an end. While previously in history a particular condition always appeared as accidental, now the isolation of individuals and the particular private gain of each man have themselves become accidental. . . .

Finally, from the conception of history we have sketched we obtain these further conclusions: (1) In the development of productive forces there comes a stage when productive forces and means of intercourse are brought into being which, under the existing relationships, only cause mischief, and are no longer productive but destructive forces (machinery and money); and connected with this a class is called forth which has to bear all the burdens of society without enjoying its advantages, which, ousted from society, is forced into the most decided antagonism to all other classes; a class which forms the majority of all members of society, and from which emanates the consciousness of the necessity of a fundamental revolution, the Communist consciousness, which may, of course, arise among the other classes too through the contemplation of the situation of this class. (2) The conditions under which definite productive forces can be applied are the conditions of the rule of a definite class of society, whose social power, deriving from its property, has its practical-idealistic expression in each case in the form of the State; and, therefore, every revolutionary struggle is directed against a class which till then has been in power. (3) In all revolutions up till now the mode of activity always remained unscathed and it was only a question of a different distribution of this activity, a new distribution of labour to other persons, while the Communist revolution is directed against the preceding mode of activity, does away with labour, and abolishes the rule of all classes with the classes themselves, because it is carried through by the class which no longer counts as a class in society, is not recognized as a class, and is in itself the expression of the dissolution of all classes, nationalities, etc. within present society; and (4) both for the production on a mass scale of this Communist consciousness, and for the success of the cause itself, the alteration of men on a mass scale is necessary, an alteration which can only take place in a practical movement, a revolution; this revolution is necessary, therefore, not only because the ruling class cannot be overthrown in any other way, but also because the class overthrowing it can only in a revolution succeed in ridding itself of all the muck of ages and become fitted to found society anew....

(b) FROM THE PREFACE TO *A Critique of Political Economy*

My investigation led to the result that legal relations as well as forms of state are to be grasped neither from themselves nor from the so-called general development of the human mind, but rather have their roots in the material conditions of life, the sum total of which Hegel, following the example of the Englishmen and Frenchmen of the eighteenth century, combines under the name of 'civil society', that, however, the anatomy of civil society is to be sought in political economy. The investigation of the latter, which I began in Paris, I continued in Brussels, whither I had emigrated in consequence of an expulsion order of M. Guizot. The general result at which I arrived and which, once won, served as a guiding thread for my studies, can be briefly formulated as follows: In the social production of their life, men enter into definite relations that are indispensable and independent of their will, relations of production which correspond to a definite stage of development of their material productive forces. The sum total of these relations of production constitutes the economic structure of society, the real foundation, on which rises a legal and political superstructure and to which correspond definite forms of social consciousness. The mode of production of material life conditions the social, political, and intellectual life-process in general. It is not the consciousness of men that determines their being, but, on the contrary, their social being that determines their consciousness. At a certain stage of their development, the material productive forces of society come in conflict with the existing relations of production, or—what is but a legal expression for the same thing—with the property relations within which they have been at work hitherto. From forms of development of the productive forces these relations turn into their fetters. Then begins an epoch of social revolution. With the change of the economic foundations the entire immense superstructure is more or less rapidly transformed. In considering such transformations a distinction should always be made between the material transformation of the economic conditions of production, which can be determined with the precision of natural science, and the legal, political, religious, aesthetic, or philosophic—in short, ideological forms in which men become conscious of this conflict and fight it out. Just as our opinion of an individual is not based on what he thinks of himself, so can we not judge of such a period of transformation by its own consciousness; on the contrary, this consciousness must be explained rather from the contradictions of material life, from the existing conflict between the social productive forces and the relations of production. No social order ever perishes before all the productive forces for which there is room in it have developed; and new, higher relations of production never appear before the material conditions of their existence have matured in the womb of the old society itself. Therefore mankind always sets itself only such tasks as it can solve; since, looking at the matter more closely, it will always be found that the task itself arises only when the material conditions for its solution already exist or are at least in the process of formation. In broad outlines Asiatic, ancient, feudal, and modern bourgeois modes of production can be designated as progressive epochs in the economic formation of society. The bourgeois relations of production are the last antagonistic form of the social process of production—antagonistic not in the sense of individual antagonism, but of one arising from the social conditions of life of the individuals; at the same time the productive forces developing in the womb of bourgeois society create the material conditions for the solution of that antagonism. This social formation brings, therefore, the prehistory of human society to a close.

2.   The *Communist Manifesto*

(a)   THE 1848 TEXT

A spectre is haunting Europe—the spectre of Communism. All the Powers of old Europe have entered into a holy alliance to exorcize this spectre: pope and tsar, Metternich and Guizot, French Radicals and German police spies.

Where is the party in opposition that has not been decried as Communistic by its opponents in power? Where the opposition that has not hurled back the branding reproach of Communism, against the more advanced opposition parties, as well as against its reactionary adversaries?

Two things result from this fact.

i. Communism is already acknowledged by all European Powers to be itself a Power.

ii. It is high time that Communists should openly, in the face of the whole world, publish their views, their aims, their tendencies, and meet this nursery tale of the Spectre of Communism with a Manifesto of the party itself.

To this end, Communists of various nationalities have assembled in London, and sketched the following Manifesto, to be published in the English, French, German, Italian, Flemish, and Danish languages.

(i) *Bourgeois and Proletarians*

The history of all hitherto existing society is the history of class struggles.

Freeman and slave, patrician and plebeian, lord and serf, guild-master and journeyman—in a word, oppressor and oppressed, stood in constant opposition to one another, carried on an uninterrupted, now hidden, now open fight, a fight that each time ended either in a revolutionary reconstitution of society at large or in the common ruin of the contending classes.

In the earlier epochs of history, we find almost everywhere a complicated arrangement of society into various orders, a manifold gradation of social rank. In ancient Rome we have patricians, knights, plebeians, slaves; in the Middle Ages, feudal lords, vassals, guild-masters, journeymen, apprentices, serfs; in almost all of these classes, again, subordinate gradations.

The modern bourgeois society that has sprouted from the ruins of feudal society has not done away with class antagonisms. It has but established new classes, new conditions of oppression, new forms of struggle in place of the old ones.

Our epoch, the epoch of the bourgeoisie, possesses, however, this distinctive feature: it has simplified the class antagonisms. Society as a whole is more and more splitting up into two great hostile camps, into two great classes directly facing each other: Bourgeoisie and Proletariat.

From the serfs of the Middle Ages sprang the chartered burghers of the earliest towns. From these burgesses the first elements of the bourgeoisie were developed.

The discovery of America, the rounding of the Cape, opened up fresh ground for the rising bourgeoisie. The East Indian and Chinese markets, the colonization of America, trade with the colonies, the increase in the means of exchange and in commodities generally, gave to commerce, to navigation, to industry, an impulse never before known, and thereby, to the revolutionary element in the tottering feudal society, a rapid development.

The feudal system of industry, under which industrial production was monopolized by closed guilds, now no longer sufficed for the growing wants of the new markets. The manufacturing system took its place. The guild-masters were pushed on one side by the manufacturing middle class; division of labour between the different corporate guilds vanished in the face of division of labour in each single workshop.

Meantime the markets kept ever growing, the demand ever rising. Even manufacture no longer sufficed. Thereupon, steam and machinery revolutionized industrial production. The place of manufacture was taken by the giant Modern Industry, the place of the industrial middle class by industrial millionaires, the leaders of whole industrial armies, the modern bourgeois.

Modern industry has established the world market, for which the discovery of America paved the way. This market has given an immense development to commerce, to navigation, to communication by land. This development has, in its turn, reacted on the extension of industry; and in proportion as industry, commerce, navigation, railways extended, in the same proportion the bourgeoisie developed, increased its capital, and pushed into the background every class handed down from the Middle Ages.

We see, therefore, how the modern bourgeoisie is itself the product of a long course of development, of a series of revolutions in the modes of production and of exchange.

Each step in the development of the bourgeoisie was accompanied by a corresponding political advance of that class. An oppressed class under the sway of the feudal nobility, an armed and self-governing association in the medieval commune; here independent urban republic (as in Italy and Germany), there taxable 'third estate' of the monarchy (as in France), afterwards, in the period of manufacture proper, serving either the semi-feudal or the absolute monarchy as a counterpoise against the nobility, and, in fact, corner-stone of the great monarchies in general, the bourgeoisie has at last, since the establishment of Modern Industry and of the world market, conquered for itself, in the modern representative State, exclusive political sway. The executive of the modern State is but a committee for managing the common affairs of the whole bourgeoisie.

The bourgeoisie, historically, has played a most revolutionary part.

The bourgeoisie, wherever it has got the upper hand, has put an end to all feudal, patriarchal, idyllic relations. It has pitilessly torn asunder the motley feudal ties that bound man to his 'natural superiors', and has left remaining no other nexus between man and man than naked self-interest, than callous 'cash payment'. It has drowned the most heavenly ecstasies of religious fervour, of chivalrous enthusiasm, of philistine sentimentalism, in the icy water of egotistical calculation. It has resolved personal worth into exchange value, and in place of the numberless indefeasible chartered freedoms, has set up that single, unconscionable freedom—Free Trade. In one word, for exploitation, veiled by religious and political illusions, it has substituted naked, shameless, direct, brutal exploitation.

The bourgeoisie has stripped of its halo every occupation hitherto honoured and looked up to with reverent awe. It has converted the physician, the lawyer, the priest, the poet, the man of science into its paid wage-labourers.

The bourgeoisie has torn away from the family its sentimental veil, and has reduced the family relation to a mere money relation.

The bourgeoisie has disclosed how it came to pass that the brutal display of vigour in the Middle Ages, which reactionists so much admire, found its fitting complement in the most slothful indolence. It has been the first to show what man's activity can bring about. It has accomplished wonders far surpassing Egyptian pyramids, Roman aqueducts, and Gothic cathedrals; it has conducted expeditions that put in the shade all former exoduses of nations and crusades.

The bourgeoisie cannot exist without constantly revolutionizing the instruments of production, and thereby the relations of production, and with them the whole relations of society. Conservation of the old modes of production in unaltered form, was, on the contrary, the first condition of existence for all earlier industrial classes. Constant revolutionizing of production, uninterrupted disturbance of all social conditions, everlasting uncertainty and agitation distinguish the bourgeois epoch from all earlier ones. All fixed, fast-frozen relations, with their train of ancient and venerable prejudices and opinions, are swept away, all new-formed ones become antiquated before they can ossify. All that is solid melts into air, all that is holy is profaned, and man is at last compelled to face with sober senses his real conditions of life, and his relations with his kind.

The need of a constantly expanding market for its products chases the bourgeoisie over the whole surface of the globe. It must nestle everywhere, settle everywhere, establish connections everywhere.

The bourgeoisie has through its exploitation of the world market given a cosmopolitan character to production and consumption in every country. To the great chagrin of reactionists, it has drawn from under the feet of industry the national ground on which it stood. All old-established national industries have been destroyed or are daily being destroyed. They are dislodged by new industries, whose introduction becomes a life-and-death question for all civilized nations, by industries that no longer work up indigenous raw material, but raw material drawn from the remotest zones; industries whose products are consumed, not only at home, but in every quarter of the globe. In place of the old wants, satisfied by the productions of the country, we find new wants, requiring for their satisfaction the products of distant lands and climes. In place of the old local and national seclusion and self-sufficiency, we have intercourse in every direction, universal interdependence of nations. And as in material, so also in intellectual production. The intellectual creations of individual nations become common property. National one-sidedness and narrow-mindedness become more and more impossible, and from the numerous national and local literatures, there arises a world literature.

The bourgeoisie, by the rapid improvement of all instruments of production, by the immensely facilitated means of communication, draws all, even the most barbarian, nations into civilization. The cheap prices of its commodities are the heavy artillery with which it batters down all Chinese walls, with which it forces the barbarians' intensely obstinate hatred of foreigners to capitulate. It compels all nations, on pain of extinction, to adopt the bourgeois mode of production; it compels them to introduce what it calls civilization into their midst, i.e., to become bourgeois themselves. In one word, it creates a world after its own image.

The bourgeoisie has subjected the country to the rule of the towns. It has created enormous cities, has greatly increased the urban population as compared with the rural, and has thus rescued a considerable part of the population from the idiocy of rural life. Just as it has made the country dependent on the towns, so it has made barbarian and semi-barbarian countries dependent on the civilized ones, nations of peasants on nations of bourgeois, the East on the West.

The bourgeoisie keeps more and more doing away with the scattered state of the population, of the means of production, and of property. It has agglomerated population, centralized means of production, and has concentrated property in a few hands. The necessary consequence of this was political centralization. Independent or but loosely connected provinces, with separate interests, laws, governments, and systems of taxation, became lumped together into one nation, with one government, one code of laws, one national class-interest, one frontier, and one customs-tariff.

The bourgeoisie, during its rule of scarcely one hundred years, has created more massive and more colossal productive forces than have all preceding generations together. Subjection of Nature's forces to man, machinery, application of chemistry to industry and agriculture, steam-navigation, railways, electric telegraphs, clearing of whole continents for cultivation, canalization of rivers, whole populations conjured out of the ground—what earlier century had even a presentiment that such productive forces slumbered in the lap of social labour?

We see then that the means of production and of exchange, on whose foundation the bourgeoisie built itself up, were generated in feudal society. At a certain stage in the development of these means of production and of exchange, the conditions under which feudal society produced and exchanged, the feudal organization of agriculture and manufacturing industry, in one word, the feudal relations of property become no longer compatible with the already developed productive forces; they became so many fetters. They had to be burst asunder; they were burst asunder.

Into their place stepped free competition, accompanied by a social and political constitution adapted to it, and by the economical and political sway of the bourgeois class.

A similar movement is going on before our own eyes. Modern bourgeois society with its relations of production, of exchange, and of property, a society that has conjured up such gigantic means of production and of exchange, is like the sorcerer who is no longer able to control the powers of the nether world which he has called up by his spells. The history of industry and commerce for many a decade past is but the history of the revolt of modern productive forces against modern conditions of production, against the property relations that are the conditions for the existence of the bourgeoisie and of its rule. It is enough to mention the commercial crises that by their periodical return put on trial, each time more threateningly, the existence of the entire bourgeois society. In these crises a great part not only of the existing products, but also of the previously created productive forces, are periodically destroyed. In these crises there breaks out an epidemic that, in all earlier epochs, would have seemed an absurdity—the epidemic of overproduction. Society suddenly finds itself put back into a state of momentary barbarism; it appears as if a famine, a universal war of devastation, has cut off the supply of every means of subsistence; industry and commerce seem to be destroyed; and why? Because there is too much civilization, too much means of subsistence, too much industry, too much commerce. The productive forces at the disposal of society no longer tend to further the development of the conditions of bourgeois property; on the contrary, they have become too powerful for these conditions, by which they are fettered, and so soon as they overcome these fetters, they bring disorder into the whole of bourgeois society, endanger the existence of bourgeois property. The conditions of bourgeois society are too narrow to comprise the wealth created by them. And how does the bourgeoisie get over these crises? On the one hand by enforced destruction of a mass of productive forces; on the other, by the conquest of new markets, and by the more thorough exploitation of the old ones. That is to say, by paving the way for more extensive and more destructive crises, and by diminishing the means whereby crises are prevented.

The weapons with which the bourgeoisie felled feudalism to the ground are now turned against the bourgeoisie itself.

But not only has the bourgeoisie forged the weapons that bring death to itself; it has also called into existence the men who are to wield those weapons—the modern working class—the proletarians.

In proportion as the bourgeoisie, i.e., capital, is developed, in the same proportion is the proletariat, the modern working class, developed—a class of labourers, who live only so long as they find work, and who find work only as long as their labour increases capital. These labourers, who must sell themselves piecemeal, are a commodity, like every other article of commerce, and are consequently exposed to all the vicissitudes of competition, to all the fluctuations of the market.

Owing to the extensive use of machinery and to division of labour, the work of the proletarians has lost all individual character, and, consequently, all charm for the workman. He becomes an appendage of the machine, and it is only the most simple, most monotonous, and most easily acquired knack, that is required of him. Hence, the cost of production of a workman is restricted, almost entirely, to the means of subsistence that he requires for his maintenance, and for the propagation of his race. But the price of a commodity, and therefore also of labour, is equal to its cost of production. In proportion, therefore, as the repulsiveness of the work increases, the wage decreases. Nay more, in proportion as the use of machinery and division of labour increases, in the same proportion the burden of toil also increases, whether by prolongation of the working hours, by increase of the work exacted in a given time or by increased speed of the machinery, etc.

Modern industry has converted the little workshop of the patriarchal master into the great factory of the industrial capitalist. Masses of labourers, crowded into the factory, are organized like soldiers. As privates of the industrial army they are placed under the command of a perfect hierarchy of officers and sergeants. Not only are they slaves of the bourgeois class, and of the bourgeois State; they are daily and hourly enslaved by the machine, by the overlooker, and, above all, by the individual bourgeois manufacturer himself. The more openly this despotism proclaims gain to be its end and aim, the more petty, the more hateful, and the more embittering it is.

The less the skill and exertion of strength implied in manual labour, in other words, the more modern industry becomes developed, the more is the labour of men superseded by that of women. Differences of age and sex have no longer any distinctive social validity for the working class. All are instruments of labour, more or less expensive to use, according to their age and sex.

No sooner is the exploitation of the labourer by the manufacturer, so far, at an end, and he receives his wages in cash, than he is set upon by the other portions of the bourgeoisie, the landlord, the shopkeeper, the pawnbroker, etc.

The lower strata of the middle class—the small tradespeople, shopkeepers, and retired tradesmen generally, the handicraftsmen and peasants—all these sink gradually into the proletariat, partly because their diminutive capital does not suffice for the scale on which Modern Industry is carried on, and is swamped in the competition with the large capitalists, partly because their specialized skill is rendered worthless by new methods of production. Thus the proletariat is recruited from all classes of the population.

The proletariat goes through various stages of development. With its birth begins its struggle with the bourgeoisie. At first the contest is carried on by individual labourers, then by the workpeople of a factory, then by the operatives of one trade, in one locality, against the individual bourgeois who directly exploits them. They direct their attacks not against the bourgeois conditions of production, but against the instruments of production themselves; they destroy imported wares that compete with their labour, they smash to pieces machinery, they set factories ablaze, they seek to restore by force the vanished status of the workman of the Middle Ages.

At this stage the labourers still form an incoherent mass scattered over the whole country, and broken up by their mutual competition. If anywhere they unite to form more compact bodies, this is not yet the consequence of their own active union, but of the union of the bourgeoisie, which class, in order to attain its own political ends, is compelled to set the whole proletariat in motion, and is moreover yet, for a time, able to do so. At this stage, therefore, the proletarians do not fight their enemies, but the enemies of their enemies, the remnants of absolute monarchy, the landowners, the non-industrial bourgeois, the petty bourgeoisie. Thus the whole historical movement is concentrated in the hands of the bourgeoisie; every victory so obtained is a victory for the bourgeoisie.

But with the development of industry the proletariat not only increases in number; it becomes concentrated in greater masses, its strength grows, and it feels that strength more. The various interests and conditions of life within the ranks of the proletariat are more and more equalized, in proportion as machinery obliterates all distinctions of labour, and nearly everywhere reduces wages to the same low level. The growing competition among the bourgeois, and the resulting commercial crises, make the wages of the workers ever more fluctuating. The unceasing improvement of machinery, ever more rapidly developing, makes their livelihood more and more precarious; the collisions between individual workmen and individual bourgeois take more and more the character of collisions between two classes. Thereupon the workers begin to form combinations (trades unions) against the bourgeois; they club together in order to keep up the rate of wages; they found permanent associations in order to make provision beforehand for these occasional revolts. Here and there the contest breaks out into riots.

Now and then the workers are victorious, but only for a time. The real fruit of their battles lies, not in the immediate result, but in the ever-expanding union of the workers. This union is helped on by the improved means of communication that are created by modern industry and that place the workers of different localities in contact with one another. It was just this contact that was needed to centralize the numerous local struggles, all of the same character, into one national struggle between classes. But every class struggle is a political struggle. And that union, to attain which the burghers of the Middle Ages, with their miserable highways, required centuries, the modern proletarians, thanks to railways, achieve in a few years.

This organization of the proletarians into a class, and consequently into a political party, is continually being upset again by the competition between the workers themselves. But it ever rises up again, stronger, firmer, mightier. It compels legislative recognition of particular interest of the workers, by taking advantage of the divisions among the bourgeoisie itself. Thus the ten-hours' bill in England was carried.

Altogether, collisions between the classes of the old society further in many ways the course of development of the proletariat. The bourgeoisie finds itself involved in a constant battle. At first with the aristocracy; later on, with those portions of the bourgeoisie itself whose interests have become antagonistic to the progress of industry; at all times, with the bourgeoisie of foreign countries. In all these battles it sees itself compelled to appeal to the proletariat, to ask for its help, and thus to drag it into the political arena. The bourgeoisie itself, therefore, supplies the proletariat with its own elements of political and general education, in other words, it furnishes the proletariat with weapons for fighting the bourgeoisie.

Further, as we have already seen, entire sections of the ruling classes are, by the advance of industry, precipitated into the proletariat, or are at least threatened in their conditions of existence. These also supply the proletariat with fresh elements of enlightenment and progress.

Finally, in times when the class struggle nears the decisive hour, the process of dissolution going on within the ruling class, in fact within the whole range of old society, assumes such a violent, glaring character, that a small section of the ruling class cuts itself adrift, and joins the revolutionary class, the class that holds the future in its hands. Just as, therefore, at an earlier period, a section of the nobility went over to the bourgeoisie, so now a portion of the bourgeoisie goes over to the proletariat, and in particular, a portion of the bourgeois ideologists, who have raised themselves to the level of comprehending theoretically the historical movement as a whole.

Of all the classes that stand face to face with the bourgeoisie today, the proletariat alone is a really revolutionary class. The other classes decay and finally disappear in the face of Modern Industry; the proletariat is its special and essential product.

The lower middle class, the small manufacturer, the shopkeeper, the artisan, the peasant, all these fight against the bourgeoisie, to save from extinction their existence as fractions of the middle class. They are therefore not revolutionary, but conservative. Nay more, they are reactionary, for they try to roll back the wheel of history. If by chance they are revolutionary, they are so only in view of their impending transfer into the proletariat; they thus defend not their present, but their future interests, they desert their own standpoint to place themselves at that of the proletariat.

The 'dangerous class', the social scum, that passively rotting mass thrown off by the lowest layers of old society, may, here and there, be swept into the movement by a proletarian revolution; its conditions of life, however, prepare it far more for the part of a bribed tool of reactionary intrigue.

In the conditions of the proletariat, those of old society at large are already virtually swamped. The proletarian is without property; his relation to his wife and children has no longer anything in common with the bourgeois family relations; modern industrial labour, modern subjection to capital, the same in England as in France, in America as in Germany, has stripped him of every trace of national character. Law, morality, religion are to him so many bourgeois prejudices, behind which lurk in ambush just as many bourgeois interests.

All the preceding classes that got the upper hand sought to fortify their already acquired status by subjecting society at large to their conditions of appropriation. The proletarians cannot become masters of the productive forces of society, except by abolishing their own previous mode of appropriation, and thereby also every other previous mode of appropriation. They have nothing of their own to secure and to fortify; their mission is to destroy all previous securities for, and insurances of, individual property.

All previous historical movements were movements of minorities, or in the interests of minorities. The proletarian movement is the self-conscious, independent movement of the immense majority, in the interests of the immense majority. The proletariat, the lowest stratum of our present society, cannot stir, cannot raise itself up, without the whole super-incumbent strata of official society being sprung into the air.

Though not in substance, yet in form, the struggle of the proletariat with the bourgeoisie is at first a national struggle. The proletariat of each country must, of course, first of all settle matters with its own bourgeoisie.

In depicting the most general phases of the development of the proletariat, we traced the more or less veiled civil war, raging within existing society, up to the point where that war breaks out into open revolution, and where the violent overthrow of the bourgeoisie lays the foundation for the sway of the proletariat.

Hitherto, every form of society has been based, as we have already seen, on the antagonism of oppressing and oppressed classes. But in order to oppress a class, certain conditions must be assured to it under which it can, at least, continue its slavish existence. The serf, in the period of serfdom, raised himself to membership in the commune, just as the petty bourgeois, under the yoke of feudal absolutism, managed to develop into a bourgeois. The modern labourer, on the contrary, instead of rising with the progress of industry, sinks deeper and deeper below the conditions of existence of his own class. He becomes a pauper, and pauperism develops more rapidly than population and wealth. And here it becomes evident that the bourgeoisie is unfit any longer to be the ruling class in society, and to impose its conditions of existence upon society as an overriding law. It is unfit to rule because it is incompetent to assure an existence to its slave within his slavery, because it cannot help letting him sink into such a state that it has to feed him, instead of being fed by him. Society can no longer live under this bourgeoisie, in other words, its existence is no longer compatible with society.

The essential condition for the existence, and for the sway, of the bourgeois class is the formation and augmentation of capital; the condition for capital is wage labour. Wage labour rests exclusively on competition between the labourers. The advance of industry, whose involuntary promoter is the bourgeoisie, replaces the isolation of the labourers, due to competition, by their revolutionary combination, due to association. The development of Modern Industry, therefore, cuts from under its feet the very foundation on which the bourgeoisie produces and appropriates products. What the bourgeoisie, therefore, produces, above all, is its own grave-diggers. Its fall and the victory of the proletariat are equally inevitable.

(ii) *Proletarians and Communists*

In what relation do the Communists stand to the proletarians as a whole?

The Communists do not form a separate party opposed to other working-class parties.

They have no interests separate and apart from those of the proletariat as a whole.

They do not set up any sectarian principles of their own, by which to shape and mould the proletarian movement.

The Communists are distinguished from the other working-class parties by this only: (1) In the national struggles of the proletarians of the different countries, they point out and bring to the front the common interests of the entire proletariat, independently of all nationality. (2) In the various stages of development which the struggle of the working class against the bourgeoisie has to pass through, they always and everywhere represent the interests of the movement as a whole.

The Communists, therefore, are on the one hand, practically, the most advanced and resolute section of the working-class parties of every country, that section which pushes forward all others; on the other hand, theoretically, they have over the great mass of the proletariat the advantage of clearly understanding the line of march, the conditions, and the ultimate general results of the proletarian movement.

The immediate aim of the Communists is the same as that of all the other proletarian parties: formation of the proletariat into a class, overthrow of the bourgeois supremacy, conquest of political power by the proletariat.

The theoretical conclusions of the Communists are in no way based on ideas or principles that have been invented, or discovered, by this or that would-be universal reformer.

They merely express, in general terms, actual relations springing from an existing class struggle, from a historical movement going on under our very eyes. The abolition of existing property relations is not at all a distinctive feature of Communism.

All property relations in the past have continually been subject to historical change consequent upon the change in historical conditions.

The French Revolution, for example, abolished feudal property in favour of bourgeois property.

The distinguishing feature of Communism is not the abolition of property generally, but the abolition of bourgeois property. But modern bourgeois private property is the final and most complete expression of the system of producing and appropriating products, that is based on class antagonisms, on the exploitation of the many by the few.

In this sense, the theory of the Communists may be summed up in the single sentence: Abolition of private property.

We Communists have been reproached with the desire of abolishing the right of personally acquiring property as the fruit of a man's own labour, which property is alleged to be the groundwork of all personal freedom, activity, and independence.

Hard-won, self-acquired, self-earned property! Do you mean the property of the petty artisan and of the small peasant, a form of property that preceded the bourgeois form? There is no need to abolish that; the development of industry has to a great extent already destroyed it, and is still destroying it daily.

Or do you mean modern bourgeois private property?

But does wage labour create any property for the labourer? Not a bit. It creates capital, i.e., that kind of property which exploits wage labour, and which cannot increase except upon condition of begetting a new supply of wage labour for fresh exploitation. Property, in its present form, is based on the antagonism of capital and wage labour. Let us examine both sides of this antagonism.

To be a capitalist, is to have not only a purely personal, but a social, status in production. Capital is a collective product, and only by the united action of many members, nay, in the last resort, only by the united action of all members of society, can it be set in motion.

Capital is, therefore, not a personal, it is a social power.

When, therefore, capital is converted into common property, into the property of all members of society, personal property is not thereby transformed into social property. It is only the social character of the property that is changed. It loses its class-character.

Let us now take wage labour.

The average price of wage labour is the minimum wage, i.e., that quantum of the means of subsistence which is absolutely requisite to keep the labourer in bare existence as a labourer. What, therefore, the wage-labourer appropriates by means of his labour merely suffices to prolong and reproduce a bare existence. We by no means intend to abolish this personal appropriation of the products of labour, an appropriation that is made for the maintenance and reproduction of human life, and that leaves no surplus wherewith to command the labour of others. All that we want to do away with is the miserable character of this appropriation, under which the labourer lives merely to increase capital, and is allowed to live only in so far as the interest of the ruling class requires it.

In bourgeois society, living labour is but a means to increase accumulated labour. In Communist society, accumulated labour is but a means to widen, to enrich, to promote the existence of the labourer.

In bourgeois society, therefore, the past dominates the present; in Communist society, the present dominates the past. In bourgeois society capital is independent and has individuality, while the living person is dependent and has no individuality.

And the abolition of this state of things is called by the bourgeois abolition of individuality and freedom! And rightly so. The abolition of bourgeois individuality, bourgeois independence, and bourgeois freedom is undoubtedly aimed at.

By freedom is meant, under the present bourgeois conditions of production, free trade, free selling and buying.

But if selling and buying disappears, free selling and buying disappears also. This talk about free selling and buying, and all the other 'brave words' of our bourgeoisie about freedom in general, have a meaning, if any, only in contrast with restricted selling and buying, with the fettered traders of the Middle Ages, but have no meaning when opposed to the Communistic abolition of buying and selling, of the bourgeois conditions of production, and of the bourgeoisie itself.

You are horrified at our intending to do away with private property. But in your existing society, private property is already done away with for nine-tenths of the population; its existence for the few is solely due to its non-existence in the hands of those nine-tenths. You reproach us, therefore, with intending to do away with a form of property, the necessary condition for whose existence is the non-existence of any property for the immense majority of society.

In one word, you reproach us with intending to do away with your property. Precisely so; that is just what we intend.

From the moment when labour can no longer be converted into capital, money, or rent, into a social power capable of being monopolized, i.e., from the moment when individual property can no longer be transformed into bourgeois property, into capital, from that moment, you say, individuality vanishes.

You must, therefore, confess that by 'individual' you mean no other person than the bourgeois, than the middle-class owner of property. This person must, indeed, be swept out of the way, and made impossible.

Communism deprives no man of the power to appropriate the products of society; all that it does is to deprive him of the power to subjugate the labour of others by means of such appropriation.

It has been objected that upon the abolition of private property all work will cease, and universal laziness will overtake us.

According to this, bourgeois society ought long ago to have gone to the dogs through sheer idleness; for those of its members who work acquire nothing, and those who acquire anything do not work. The whole of this objection is but another expression of the tautology: that there can no longer be any wage labour when there is no longer any capital.

All objections urged against the Communistic mode of producing and appropriating material products have, in the same way, been urged against the Communistic modes of producing and appropriating intellectual products. Just as, to the bourgeois, the disappearance of class property is the disappearance of production itself, so the disappearance of class culture is to him identical with the disappearance of all culture.

That culture, the loss of which he laments, is, for the enormous majority, a mere training to act as a machine.

But don't wrangle with us so long as you apply, to our intended abolition of bourgeois property, the standard of your bourgeois notions of freedom, culture, law, etc. Your very ideas are but the outgrowth of the conditions of your bourgeois production and bourgeois property, just as your jurisprudence is but the will of your class made into a law for all, a will whose essential character and direction are determined by the economical conditions of existence of your class.

The selfish misconception that induces you to transform into eternal laws of nature and of reason the social forms springing from your present mode of production and form of property—historical relations that rise and disappear in the progress of production—this misconception you share with every ruling class that has preceded you. What you see clearly in the case of ancient property, what you admit in the case of feudal property, you are of course forbidden to admit in the case of your own bourgeois form of property.

Abolition of the family! Even the most radical flare up at this infamous proposal of the Communists.

On what foundation is the present family, the bourgeois family, based? On capital, on private gain. In its completely developed form this family exists only among the bourgeoisie. But this state of things finds its complement in the practical absence of the family among the proletarians, and in public prostitution.

The bourgeois family will vanish as a matter of course when its complement vanishes, and both will vanish with the vanishing of capital.

Do you charge us with wanting to stop the exploitation of children by their parents? To this crime we plead guilty.

But, you will say, we destroy the most hallowed of relations, when we replace home education by social.

And your education! Is not that also social, and determined by the social conditions under which you educate, by the intervention, direct or indirect, of society, by means of schools, etc.? The Communists have not invented the intervention of society in education; they do but seek to alter the character of that intervention, and to rescue education from the influence of the ruling class.

The bourgeois claptrap about the family and education, about the hallowed co-relation of parent and child, becomes all the more disgusting, the more, by the action of Modern Industry, all family ties among the proletarians are torn asunder, and their children transformed into simple articles of commerce and instruments of labour.

But you Communists would introduce community of women, screams the whole bourgeoisie in chorus.

The bourgeois sees in his wife a mere instrument of production. He hears that the instruments of production are to be exploited in common, and naturally, can come to no other conclusion than that the lot of being common to all will likewise fall to the women.

He has not even a suspicion that the real point aimed at is to do away with the status of women as mere instruments of production.

For the rest, nothing is more ridiculous than the virtuous indignation of our bourgeois at the community of women which, they pretend, is to be openly and officially established by the Communists. The Communists have no need to introduce community of women; it has existed almost from time immemorial.

Our bourgeois, not content with having the wives and daughters of their proletarians at their disposal, not to speak of common prostitutes, take the greatest pleasure in seducing each other's wives.

Bourgeois marriage is in reality a system of wives in common and thus, at the most, what the Communists might possibly be reproached with, is that they desire to introduce, in substitution for a hypocritically concealed, an openly legalized, community of women. For the rest, it is self-evident that the abolition of the present system of production must bring with it the abolition of the community of women springing from that system, i.e., of prostitution both public and private.

The Communists are further reproached with desiring to abolish countries and nationality.

The working men have no country. We cannot take from them what
they have not got. Since the proletariat must first of all acquire political
supremacy, must rise to be the leading class of the nation, must constitute
itself *the* nation, it is, so far, itself national, though not in the bourgeois
sense of the world.

National differences and antagonisms between peoples are daily more
and more vanishing, owing to the development of the bourgeoisie, to
freedom of commerce, to the world market, to uniformity in the mode of
production and in the conditions of life corresponding thereto.

The supremacy of the proletariat will cause them to vanish still faster.
United action, of the leading civilized countries at least, is one of the first
conditions for the emancipation of the proletariat.

In proportion as the exploitation of one individual by another is put an
end to, the exploitation of one nation by another will also be put an end
to. In proportion as the antagonism between classes within the nation
vanishes, the hostility of one nation to another will come to an end.

The charges against Communism made from a religious, a philosophical,
and, generally, from an ideological standpoint are not deserving of serious
examination.

Does it require deep intuition to comprehend that man's ideas, views,
and conceptions, in one word, man's consciousness, changes with every
change in the conditions of his material existence, in his social relation,
and in his social life?

What else does the history of ideas prove, than that intellectual production changes its character in proportion as material production is
changed? The ruling ideas of each age have ever been the ideas of its ruling
class.

When people speak of ideas that revolutionize society, they do but express
the fact, that within the old society, the elements of a new one have been
created, and that the dissolution of the old ideas keeps even pace with the
dissolution of the old conditions of existence.

When the ancient world was in its last throes, the ancient religions
were overcome by Christianity. When Christian ideas succumbed in the
eighteenth century to rationalist ideas, feudal society fought its death battle
with the then revolutionary bourgeoisie. The ideas of religious liberty
and freedom of conscience merely gave expression to the sway of free
competition within the domain of knowledge.

'Undoubtedly,' it will be said, 'religious, moral, philosophical, and juridical ideas have been modified in the course of historical development. But

religion, morality, philosophy, political science, and law constantly survived this change.'

'There are, besides, eternal truths, such as Freedom, Justice, etc., that are common to all states of society. But Communism abolishes eternal truths, it abolishes all religion and all morality, instead of constituting them on a new basis; it therefore acts in contradiction to all past historical experience.'

What does this accusation reduce itself to? The history of all past society has consisted in the development of class antagonisms, antagonisms that assumed different forms at different epochs.

But whatever form they may have taken, one fact is common to all past ages, viz., the exploitation of one part of society by the other. No wonder, then, that the social consciousness of past ages, despite all the multiplicity and variety it displays, moves within certain common forms, or general ideas, which cannot completely vanish except with the total disappearance of class antagonisms.

The Communist revolution is the most radical rupture with traditional property relations; no wonder that its development involves the most radical rupture with traditional ideas.

But let us have done with the bourgeois objections to Communism.

We have seen above, that the first step in the revolution by the working class is to raise the proletariat to the position of ruling class, to win the battle of democracy.

The proletariat will use its political supremacy to wrest, by degrees, all capital from the bourgeoisie, to centralize all instruments of production in the hands of the State, i.e., of the proletariat organized as the ruling class; and to increase the total of productive forces as rapidly as possible.

Of course, in the beginning this cannot be effected except by means of despotic inroads on the rights of property, and on the conditions of bourgeois production; by means of measures, therefore, which appear economically insufficient and untenable, but which, in the course of the movement, outstrip themselves, necessitate further inroads upon the old social order, and are unavoidable as a means of entirely revolutionizing the mode of production.

These measures will of course be different in different countries.

Nevertheless, in the most advanced countries, the following will be pretty generally applicable.

1. Abolition of property in land and application of all rents of land to public purposes.
2. A heavy progressive or graduated income tax.

3. Abolition of all right of inheritance.
4. Confiscation of the property of all emigrants and rebels.
5. Centralization of credit in the hands of the State, by means of a national bank with State capital and an exclusive monopoly.
6. Centralization of the means of communication and transport in the hands of the State.
7. Extension of factories and instruments of production owned by the State; the bringing into cultivation of wastelands, and the improvement of the soil generally in accordance with a common plan.
8. Equal liability of all to labour. Establishment of industrial armies, especially for agriculture.
9. Combination of agriculture with manufacturing industries; gradual abolition of the distinction between town and country, by a more equable distribution of the population over the country.
10. Free education for all children in public schools. Abolition of children's factory labour in its present form. Combination of education with industrial production, etc., etc.

When, in the course of development, class distinctions have disappeared, and all production has been concentrated in the hands of associated individuals, the public power will lose its political character. Political power, properly so called, is merely the organized power of one class for oppressing another. If the proletariat during its contest with the bourgeoisie is compelled, by the force of circumstances, to organize itself as a class, if, by means of a revolution, it makes itself the ruling class, and, as such, sweeps away by force the old conditions of production, then it will, along with these conditions, have swept away the conditions for the existence of class antagonisms and of classes generally, and will thereby have abolished its own supremacy as a class.

In place of the old bourgeois society, with its classes and class antagonisms, we shall have an association, in which the free development of each is the condition for the free development of all.

(iii) Socialist and Communist Literature

1. Reactionary Socialism

(a) Feudal Socialism. Owing to their historical position, it became the vocation of the aristocracies of France and England to write pamphlets against modern bourgeois society. In the French revolution of July 1830, and in the English reform agitation, these aristocracies again succumbed to the hateful upstart. Thenceforth, a serious political contest was altogether out of question. A literary battle alone remained possible. But even in the domain of literature the old cries of the restoration period had become impossible.

In order to arouse sympathy, the aristocracy were obliged to lose sight, apparently, of their own interests, and to formulate their indictment against the bourgeoisie in the interest of the exploited working class alone. Thus the aristocracy took their revenge by singing lampoons on their new master, and whispering in his ears sinister prophecies of coming catastrophe.

In this way arose Feudal Socialism: half lamentation, half lampoon; half echo of the past, half menace of the future; at times, by its bitter, witty, and incisive criticism, striking the bourgeoisie to the very heart's core; but always ludicrous in its effect, through total incapacity to comprehend the march of modern history.

The aristocracy, in order to rally the people to them, waved the proletarian alms-bag in front for a banner. But the people, so often as it joined them, saw on their hindquarters the old feudal coats of arms, and deserted with loud and irreverent laughter.

One section of the French Legitimists and 'Young England' exhibited this spectacle.

In pointing out that their mode of exploitation was different to that of the bourgeoisie, the feudalists forget that they exploited under circumstances and conditions that were quite different, and that are now antiquated. In showing that, under their rule, the modern proletariat never existed, they forget that the modern bourgeoisie is the necessary offspring of their own form of society.

For the rest, so little do they conceal the reactionary character of their criticism that their chief accusation against the bourgeoisie amounts to this, that under the bourgeois regime a class is being developed which is destined to cut up root and branch the old order of society.

What they upbraid the bourgeoisie with is not so much that it creates a proletariat, as that it creates a revolutionary proletariat.

In political practice, therefore, they join in all coercive measures against the working class; and in ordinary life, despite their high-falutin phrases, they stoop to pick up the golden apples dropped from the tree of industry, and to barter truth, love, and honour for traffic in wool, sugar-beet, and potato spirits.

As the parson has ever gone hand in hand with the landlord, so has Clerical Socialism with Feudal Socialism.

Nothing is easier than to give Christian asceticism a Socialist tinge. Has not Christianity declaimed against private property, against marriage, against the State? Has it not preached in the place of these charity and poverty, celibacy and mortification of the flesh, monastic life and Mother

Church? Christian Socialism is but the holy water with which the priest consecrates the heart-burnings of the aristocrat.

(b) *Petty-bourgeois Socialism.* The feudal aristocracy was not the only class that was ruined by the bourgeoisie, not the only class whose conditions of existence pined and perished in the atmosphere of modern bourgeois society. The medieval burgesses and the small peasant proprietors were the precursors of the modern bourgeoisie. In those countries which are but little developed, industrially and commercially, these two classes still vegetate side by side with the rising bourgeoisie.

In countries where modern civilization has become fully developed, a new class of petty bourgeois has been formed, fluctuating between proletariat and bourgeoisie and ever renewing itself as a supplementary part of bourgeois society. The individual members of this class, however, are being constantly hurled down into the proletariat by the action of competition, and, as modern industry develops, they even see the moment approaching when they will completely disappear as an independent section of modern society, to be replaced, in manufactures, agriculture, and commerce, by overlookers, bailiffs, and shopmen.

In countries like France, where the peasants constitute far more than half of the population, it was natural that writers who sided with the proletariat against the bourgeoisie should use, in their criticism of the bourgeois regime, the standard of the peasant and petty bourgeois, and from the standpoint of these intermediate classes should take up the cudgels for the working class. Thus arose petty-bourgeois Socialism. Sismondi was the head of this school, not only in France but also in England.

This school of Socialism dissected with great acuteness the contradictions in the conditions of modern production. It laid bare the hypocritical apologies of economists. It proved, incontrovertibly, the disastrous effects of machinery and division of labour; the concentration of capital and land in a few hands; overproduction and crises; it pointed out the inevitable ruin of the petty bourgeois and peasant, the misery of the proletariat, the anarchy in production, the crying inequalities in the distribution of wealth, the industrial war of extermination between nations, the dissolution of old moral bonds, of the old family relations, of the old nationalities.

In its positive aims, however, this form of socialism aspires either to restoring the old means of production and of exchange, and with them the old property relations, and the old society, or to cramping the modern means of production and of exchange, within the framework of the old property relations that have been, and were bound to be, exploded by those means. In either case, it is both reactionary and Utopian.

Its last words are: corporate guilds for manufacture, patriarchal relations in agriculture.

Ultimately, when stubborn historical facts had dispersed all intoxicating effects of self-deception, this form of Socialism ended in a miserable fit of the blues.

(c) *German, or 'True', Socialism.* The Socialist and Communist literature of France, a literature that originated under the pressure of a bourgeoisie in power, and that was the expression of the struggle against this power, was introduced into Germany at a time when the bourgeoisie, in that country, had just begun its contest with feudal absolutism.

German philosophers, would-be philosophers, and *beaux esprits* eagerly seized on this literature, only forgetting that when these writings immigrated from France into Germany, French social conditions had not immigrated along with them. In contact with German social conditions, this French literature lost all its immediate practical significance, and assumed a purely literary aspect. Thus, to the German philosophers of the eighteenth century, the demands of the first French Revolution were nothing more than the demands of 'Practical Reason' in general, and the utterance of the will of the revolutionary French bourgeoisie signified in their eyes the laws of pure Will, of Will as it was bound to be, of true human Will generally.

The work of the German literati consisted solely in bringing the new French ideas into harmony with their ancient philosophical conscience, or rather, in annexing the French ideas without deserting their own philosophic point of view.

This annexation took place in the same way in which a foreign language is appropriated, namely, by translation.

It is well known how the monks wrote silly lives of Catholic saints over the manuscripts on which the classical works of ancient heathendom had been written. The German literati reversed this process with the profane French literature. They wrote their philosophical nonsense beneath the French original. For instance, beneath the French criticism of the economic functions of money, they wrote 'Alienation of Humanity', and beneath the French criticism of the bourgeois State they wrote 'Dethronement of the Category of the General', and so forth.

The introduction of these philosophical phrases at the back of the French historical criticisms they dubbed 'Philosophy of Actions', 'True Socialism', 'German Science of Socialism', 'Philosophical Foundation of Socialism', and so on.

The French Socialist and Communist literature was thus completely emasculated. And, since it ceased in the hands of the German to express the struggle of one class with the other, he felt conscious of having overcome 'French one-sidedness' and of representing, not true requirements, but the requirements of Truth; not the interests of the proletariat, but the interests of Human Nature, of Man in general, who belongs to no class, has no reality, who exists only in the misty realm of philosophical fantasy.

This German Socialism, which took its schoolboy task so seriously and solemnly, and extolled its poor stock-in-trade in such mountebank fashion, meanwhile gradually lost its pedantic innocence.

The fight of the German, and, especially, of the Prussian, bourgeoisie against feudal aristocracy and absolute monarchy, in other words, the liberal movement, became more earnest.

By this, the long wished-for opportunity was offered to 'True' Socialism of confronting the political movement with the Socialist demands, of hurling the traditional anathemas against liberalism, against representative government, against bourgeois competition, bourgeois freedom of the press, bourgeois legislation, bourgeois liberty and equality, and of preaching to the masses that they had nothing to gain, and everything to lose, by this bourgeois movement. German Socialism forgot, in the nick of time, that the French criticism, whose silly echo it was, presupposed the existence of modern bourgeois society, with its corresponding economic conditions of existence, and the political constitution adapted thereto, the very things whose attainment was the object of the pending struggle in Germany.

To the absolute governments, with their following of parsons, professors, country squires, and officials, it served as a welcome scarecrow against the threatening bourgeoisie.

It was a sweet finish after the bitter pills of floggings and bullets with which these same governments, just at that time, dosed the German working-class risings.

While this 'True' Socialism thus served the governments as a weapon for fighting the German bourgeoisie, it at the same time, directly represented a reactionary interest, the interest of the German Philistines. In Germany the petty-bourgeois class, a relic of the sixteenth century, and since then constantly cropping up again under various forms, is the real social basis of the existing state of things.

To preserve this class is to preserve the existing state of things in Germany. The industrial and political supremacy of the bourgeoisie threatens it with certain destruction; on the one hand, from the concentration of capital; on the other, from the rise of a revolutionary proletariat. 'True' Socialism appeared to kill these two birds with one stone. It spread like an epidemic.

The robe of speculative cobwebs, embroidered with flowers of rhetoric, steeped in the dew of sickly sentiment, this transcendental robe in which the German Socialists wrapped their sorry 'eternal truths', all skin and bone, served wonderfully to increase the sale of their goods among such a public.

And on its part, German Socialism recognized, more and more, its own calling as the bombastic representative of the petty-bourgeois Philistine.

It proclaimed the German nation to be the model nation, and the German petty Philistine to be the typical man. To every villainous meanness of this model man it gave a hidden, higher, Socialistic interpretation, the exact contrary of its real character. It went to the extreme length of directly opposing the 'brutally destructive' tendency of Communism, and of proclaiming its supreme and impartial contempt of all class struggles. With very few exceptions, all the so-called Socialist and Communist publications that now (1847) circulate in Germany belong to the domain of this foul and enervating literature.

2. *Conservative, or Bourgeois, Socialism*

A part of the bourgeoisie is desirous of redressing social grievances, in order to secure the continued existence of bourgeois society.

To this section belong economists, philanthropists, humanitarians, improvers of the condition of the working class, organizers of charity, members of societies for the prevention of cruelty to animals, temperance fanatics, hole-and-corner reformers of every imaginable kind. This form of Socialism has, moreover, been worked out into complete systems.

We may cite Proudhon's *Philosophie de la misère* as an example of this form.

The Socialistic bourgeois want all the advantages of modern social conditions without the struggles and dangers necessarily resulting from them. They desire the existing state of society minus its revolutionary and disintegrating elements. They wish for a bourgeoisie without a proletariat. The bourgeoisie naturally conceives the world in which it is supreme to be the best; and bourgeois Socialism develops this comfortable conception into various more or less complete systems. In requiring the proletariat to carry out such a system, and thereby to march straightaway into the social New Jerusalem, it but requires in reality that the proletariat should remain within the bounds of existing society, but should cast away all its hateful ideas concerning the bourgeoisie.

A second and more practical, but less systematic, form of this Socialism sought to depreciate every revolutionary movement in the eyes of the working class, by showing that no mere political reform, but only a change in the material conditions of existence, in economical relations, could be of any advantage to them. By changes in the material conditions of existence, this form of Socialism, however, by no means understands abolition of the bourgeois relations of production, an abolition that can be effected only by a revolution, but administrative reforms, based on the continued existence of these relations; reforms, therefore, that in no respect affect the relations between capital and labour, but, at the best, lessen the cost, and simplify the administrative work, of bourgeois government.

Bourgeois Socialism attains adequate expression when, and only when, it becomes a mere figure of speech.

Free trade: for the benefit of the working class. Protective duties: for the benefit of the working class. Prison reform: for the benefit of the working class. This is the last word and the only seriously meant word of bourgeois Socialism.

It is summed up in the phrase: the bourgeois is a bourgeois—for the benefit of the working class.

3. Critical-Utopian Socialism and Communism

We do not here refer to that literature which, in every great modern revolution, has always given voice to the demands of the proletariat, such as the writings of Babeuf and others.

The first direct attempts of the proletariat to attain its own ends, made in times of universal excitement, when feudal society was being overthrown, these attempts necessarily failed, owing to the then undeveloped state of the proletariat, as well as to the absence of the economic conditions for its emancipation, conditions that had yet to be produced, and could be produced by the impending bourgeois epoch alone. The revolutionary literature that accompanied these first movements of the proletariat had necessarily a reactionary character. It inculcated universal asceticism and social levelling in its crudest form.

The Socialist and Communist systems properly so called, those of Saint-Simon, Fourier, Owen, and others, spring into existence in the early un-developed period, described above, of the struggle between proletariat and bourgeoisie (see Section I. Bourgeoisie and Proletariat).

The founders of these systems see, indeed, the class antagonisms, as well as the action of the decomposing elements, in the prevailing form of society. But the proletariat, as yet in its infancy, offers to them the spectacle of a class without any historical initiative or any independent political movement.

Since the development of class antagonism keeps even pace with the development of industry, the economic situation, as they find it, does not as yet offer to them the material conditions for the emancipation of the proletariat. They therefore search after a new social science, after new social laws, that are to create these conditions.

Historical action is to yield to their personal inventive action, historically created conditions of emancipation to fantastic ones, and the gradual, spontaneous class-organization of the proletariat to an organization of society specially contrived by these inventors. Future history resolves itself, in their eyes, into the propaganda and the practical carrying out of their social plans.

In the formation of their plans they are conscious of caring chiefly for the interests of the working class, as being the most suffering class. Only from the point of view of being the most suffering class does the proletariat exist for them.

The undeveloped state of the class struggle, as well as their own sur-roundings, causes Socialists of this kind to consider themselves far superior to all class antagonisms. They want to improve the condition of every member of society, even that of the most favoured. Hence, they habitually appeal to society at large, without distinction of class; nay, by preference, to the ruling class. For how can people, when once they understand their system, fail to see in it the best possible plan of the best possible state of society?

Hence, they reject all political, and especially all revolutionary, action; they wish to attain their ends by peaceful means, and endeavour, by small experiments, necessarily doomed to failure, and by the force of example, to pave the way for the new social gospel.

Such fantastic pictures of future society, painted at a time when the proletariat is still in a very undeveloped state and has but a fantastic conception of its own position, correspond with the first instinctive yearn-ings of that class for a general reconstruction of society.

But these Socialist and Communist publications contain also a critical element. They attack every principle of existing society. Hence they are full of the most valuable materials for the enlightenment of the working class. The practical measures proposed in them—such as the abolition of the distinction between town and country, of the family, of the carrying on of industries for the account of private individuals, and of the wage system, the proclamation of social harmony, the conversion of the functions of the State into a mere superintendence of production, all these proposals point solely to the disappearance of class antagonisms which were, at that time, only just cropping up, and which in these publications, are recognized in their earliest, indistinct, and undefined forms only. These proposals, therefore, are of a purely Utopian character.

The significance of Critical-Utopian Socialism and Communism bears an inverse relation to historical development. In proportion as the modern class struggle develops and takes definite shape, this fantastic standing apart from the contest, these fantastic attacks on it, lose all practical value and all theoretical justification. Therefore, although the originators of these systems were, in many respects, revolutionary, their disciples have, in every case, formed mere reactionary sects. They hold fast by the original views of their masters, in opposition to the progressive historical development of the proletariat. They, therefore, endeavour, and that consistently, to deaden the class struggle and to reconcile the class antagonisms. They still dream of experimental realization of their social Utopias, of founding isolated '*phalanstères*', of establishing 'Home Colonies', of setting up a 'Little Icaria'—duodecimo editions of the New Jerusalem—and to realize all these castles in the air they are compelled to appeal to the feelings and purses of the bourgeois. By degrees they sink into the category of the reactionary conservative Socialists depicted above, differing from these only by more systematic pedantry, and by their fanatical and superstitious belief in the miraculous effects of their social science.

They, therefore, violently oppose all political action on the part of the working class; such action, according to them, can only result from blind unbelief in the new gospel.

The Owenites in England and the Fourierists in France, respectively, oppose the Chartists and the *Réformistes*.

(iv) *Position of the Communists in Relation to the Various Existing Opposition Parties*

Section II has made clear the relations of the Communists to the existing working-class parties, such as the Chartists in England and the Agrarian Reformers in America.

The Communists fight for the attainment of the immediate aims, for the enforcement of the momentary interests of the working class; but in the movement of the present, they also represent and take care of the future of that movement. In France the Communists ally themselves with the Social Democrats, against the conservative and radical bourgeoisie, reserving, however, the right to take up a critical position in regard to phrases and illusions traditionally handed down from the great Revolution.

In Switzerland they support the Radicals, without losing sight of the fact that this party consists of antagonistic elements, partly of Democratic Socialists, in the French sense, partly of radical bourgeois.

In Poland they support the party that insists on an agrarian revolution as the prime condition for national emancipation, that party which fomented the insurrection of Cracow in 1846.

In Germany they fight with the bourgeoisie whenever it acts in a revolutionary way, against the absolute monarchy, the feudal squirearchy, and the petty bourgeoisie.

But they never cease, for a single instant, to instil into the working class the clearest possible recognition of the hostile antagonism between bourgeoisie and proletariat, in order that the German workers may straightaway use, as so many weapons against the bourgeoisie, the social and political conditions that the bourgeoisie must necessarily introduce along with its supremacy, and in order that, after the fall of the reactionary classes in Germany, the fight against the bourgeoisie itself may immediately begin.

The Communists turn their attention chiefly to Germany, because that country is on the eve of a bourgeois revolution that is bound to be carried out under more advanced conditions of European civilization, and with a much more developed proletariat, than that of England was in the seventeenth, and of France in the eighteenth century, and because the bourgeois revolution in Germany will be but the prelude to an immediately following proletarian revolution.

In short, the Communists everywhere support every revolutionary movement against the existing social and political order of things.

In all these movements they bring to the front, as the leading question in each, the property question, no matter what its degree of development at the time.

Finally, they labour everywhere for the union and agreement of the democratic parties of all countries.

The Communists disdain to conceal their views and aims. They openly declare that their ends can be attained only by the forcible overthrow of all existing social conditions. Let the ruling classes tremble at a Communistic revolution. The proletarians have nothing to lose but their chains. They have a world to win.

WORKING MEN OF ALL COUNTRIES, UNITE!

(*b*) FROM THE 1872 PREFACE

However much the state of things may have altered during the last twenty-five years, the general principles laid down in this Manifesto are, on the whole, as correct today as ever. Here and there some detail might be improved. The practical application of the principles will depend, as the Manifesto itself states, everywhere and at all times, on the historical conditions for the time being existing, and, for that reason, no special stress is laid on the revolutionary measures proposed at the end of Section II. That passage would, in many respects, be very differently worded today. In view of the gigantic strides of Modern Industry in the last twenty-five years, and of the accompanying improved and extended party organization of the working class, in view of the practical experience gained, first in the February Revolution, and then, still more, in the Paris Commune, where the proletariat for the first time held political power for two whole months, this programme has in some details become antiquated. One thing especially was proved by the Commune, viz., that 'the working class cannot simply lay hold of the ready-made State machinery, and wield it for its own purposes'. (See *The Civil War in France; Address of the General Council of the International Working Men's Association*, London, Truelove, 1871, p. 15, where this point is further developed.) Further, it is self-evident that the criticism of socialist literature is deficient in relation to the present time, because it comes down only to 1847; also, that the remarks on the relation of the Communists to the various opposition parties (Section IV), although in principle still correct, yet in practice are antiquated, because the political situation has been entirely changed, and the progress of history has swept from off the earth the greater portion of the political parties there enumerated.

But, then, the Manifesto has become a historical document which we have no longer any right to alter. A subsequent edition may perhaps appear with an introduction bridging the gap from 1847 to the present day; this reprint was too unexpected to leave us time for that. . . .

(c) FROM THE 1882 PREFACE

What a limited field the proletarian movement still occupied at that time (December 1847) is most clearly shown by the last section of the Manifesto: the position of the Communists in relation to the various opposition parties in the various countries. Precisely Russia and the United States are missing here. It was the time when Russia constituted the last great reserve of all European reaction, when the United States absorbed the surplus proletarian forces of Europe through immigration. Both countries provided Europe with raw materials and were at the same time markets for the sale of its industrial products. At that time both were, therefore, in one way or another, pillars of the existing European order.

How very different today! Precisely European immigration fitted North America for a gigantic agricultural production, whose competition is shaking the very foundations of European landed property—large and small. In addition it enabled the United States to exploit its tremendous industrial resources with an energy and on a scale that must shortly break the industrial monopoly of Western Europe, and especially of England, existing up to now. Both circumstances react in revolutionary manner upon America itself. Step by step the small and middle landownership of the farmers, the basis of the whole political constitution, is succumbing to the competition of giant farms; simultaneously, a mass proletariat and a fabulous concentration of capitals are developing for the first time in the industrial regions.

And now Russia! During the Revolution of 1848–9 not only the European princes, but the European bourgeois as well, found their only salvation from the proletariat, just beginning to awaken, in Russian intervention. The tsar was proclaimed the chief of European reaction. Today he is a prisoner of war of the revolution, in Gatchina, and Russia forms the vanguard of revolutionary action in Europe.

The Communist Manifesto had as its object the proclamation of the inevitably impending dissolution of modern bourgeois property. But in Russia we find, face to face with the rapidly developing capitalist swindle and bourgeois landed property, just beginning to develop, more than half the land owned in common by the peasants. Now the question is: Can the Russian *obshchina*, though greatly undermined, yet a form of the primeval common ownership of land, pass directly to the higher form of Communist common ownership? Or, on the contrary, must it first pass through the same process of dissolution as constitutes the historical evolution of the West?

The only answer to that possible today is this: If the Russian Revolution becomes the signal for a proletarian revolution in the West, so that both complement each other, the present Russian common ownership of land may serve as the starting-point for a Communist development.

3.   The Destiny of Capitalism

In this chapter we consider the influence of the growth of capital on the lot of the labouring class. The most important factor in this inquiry is the composition of capital and the changes it undergoes in the course of the process of accumulation.

The composition of capital is to be understood in a twofold sense. On the side of value, it is determined by the proportion in which it is divided into constant capital or value of the means of production, and variable capital or value of labour-power, the sum total of wages. On the side of material, as it functions in the process of production, all capital is divided into means of production and living labour-power. This latter composition is determined by the relation between the mass of the means of production employed, on the one hand, and the mass of labour necessary for their employment on the other. I call the former the value-composition, the latter the technical composition of capital. Between the two there is a strict correlation. To express this, I call the value-composition of capital, in so far as it is determined by its technical composition and mirrors the changes of the latter, the organic composition of capital. Wherever I refer to the composition of capital, without further qualification, its organic composition is always understood.

The many individual capitals invested in a particular branch of production have, one with another, more or less different compositions. The average of their individual compositions gives us the composition of the total capital in this branch of production. Lastly, the average of these averages, in all branches of production, gives us the composition of the total social capital of a country, and with this alone are we, in the last resort, concerned in the following investigation.

Growth of capital involves growth of its variable constituent or of the part invested in labour power. A part of the surplus value turned into additional capital must always be re-transformed into variable capital, or additional labour-fund. If we suppose that, all other circumstances remaining the same, the composition of capital also remains constant (i.e., that a definite mass of means of production constantly needs the same mass of labour-power to set it in motion), then the demand for labour and the subsistence-fund of the labourers clearly increase in the same proportion as the capital, and the more rapidly, the more rapidly the capital increases. Since the capital produces yearly a surplus value, of which one part is yearly added to the original capital; since this increment itself grows yearly along with the augmentation of the capital already functioning; since lastly, under special stimulus to enrichment, such as the opening of new markets, or of new spheres for the outlay of capital in consequence of newly developed social wants, etc., the scale of accumulation may be suddenly extended, merely by a change in the division of the surplus value or surplus product into capital and revenue, the requirements of accumulating capital may exceed the increase of labour-power or of the number of labourers; the demand for labourers may exceed the supply, and, therefore, wages may rise. This must, indeed, ultimately be the case if the conditions supposed above continue. For since in each year more labourers are employed than in its predecessor, sooner or later a point must be reached at which the requirements of accumulation begin to surpass the customary supply of labour, and, therefore, a rise of wages takes place. A lamentation on this score was heard in England during the whole of the fifteenth, and the first half of the eighteenth, centuries. The more or less favourable circumstances in which the wage-working class supports and multiplies itself, in no way alter the fundamental character of capitalist production. As simple reproduction constantly reproduces the capital-relation itself, i.e., the relation of capitalists on the one hand, and wage-workers on the other, so reproduction on a progressive scale, i.e., accumulation, reproduces the capital-relation on a progressive scale, more capitalists or larger capitalists at this pole, more wage-workers at that. The reproduction of a mass of labour-power, which must incessantly reincorporate itself with capital for the capital's self-expansion; which cannot get free from capital, and whose enslavement to capital is only concealed by the variety of individual capitalists to whom it sells itself, this reproduction of labour power forms, in fact, an essential of the reproduction of capital itself. Accumulation of capital is, therefore, increase of the proletariat.

The law of capitalist production, that is at the bottom of the pretended 'natural law of population', reduces itself simply to this: The correlation between accumulation of capital and rate of wages is nothing else than the correlation between the unpaid labour transformed into capital, and the additional paid labour necessary for the setting in motion of this additional capital. It is therefore in no way a relation between two magnitudes, independent one of the other: on the one hand, the magnitude of the capital; on the other, the number of the labouring population; it is rather, at bottom, only the relation between the unpaid and the paid labour of the same labouring population. If the quantity of unpaid labour supplied by the working class, and accumulated by the capitalist class, increases so rapidly that its conversion into capital requires an extraordinary addition of paid labour, then wages rise, and, all other circumstances remaining equal, the unpaid labour diminishes in proportion. But as soon as this diminution touches the point at which the surplus labour that nourishes capital is no longer supplied in normal quantity, a reaction sets in: a smaller part of revenue is capitalized, accumulation lags, and the movement of rise in wages receives a check. The rise of wages therefore is confined within limits that not only leave intact the foundations of the capitalistic system, but also secure its reproduction on a progressive scale. The law of capitalistic accumulation, metamorphosed by economists into a pretended law of Nature, in reality merely states that the very nature of accumulation excludes every diminution in the degree of exploitation of labour, and every rise in the price of labour, which could seriously imperil the continual reproduction, on an ever-enlarging scale, of the capitalistic relation. It cannot be otherwise in a mode of production in which the labourer exists to satisfy the needs of self-expansion of existing values, instead of, on the contrary, material wealth existing to satisfy the needs of development on the part of the labourer. As in religion man is governed by the products of his own brain, so in capitalistic production, he is governed by the products of his own hand. . . .

But if a surplus labouring population is a necessary product of accumulation or of the development of wealth on a capitalist basis, this surplus population becomes, conversely, the lever of capitalistic accumulation, nay, a condition of existence of the capitalist mode of production. It forms a disposable industrial reserve army, that belongs to capital quite as absolutely as if the latter had bred it at its own cost. Independently of the limits of the actual increase of population, it creates, for the changing needs of the self-expansion of capital, a mass of human material always ready for exploitation. With accumulation, and the development of the productiveness of labour that accompanies it, the power of sudden expansion of capital grows also; it grows, not merely because the elasticity of the capital already functioning increases, not merely because the absolute wealth of society expands, of which capital only forms an elastic part, not merely because credit, under every special stimulus, at once places an unusual part of this wealth at the disposal of production in the form of additional capital; it grows, also, because the technical conditions of the process of production themselves—machinery, means of transport, etc.— now admit of the rapidest transformation of masses of surplus product into additional means of production. The mass of social wealth, overflowing with the advance of accumulation, and transformable into additional capital, thrusts itself frantically into old branches of production, whose market suddenly expands, or into newly formed branches, such as railways, etc., the need for which grows out of the development of the old ones. In all such cases, there must be the possibility of throwing great masses of men suddenly on the decisive points without injury to the scale of production in other spheres. Overpopulation supplies these masses. The course characteristic of modern industry, viz., a decennial cycle (interrupted by smaller oscillations) of periods of average activity, production at high pressure, crisis, and stagnation, depends on the constant formation, the greater or less absorption, and the re-formation of the industrial reserve army or surplus population. In their turn, the varying phases of the industrial cycle recruit the surplus population, and become one of the most energetic agents of its reproduction. This peculiar course of modern industry, which occurs in no earlier period of human history, was also impossible in the childhood of capitalist production. The composition of capital changed but very slowly. With its accumulation, therefore, there kept pace, on the whole, a corresponding growth in the demand for labour. Slow as was the advance of accumulation compared with that of more modern times, it found a check in the natural limits of the exploitable labouring population, limits which could only be got rid of by forcible means to be mentioned later. The expansion by fits and starts of the scale of production is the preliminary to its equally sudden contraction; the latter again evokes the former, but the former is impossible without disposable human material, without an increase in the number of labourers independently of the absolute growth of the population. This increase is effected by the simple process that constantly 'sets free' a part of the labourers; by methods which lessen the number of labourers employed in proportion to the increased production. The whole form of the movement of modern industry depends, therefore, upon the constant transformation of a part of the labouring population into unemployed or half-employed hands. The superficiality of Political Economy shows itself in the fact that it looks upon the expansion and contraction of credit, which is a mere symptom of the periodic changes of the industrial cycle, as their cause. As the heavenly bodies, once thrown into a certain definite motion, always repeat this, so is it with social production as soon as it is once thrown into this movement of alternate expansion and contraction. Effects, in their turn, become causes, and the varying accidents of the whole process, which always reproduces its own conditions, take on the form of periodicity. When this periodicity is once consolidated, even Political Economy then sees that the production of a relative surplus population—i.e., surplus with regard to the average needs of the self-expansion of capital—is a necessary condition of modern industry....

The industrial reserve army, during the periods of stagnation and average prosperity, weighs down the active labour-army; during the periods of overproduction and paroxysm it holds its pretensions in check. Relative surplus population is therefore the pivot upon which the law of demand and supply of labour works. It confines the field of action of this law within the limits absolutely convenient to the activity of exploitation and to the domination of capital....

The relative surplus population exists in every possible form. Every labourer belongs to it during the time when he is only partially employed or wholly unemployed. Not taking into account the great periodically recurring forms that the changing phases of the industrial cycle impress on it, now an acute form during the crisis, then again a chronic form during dull times—it has always three forms, the floating, the latent, the stagnant....

The lowest sediment of the relative surplus population finally dwells in the sphere of pauperism. Exclusive of vagabonds, criminals, prostitutes, in a word, the 'dangerous' classes, this layer of society consists of three categories. First, those able to work. One need only glance superficially at the statistics of English pauperism to find that the quantity of paupers increases with every crisis, and diminishes with every revival of trade. Second, orphans and pauper children. These are candidates for the industrial reserve army, and are, in times of great prosperity, as 1860, e.g., speedily and in large numbers enrolled in the active army of labourers. Third, the demoralized and ragged, and those unable to work, chiefly people who succumb to their incapacity for adaptation, due to the division of labour; people who have passed the normal age of the labourer; the victims of industry, whose number increases with the increase of dangerous machinery, of mines, chemical works, etc., the mutilated, the sickly, the widows, etc. Pauperism is the hospital of the active labour-army and the dead weight of the industrial reserve army. Its production is included in that of the relative surplus population, its necessity in theirs; along with the surplus population, pauperism forms a condition of capitalist production, and of the capitalist development of wealth. It enters into the *faux frais* [unnecessary expenditure] of capitalist production; but capital knows how to throw these, for the most part, from its own shoulders onto those of the working class and the lower middle class.

The greater the social wealth, the functioning capital, the extent and energy of its growth, and, therefore, also the absolute mass of the proletariat and the productiveness of its labour, the greater is the industrial reserve army. The same causes which develop the expansive power of capital develop also the labour-power at its disposal. The relative mass of the industrial reserve army increases therefore with the potential energy of wealth. But the greater this reserve army in proportion to the active labour-army, the greater is the mass of a consolidated surplus population, whose misery is in inverse ratio to its torment of labour. The more extensive, finally, the lazarus-layers of the working class, and the industrial reserve army, the greater is official pauperism. *This is the absolute general law of capitalist accumulation.* Like all other laws it is modified in its working by many circumstances, the analysis of which does not concern us here.

The folly of the economic wisdom that preaches to the labourers the accommodation of their number to the requirements of capital is now patent. The mechanism of capitalist production and accumulation constantly effects this adjustment. The first word of this adaptation is the creation of a relative surplus population, or industrial reserve army. Its last word is the misery of constantly extending strata of the active army of labour, and the dead weight of pauperism.

The law by which a constantly increasing quantity of means of production, thanks to the advance in the productiveness of social labour, may be set in movement by a progressively diminishing expenditure of human power, this law, in a capitalist society—where the labourer does not employ the means of production, but the means of production employ the labourer—undergoes a complete inversion and is expressed thus: the higher the productiveness of labour, the greater is the pressure of the labourers on the means of employment, the more precarious, therefore, becomes their condition of existence, viz., the sale of their own labour-power for the increasing of another's wealth, or for the self-expansion of capital. The fact that the means of production, and the productiveness of labour, increase more rapidly than the productive population, expresses itself, therefore, capitalistically in the inverse form that the labouring population always increases more rapidly than the conditions under which capital can employ this increase for its own self-expansion. . . .

Within the capitalist system all methods for raising the social productiveness of labour are brought about at the cost of the individual labourer; all means for the development of production transform themselves into means of domination over, and exploitation of, the producers; they mutilate the labourer into a fragment of a man, degrade him to the level of an appendage of a machine, destroy every remnant of charm in his work and turn it into a hated toil; they estrange from him the intellectual potentialities of the labour process in the same proportion as science is incorporated in it as an independent power; they distort the conditions under which he works, subject him during the labour process to a despotism the more hateful for its meanness; they transform his lifetime into working-time, and drag his wife and child beneath the wheels of the Juggernaut of capital. But all methods for the production of surplus value are at the same time methods of accumulation; and every extension of accumulation becomes again a means for the development of those methods. It follows therefore that in proportion as capital accumulates, the lot of the labourer, be his payment high or low, must grow worse. The law, finally, that always equilibrates the relative surplus population, or industrial reserve army, to the extent and energy of accumulation, this law rivets the labourer to capital more firmly than the wedges of Vulcan did Prometheus to the rock. It establishes an accumulation of misery, corresponding with accumulation of capital. Accumulation of wealth at one pole is, therefore, at the same time accumulation of misery, agony of toil, slavery, ignorance, brutality, mental degradation, at the opposite pole, i.e., on the side of the class that produces its own product in the form of capital. . . .

Primitive Accumulation

We have seen how money is changed into capital; how through capital surplus value is made, and from surplus value more capital. But the accumulation of capital presupposes surplus value; surplus value presupposes capitalistic production; capitalistic production presupposes the pre-existence of considerable masses of capital and of labour power in the hands of producers of commodities. The whole movement, therefore, seems to turn in a vicious circle, out of which we can only get by supposing a primitive accumulation (previous accumulation of Adam Smith) preceding capitalistic accumulation; an accumulation not the result of the capitalist mode of production, but its starting-point.

This primitive accumulation plays in Political Economy about the same part as original sin in theology. Adam bit the apple, and thereupon sin fell on the human race. Its origin is supposed to be explained when it is told as an anecdote of the past. In times long gone by there were two sorts of people; one, the diligent, intelligent, and, above all, frugal élite; the other, lazy rascals, spending their substance, and more, in riotous living. The legend of theological original sin tells us certainly how man came to be condemned to eat his bread in the sweat of his brow; but the history of economic original sin reveals to us that there are people to whom this is by no means essential. Never mind! Thus it came to pass that the former sort accumulated wealth, and the latter sort had at last nothing to sell except their own skins. And from this original sin dates the poverty of the great majority that, despite all its labour, has up to now nothing to sell but itself, and the wealth of the few that increases constantly although they have long ceased to work. Such insipid childishness is every day preached to us in the defence of property. M. Thiers, for example, had the assurance to repeat it with all the solemnity of a statesman, to the French people, once so *spirituel*. But as soon as the question of property crops up, it becomes a sacred duty to proclaim the intellectual food of the infant as the one thing fit for all ages and for all stages of development. In actual history it is notorious that conquest, enslavement, robbery, murder, briefly force, play the great part. In the tender annals of Political Economy, the idyllic reigns from time immemorial. Right and 'labour' were from all time the sole means of enrichment, the present year of course always excepted. As a matter of fact, the methods of primitive accumulation are anything but idyllic.

In themselves money and commodities are no more capital than are the means of production and of subsistence. They want transforming into capital. But this transformation itself can only take place under certain circumstances that centre in this, viz., that two very different kinds of commodity-possessors must come face to face and into contact; on the one hand, the owners of money, means of production, means of subsistence, who are eager to increase the sum of values they possess, by buying other people's labour-power; on the other hand, free labourers, the sellers of their own labour-power, and therefore the sellers of labour. Free labourers, in the double sense that neither they themselves form part and parcel of the means of production, as in the case of slaves, bondsmen, etc., nor do the means of production belong to them, as in the case of peasant-proprietors; they are therefore, free from, unencumbered by, any means of production of their own. With this polarization of the market for commodities, the fundamental conditions of capitalist production are given. The capitalist system presupposes the complete separation of the labourers from all property in the means by which they can realize their labour. As soon as capitalist production is once on its own legs, it not only maintains this separation, but reproduces it on a continually extending scale. The process, therefore, that clears the way for the capitalist system, can be none other than the process which takes away from the labourer the possession of his means of production; a process that transforms, on the one hand, the social means of subsistence and of production into capital, on the other, the immediate producers into wage-labourers. The so-called primitive accumulation, therefore, is nothing else than the historical process of divorcing the producer from the means of production. It appears as primitive, because it forms the prehistoric stage of capital and of the mode of production corresponding with it.

The economic structure of capitalistic society has grown out of the economic structure of feudal society. The dissolution of the latter set free the elements of the former.

The immediate producer, the labourer, could only dispose of his own person after he had ceased to be attached to the soil and ceased to be the slave, serf, or bondman of another. To become a free seller of labour power, who carries his commodity wherever he finds a market, he must further have escaped from the regime of the guilds, their rules for apprentices and journeymen, and the impediments of their labour regulations. Hence, the historical movement which changes the producers into wage-workers appears, on the one hand, as their emancipation from serfdom and from the fetters of the guilds, and this side alone exists for our bourgeois historians. But, on the other hand, these new freedmen became sellers of themselves only after they had been robbed of all their own means of production, and of all the guarantees of existence afforded by the old feudal arrangements. And the history of this, their expropriation, is written in the annals of mankind in letters of blood and fire.

The industrial capitalists, these new potentates, had on their part not only to displace the guild masters of handicrafts, but also the feudal lords, the possessors of the sources of wealth. In this respect their conquest of social power appears as the fruit of a victorious struggle both against feudal lordship and its revolting prerogatives, and against the guilds and the fetters they laid on the free development of production and the free exploitation of man by man. The *chevaliers d'industrie*, however, only succeeded in supplanting the chevaliers of the sword by making use of events of which they themselves were wholly innocent. They have risen by means as vile as those by which the Roman freedman once on a time made himself the master of his *patronus*.

The starting-point of the development that gave rise to the wage-labourer as well as to the capitalist was the servitude of the labourer. The advance consisted in a change of form of this servitude, in the transformation of feudal exploitation into capitalist exploitation. To understand its march, we need not go back very far. Although we come across the first beginnings of capitalist production as early as the fourteenth or fifteenth century, sporadically, in certain towns of the Mediterranean, the capitalistic era dates from the sixteenth century. Wherever it appears, the abolition of serfdom has been long effected, and the highest development of the Middle Ages, the existence of sovereign towns, has been long on the wane.

In the history of primitive accumulation, all revolutions are epoch-making that act as levers for the capitalist class in course of formation; but, above all, those moments when great masses of men are suddenly and forcibly torn from their means of subsistence, and hurled as free and 'unattached' proletarians on the labour-market. The expropriation of the agricultural producer, of the peasant, from the soil, is the basis of the whole process. The history of this expropriation, in different countries, assumes different aspects, and runs through its various phases in different orders of succession, and at different periods. In England alone, which we take as our example, has it the classic form. . . .

The Historical Tendency of Capitalist Accumulation

What does the primitive accumulation of capital, i.e., its historical genesis, resolve itself into? In so far as it is not immediate transformation of slaves and serfs into wage-labourers, and therefore a mere change of form, it only means the expropriation of the immediate producers, i.e., the dissolution of private property based on the labour of its owner. Private property, as the antithesis to social, collective property, exists only where the means of labour and the external conditions of labour belong to private individuals. But according as these private individuals are labourers or not labourers, private property has a different character. The numberless shades that it at first sight presents correspond to the intermediate stages lying between these two extremes. The private property of the labourer in his means of production is the foundation of petty industry, whether agricultural, manufacturing, or both; petty industry, again, is an essential condition for the development of social production and of the free individuality of the labourer himself. Of course, this petty mode of production exists also under slavery, serfdom, and other states of dependence. But it flourishes, it lets loose its whole energy, it attains its adequate classical form, only where the labourer is the private owner of his own means of labour set in action by himself: the peasant of the land which he cultivates, the artisan of the tool which he handles as a virtuoso. This mode of production presupposes parcelling of the soil, and scattering of the other means of production. As it excludes the concentration of these means of production, so also it excludes co-operation, division of labour within each separate process of production, the control over, and the productive application of the forces of Nature by society, and the free development of the social productive powers. It is compatible only with a system of production, and a society, moving within narrow and more or less primitive bounds. To perpetuate it would be, as Pecqueur rightly says, 'to decree universal mediocrity'. At a certain stage of development it brings forth the material agencies for its own dissolution. From that moment new forces and new passions spring up in the bosom of society; but the old social organization fetters them and keeps them down. It must be annihilated; it is annihilated. Its annihilation, the transformation of the individualized and scattered means of production into socially concentrated ones, of the pigmy property of the many into the huge property of the few, the expropriation of the great mass of the people from the soil, from the means of subsistence, and from the means of labour, this fearful and painful expropriation of the mass of the people forms the prelude to the history of capital. It comprises a series of forcible methods, of which we have passed in review only those that have been epoch-making as methods of the primitive accumulation of capital. The expropriation of the immediate producers was accomplished with merciless vandalism, and under the stimulus of passions the most infamous, the most sordid, the pettiest, the most meanly odious. Self-earned private property, that is based, so to say, on the fusing together of the isolated, independent labouring-individual with the conditions of his labour, is supplanted by capitalistic private property, which rests on exploitation of the nominally free labour of others, i.e., on wage labour.

As soon as this process of transformation has sufficiently decomposed the old society from top to bottom, as soon as the labourers are turned into proletarians, their means of labour into capital, as soon as the capitalist mode of production stands on its own feet, then the further socialization of labour and further transformation of the land and the other means of production into socially exploited and, therefore, common means of production, as well as the further expropriation of private proprietors, takes a new form. That which is now to be expropriated is no longer the labourer working for himself, but the capitalist exploiting many labourers. This expropriation is accomplished by the action of the immanent laws of capitalistic production itself, by the centralization of capital. One capitalist always kills many. Hand in hand with this centralization, or this expropriation of many capitalists by few, develop, on an ever-extending scale, the co-operative form of the labour process, the conscious technical application of science, the methodical cultivation of the soil, the transformation of the instruments of labour into instruments of labour only usable in common, the economizing of all means of production by their use as the means of production of combined, socialized labour, the entanglement of all peoples in the net of the world market, and, with this, the international character of the capitalistic regime. Along with the constantly diminishing number of the magnates of capital, who usurp and monopolize all advantages of this process of transformation, grows the mass of misery, oppression, slavery, degradation, exploitation; but with this too grows the revolt of the working class, a class always increasing in numbers, and disciplined, united, organized by the very mechanism of the process of capitalist production itself. The monopoly of capital becomes a fetter upon the mode of production which has sprung up and flourished along with, and under it. Centralization of the means of production and socialization of labour at last reach a point where they become incompatible with their capitalist integument. The integument is burst asunder. The knell of capitalist private property sounds. The expropriators are expropriated.

The capitalist mode of appropriation, the result of the capitalist mode of production, produces capitalist private property. This is the first negation of individual private property, as founded on the labour of the proprietor. But capitalist production begets, with the inexorability of a law of Nature, its own negation. It is the negation of negation. This does not re-establish private property for the producer, but gives him individual property based on the acquisitions of the capitalist era: i.e., on co-operation and the possession in common of the land and of the means of production.

The transformation of scattered private property, arising from individual labour, into capitalist private property is, naturally, a process, incomparably more protracted, violent, and difficult, than the transformation of capitalistic private property, already practically resting on socialized production, into socialized property. In the former case, we had the expropriation of the mass of the people by a few usurpers; in the latter, we have the expropriation of a few usurpers by the mass of the people....

Source: K. Marx, *Selected Writings*, ex. D. McLellan (Oxford, 1977), pp. 160 ff., 170 ff., 221 ff., 388 ff., 477 ff., 559, 583 f.

FURTHER NOTES

WORKS

MARX, K., *Selected Writings*, ed. D. McLellan (Oxford, 1977).
——and ENGELS F., *Collected Works*, 50 vols. (New York and London, 1975–).
——— *Selected Works*, 2 vols. (Moscow, 1962).

COMMENTARIES

AVINERI, S., *The Social and Political Thought of Karl Marx* (Cambridge, 1968).
McLELLAN, D., *The Thought of Karl Marx* (New York, 1980).
McMURTRY, J., *The Structure of Marx's World-view* (Princeton, 1978).
MAZLISH, B., *The Meaning of Marx* (New York, 1984).
PLAMENATZ, J., *Karl Marx's Philosophy of Man* (Oxford, 1975).
RADER, M., *Marx's Interpretation of History* (New York, 1979).
SOWELL, T., *Marxism: Philosophy and Economics* (New York, 1986).
SUCHTING, W., *Marx: An Introduction* (Brighton, 1983).
TUCKER, R., *The Marxian Revolutionary Idea* (London, 1970).

2

FRIEDRICH ENGELS

The world is not to be comprehended as a complex of ready-made *things*, but as a complex of *processes*, in which the things are apparently stable, no less than their mental images in our heads, the concepts, go through an uninterrupted change of coming into being and passing away, in which, in spite of all seeming accidentality and all temporary retrogression, a progressive development asserts itself in the end.

Selected Works, (Moscow, 1962), ii. 387

FRIEDRICH ENGELS (1820–95), *lifelong friend and collaborator of Marx, enjoyed unrivalled prestige in the years before 1914 as an interpreter of Marx's work. The eldest son of a wealthy textile manufacturer in the Ruhr, Engels espoused Communism very early, joined forces with Marx in Paris in 1844, and emigrated with him to England, where he worked in the Manchester branch of his father's firm until moving to London on retirement in 1870. Engels survived Marx by twelve years and became responsible for the exposition and popularization of Marxism as a world view. Engels' gifts as a publicist are displayed in the first extract below taken from the pamphlet* Socialism: Utopian and Scientific *(1892). Itself an excerpt from Engels' larger work* Anti-Dühring, *it was, so he claimed, translated into more languages than even the* Communist Manifesto. *It contains a succinct summary of the materialist conception of history and a description of the Communist reorganization of society that a proletarian revolution will make possible. The second extract—from a letter of 1890 to his fellow Marxist Joseph Bloch—gives a clear account of what Engels understood by historical materialism. The third extract is from his 1895 Preface to a re-edition of Marx's* Class Struggles in France. *Here Engels explains why he considers the insurrectionary tactics of 1848 to be no longer applicable and argues a non-violent approach to political struggle, since bourgeois legality will itself provide the means for the proletariat to gain ascendancy.*

1. Socialism: Utopian and Scientific

THE materialist conception of history starts from the proposition that the production of the means to support human life and, next to production, the exchange of things produced, is the basis of all social structure; that in every society that has appeared in history, the manner in which wealth is distributed and society divided into classes or orders is dependent upon what is produced, how it is produced, and how the products are exchanged. From this point of view the final causes of all social changes and political revolutions are to be sought, not in men's brains, not in man's better insight into eternal truth and justice, but in changes in the modes of production and exchange. They are to be sought, not in the *philosophy*, but in the *economics* of each particular epoch. The growing perception that existing social institutions are unreasonable and unjust, that reason has become unreason, and right wrong, is only proof that in the modes of production and exchange changes have silently taken place with which the social order, adapted to earlier economic conditions, is no longer in keeping. From this it also follows that the means of getting rid of the incongruities that have been brought to light must also be present, in a more or less developed condition, within the changed modes of production themselves. These means are not to be invented by deduction from fundamental principles, but are to be discovered in the stubborn facts of the existing system of production.

What is, then the position of modern Socialism in this connection?

The present structure of society—this is now pretty generally conceded—is the creation of the ruling class today, of the bourgeoisie. The mode of production peculiar to the bourgeoisie, known, since Marx, as the capitalist mode of production, was incompatible with the feudal system, with the privileges it conferred upon individuals, entire social ranks, and local corporations, as well as with the hereditary ties of subordination which constituted the framework of its social organization. The bourgeoisie broke up the feudal system and built upon its ruins the capitalist order of society, the kingdom of free competition, of personal liberty, of the equality, before the law, of all commodity owners, of all the rest of the capitalist blessings. Thenceforward the capitalist mode of production could develop in freedom. Since steam, machinery, and the making of machines by machinery transformed the older manufacture into modern industry, the productive forces evolved under the guidance of the bourgeoisie developed with a rapidity and in a degree unheard of before. But just as the older manufacture, in its time, and handicraft, becoming more developed under its influence, had come into collision with the feudal trammels of the guilds, so now modern industry, in its more complete development, comes into collision with the bounds within which the capitalistic mode of production holds it confined. The new productive forces have already outgrown the capitalistic mode of using them. And this conflict between productive forces and modes of production is not a conflict engendered in the mind of man, like that between original sin and divine justice. It exists, in fact, objectively, outside us, independently of the will and actions even of the men that have brought it on. Modern Socialism is nothing but the reflex, in thought, of this conflict in fact; its ideal reflection in the minds, first, of the class directly suffering under it, the working class. . . .

Active social forces work exactly like natural forces: blindly, forcibly, destructively, so long as we do not understand and reckon with them. But when once we understand them, when once we grasp their action, their direction, their effects, it depends only upon ourselves to subject them more and more to our own will, and by means of them to reach our own ends. And this holds quite especially of the mighty productive forces of today. As long as we obstinately refuse to understand the nature and the character of these social means of action—and this understanding goes against the grain of the capitalist mode of production and its defenders—so long these forces are at work in spite of us, in opposition to us, so long they master us, as we have shown above in detail.

But when once their nature is understood, they can, in the hands of the producers working together, be transformed from master demons into willing servants. The difference is as that between the destructive force of electricity in the lightning of the storm, and electricity under command in the telegraph and the voltaic arc; the difference between a conflagration, and fire working in the service of man. With this recognition at last of the real nature of the productive forces of today, the social anarchy of production gives place to a social regulation of production upon a definite plan, according to the needs of the community and of each individual. Then the capitalist mode of appropriation, in which the product enslaves first the producer and then the appropriator, is replaced by the mode of appropriation of the products that is based upon the nature of the modern means of production; upon the one hand, direct social appropriation, as means to the maintenance and extension of production—on the other, direct individual appropriation, as means of subsistence and of enjoyment.

While the capitalist mode of production more and more completely transforms the great majority of the population into proletarians, it creates the power which, under penalty of its own destruction, is forced to accomplish this revolution. While it forces on more and more the transformation of the vast means of production, already socialized, into State property, it shows itself the way to accomplishing this revolution. *The proletariat seizes political power and turns the means of production into State property.*

But, in doing this, it abolishes itself as proletariat, abolishes all class distinctions and class antagonisms, abolishes also the State as State. Society thus far, based upon class antagonisms, had need of the State. That is, of an organization of the particular class which was *pro tempore* the exploiting class, an organization for the purpose of preventing any interference from without with the existing conditions of production, and therefore, especially, for the purpose of forcibly keeping the exploited classes in the condition of oppression corresponding with the given mode of production (slavery, serfdom, wage labour). The State was the official representative of society as a whole; the gathering of it together into a visible enbodiment. But it was this only in so far as it was the State of that class which itself represented, for the time being, society as a whole; in ancient times, the State of slave-owning citizens; in the middle ages, the feudal lords; in our own time, the bourgeoisie. When at last it becomes the real representative of the whole of society, it renders itself unnecessary. As soon as there is no longer any social class to be held in subjection; as soon as class rule, and the individual struggle for existence based upon our present anarchy in production, with the collisions and excesses arising from these, are removed, nothing more remains to be repressed, and a special repressive force, a State, is no longer necessary. The first act by virtue of which the State really constitutes itself the representative of the whole of society—the taking possession of the means of production in the name of society—this is, at the same time, its last independent act as a State. State interference in social relations becomes, in one domain after another, superfluous, and then dies out of itself; the government of persons is replaced by the administration of things and by the conduct of processes of production. The State is not 'abolished'. *It dies out.* This gives the measure of the value of the phrase 'a free State', both as to its justifiable use at times by agitators, and as to its ultimate scientific insufficiency; and also of the demands of the so-called anarchists for the abolition of the State out of hand.

Since the historical appearance of the capitalist mode of production, the appropriation by society of all the means of production has often been dreamed of, more or less vaguely, by individuals, as well as by sects, as the ideal of the future. But it could become possible, could become a historical necessity, only when the actual conditions for its realization were there. Like every other social advance, it becomes practicable, not by men understanding that the existence of classes is in contradiction to justice, equality, etc., not by the mere willingness to abolish these classes, but by virtue of certain new economic conditions. The separation of society into an exploiting and an exploited class, a ruling and an oppressed class, was the necessary consequence of the deficient and restricted development of production in former times. So long as the total social labour only yields a produce which but slightly exceeds that barely necessary for the existence of all; so long, therefore, as labour engages all or almost all the time of the great majority of the members of society—so long, of necessity, this society is divided into classes. Side by side with the great majority, exclusively bond slaves to labour, arises a class freed from directly productive labour, which looks after the general affairs of society; the direction of labour, State business, law, science, art, etc. It is, therefore, the law of division of labour that lies at the basis of the division into classes. But this does not prevent this division into classes from being carried out by means of violence and robbery, trickery and fraud. It does not prevent the ruling class, once having the upper hand, from consolidating its power at the expense of the working class, from turning their social leadership into an intensified exploitation of the masses.

But if, upon this showing, division into classes has a certain historical justification, it has this only for a given period, only under given social conditions. It was based upon the insufficiency of production. It will be swept away by the complete development of modern productive forces. And, in fact, the abolition of classes in society presupposes a degree of historical evolution, at which the existence, not simply of this or that particular ruling class, but of any ruling class at all, and, therefore, the existence of class distinction itself has become an obsolete anachronism. It presupposes, therefore, the development of production carried out to a degree at which appropriation of the means of production and of the products, and, with this, of political domination, of the monopoly of culture, and of intellectual leadership by a particular class of society, has become not only superfluous, but economically, politically, intellectually a hindrance to development.

This point is now reached. Their political and intellectual bankruptcy is scarcely any longer a secret to the bourgeoisie themselves. Their economic bankruptcy recurs regularly every ten years. In every crisis, society is suffocated beneath the weight of its own productive forces and products, which it cannot use, and stands helpless, face to face with the absurd contradiction that the producers have nothing to consume, because consumers are wanting. The expansive force of the means of production bursts the bonds that the capitalist mode of production had imposed upon them, Their deliverance from these bonds is the one pre-condition for an unbroken, constantly accelerated development of the productive forces, and therewith for a practically unlimited increase of production itself. Nor is this all. The socialized appropriation of the means of production does away, not only with the present artificial restrictions upon production, but also with the positive waste and devastation of productive forces and products that are at the present time the inevitable concomitants of production, and that reach their height in the crises. Further, it sets free for the community at large a mass of means of production and of products, by doing away with the senseless extravagance of the ruling classes of today, and their political representatives. The possibility of securing for every member of society, by means of socialized production, an existence not only fully sufficient materially, and becoming day by day more full, but an existence guaranteeing to all the free development and exercise of their physical and mental faculties—this possibility is now for the first time here, truly *it is here*.

With the seizing of the means of production by society, production of commodities is done away with, and, simultaneously the mastery of the product over the producer. Anarchy in social production is replaced by systematic, definite organization. The struggle for individual existence disappears. Then for the first time, man, in a certain sense, is finally marked off from the rest of the animal kingdom, and emerges from mere animal conditions of existence into really human ones. The whole sphere of the conditions of life which environ man, and which have hitherto ruled man, now comes under the dominion and control of man, who for the first time becomes the real, conscious lord of Nature, because he has now become master of his own social organization. The laws of his own social action, hitherto standing face to face with man as laws of Nature foreign to, and dominating, him, will then be used with full understanding, and so mastered by him. Man's own social organization, hitherto confronting him as a necessity imposed by Nature and history, now becomes the result of his own free action. The extraneous objective forces that have hitherto governed history pass under the control of man himself. Only from that time will man himself, more and more consciously, make his own history— only from that time will the social causes set in movement by him have, in the main and in a constantly growing measure, the results intended by him. It is the ascent of man from the kingdom of necessity to the kingdom of freedom.

Let us briefly sum up our sketch of historical evolution.

i. *Medieval Society*. Individual production on a small scale. Means of production adapted for individual use; hence primitive, ungainly, petty, dwarfed in action. Production for immediate consumption, either of the producer himself or of his feudal lord. Only where an excess of production over this consumption occurs is such excess offered for sale, enters into exchange. Production of commodities, therefore, only in its infancy. But already it contains within itself, in embryo, *anarchy in the production of society at large*.

ii. *Capitalist Revolution.* Transformation of industry, at first by means of simple co-operation and manufacture. Concentration of the means of production, hitherto scattered, into great workshops. As a consequence, their transformation from individual to social means of production—a transformation which does not, on the whole, affect the form of exchange. The old forms of appropriation remain in force. The capitalist appears. In his capacity as owner of the means of production, he also appropriates the products and turns them into commodities. Production has become a *social* act. Exchange and appropriation continue to be *individual* acts, the acts of individuals. *The social product is appropriated by the individual capitalist.* Fundamental contradiction, whence arise all the contradictions in which our present day society moves, and which modern industry brings to light.

(*a*) Severance of the producer from the means of production. Condemnation of the worker to wage labour for life. *Antagonism between the proletariat and the bourgeoisie.*

(*b*) Growing predominance and increasing effectiveness of the laws governing the production of commodities. Unbridled competition. *Contradiction between socialized organization in the individual factory and social anarchy in production as a whole.*

(*c*) On the one hand, perfecting of machinery, made by competition compulsory for each individual manufacturer, and complemented by a constantly growing displacement of labourers. *Industrial reserve-army.* On the other hand, unlimited extension of production, also compulsory under competition, for every manufacturer. On both sides, unheard of development of productive forces, excess of supply over demand, over-production, glutting of the markets, crises every ten years, the vicious circle: excess here, of means of production and products—excess there, of labourers, without employment and without means of existence. But these two levers of production and of social well-being are unable to work together, because the capitalist form of production prevents the productive forces from working and the products from circulating, unless they are first turned into capital—which their very superabundance prevents. The contradiction has grown into an absurdity. The mode of production rises in rebellion against the form of exchange. The bourgeoisie are convicted of incapacity further to manage their own social productive forces.

(*d*) Partial recognition of the social character of the productive forces forced upon the capitalists themselves. Taking over of the great institutions for production and communication, first by joint-stock companies, later on by trusts, then by the State. The bourgeoisie demonstrated to be a superfluous class. All its social functions are now performed by salaried employees.

iii. *Proletariat Revolution.* Solution of the contradictions. The proletariat seizes the public power, and by means of this transforms the socialized means of production, slipping from the hands of the bourgeoisie, into public property. By this act, the proletariat frees the means of production from the character of capital they have thus far borne, and gives their socialized character complete freedom to work itself out. Socialized production upon a predetermined plan becomes henceforth possible. The development of production makes the existence of different classes of society thenceforth an anachronism. In proportion as anarchy in social production vanishes, the political authority of the State dies out. Man, at last the master of his own form of social organization, becomes at the same time the lord over Nature, his own master—free.

To accomplish this act of universal emancipation is the historical mission of the modern proletariat. To thoroughly comprehend the historical conditions and thus the very nature of this act, to impart to the now oppressed proletarian class a full knowledge of the conditions and of the meaning of the momentous act it is called upon to accomplish, this is the task of the theoretical expression of the proletarian movement, scientific Socialism.

2.   The Materialist Conception of History

According to the materialist conception of history, the *ultimately* determining element in history is the production and reproduction of real life. More than this neither Marx nor I have ever asserted. Hence if somebody twists this into saying that the economic element is the *only* determining one, he transforms that proposition into a meaningless, abstract, senseless phrase. The economic situation is the basis, but the various elements of the superstructure—political forms of the class struggle and its results, to wit: constitutions established by the victorious class after a successful battle, etc., juridical forms, and even the reflexes of all these actual struggles in the brains of the participants, political, juristic, philosophical theories, religious views and their further development into systems of dogmas—also exercise their influence upon the course of the historical struggles and in many cases preponderate in determining their *form*. There is an interaction of all these elements in which, amid all the endless host of accidents (that is, of things and events whose inner interconnection is so remote or so impossible of proof that we can regard it as non-existent, as negligible) the economic movement finally asserts itself as necessary. Otherwise the application of the theory to any period of history would be easier than the solution of a simple equation of the first degree.

We make our history ourselves, but, in the first place, under very definite assumptions and conditions. Among these the economic ones are ultimately decisive. But the political ones, etc., and indeed even the traditions which haunt human minds, also play a part, although not the decisive one. The Prussian state also arose and developed from historical, ultimately economic, causes. But it could scarcely be maintained without pedantry that among the many small states of North Germany, Brandenburg was specifically determined by economic necessity to become the great power embodying the economic, linguistic and, after the Reformation, also the religious difference between North and South, and not by other elements as well (above all by its entanglement with Poland, owing to the possession of Prussia, and hence with international political relations—which were indeed also decisive in the formation of the Austrian dynastic power). Without making oneself ridiculous it would be a difficult thing to explain in terms of economics the existence of every small state in Germany, past and present, or the origin of the High German consonant permutations, which widened the geographic partition wall formed by the mountains from the Sudeten range to the Taunus to form a regular fissure across all Germany.

In the second place, however, history is made in such a way that the final result always arises from conflicts between many individual wills, of which each in turn has been made what it is by a host of particular conditions of life. Thus there are innumerable intersecting forces, an infinite series of parallelograms of forces which give rise to one resultant—the historical event. This may again itself be viewed as the product of a power which works as a whole *unconsciously* and without volition. For what each individual wills is obstructed by everyone else, and what emerges is something that no one willed. Thus history has proceeded hitherto in the manner of a natural process and is essentially subject to the same laws of motion. But from the fact that the wills of individuals—each of whom desires what he is impelled to by his physical constitution and external, in the last resort economic, circumstances (either his own personal circumstances or those of society in general)—do not attain what they want, but are merged into an aggregate mean, a common resultant, it must not be concluded that they are equal to zero. On the contrary, each contributes to the resultant and is to this extent included in it.

I would furthermore ask you to study this theory from its original sources and not at second hand; it is really much easier. Marx hardly wrote anything in which it did not play a part. But especially *The Eighteenth Brumaire of Louis Bonaparte* is a most excellent example of its application. There are also many allusions to it in *Capital*. Then may I also direct you to my writings: *Herr Eugen Dühring's Revolution in Science* and *Ludwig Feuerbach and the End of Classical German Philosophy*, in which I have given the most detailed account of historical materialism which, as far as I know, exists.

Marx and I are ourselves partly to blame for the fact that the younger people sometimes lay more stress on the economic side than is due to it. We had to emphasize the main principle *vis-à-vis* our adversaries, who denied it, and we had not always the time, the place or the opportunity to give their due to the other elements involved in the interaction. But when it came to presenting a section of history, that is, to making a practical application, it was a different matter and there no error was permissible. Unfortunately, however, it happens only too often that people think they have fully understood a new theory and can apply it without more ado from the moment they have assimilated its main principles, and even those not always correctly. And I cannot exempt many of the more recent 'Marxists' from this reproach, for the most amazing rubbish has been produced in this quarter, too....

3. Revolution: Peaceful or Violent?

All revolutions of modern times, beginning with the great English revolution of the seventeenth century, showed these features, which appeared inseparable from every revolutionary struggle. They appeared applicable, also, to the struggles of the proletariat for its emancipation; all the more applicable, since in 1848 there were few people who had any idea at all of the direction in which this emancipation was to be sought. The proletarian masses themselves, even in Paris, after the victory, were still absolutely in the dark as to the path to be taken. And yet the movement was there, instinctive, spontaneous, irrepressible. Was not this just the situation in which a revolution had to succeed, led certainly by a minority, but this time not in the interests of the minority, but in the real interests of the majority? If, in all the longer revolutionary periods, it was so easy to win the great masses of the people by the merely plausible and delusive views of the minorities thrusting themselves forward, how could they be less susceptible to ideas which were the truest reflex of their economic position, which were nothing but the clear, comprehensible expression of their needs, of needs not yet understood by themselves, but only vaguely felt? To be sure, this revolutionary mood of the masses had almost always, and usually very speedily, given way to lassitude or even to a revulsion to its opposite, so soon as illusion evaporated and disappointment set in. But here it was not a question of delusive views, but of giving effect to the very special interests of the great majority itself, interests which at that time were certainly by no means clear to this great majority, but which must soon enough become clear in the course of giving practical effect to them, by their convincing obviousness. And if now, as Marx showed in the third article, in the spring of 1850, the development of the bourgeois republic that had arisen out of the 'social' revolution of 1848 had concentrated the real power in the hands of the big bourgeoisie—monarchistically inclined as it was—and, on the other hand, had grouped all the other social classes, peasants as well as petty bourgeoisie, round the proletariat, so that, during and after the common victory, not they, but the proletariat grown wise by experience, must become the decisive factor—was there not every prospect here of turning the revolution of the minority into the revolution of the majority?

History has proved us, and all who thought like us, wrong. It has made it clear that the state of economic development on the Continent at that time was not, by a long way, ripe for the removal of capitalist production; it has proved this by the economic revolution which, since 1848, has seized the whole of the Continent, has really caused big industry for the first time to take root in France, Austria, Hungary, Poland, and, recently, in Russia, while it has made Germany positively an industrial country of the first rank—all on a capitalist basis, which in the year 1848, therefore, still had great capacity for expansion. But it is just this industrial revolution which has everywhere for the first time produced clarity in the class relationships, which has removed a number of transition forms handed down from the manufacturing period and in Eastern Europe even from guild handicraft, and has created a genuine bourgeoisie and a genuine large-scale industrial proletariat and pushed them into the foreground of social development. But, owing to this, the struggle of these two great classes, which apart from England, existed in 1848 only in Paris and, at the most, a few big industrial centres, has been spread over the whole of Europe and has reached an intensity such as was unthinkable in 1848. At that time the many obscure evangels of the sects, with their panaceas; today the one generally recognized, transparently clear theory of Marx, sharply formulating the final aims of the struggle. At that time the masses, sundered and differing according to locality and nationality, linked only by the feeling of common suffering, undeveloped, tossed to and fro in their perplexity from enthusiasm to despair; today a great international army of Socialists, marching irresistibly on and growing daily in number, organization, discipline, insight, and assurance of victory. If even this mighty army of the proletariat has still not reached its goal, if, a long way from winning victory with one mighty stroke, it has slowly to press forward from position to position in a hard tenacious struggle, this only proves, once and for all, how impossible it was in 1848 to win social reconstruction by a simple surprise attack. . . .

But whatever may happen in other countries, German Social Democracy has a special situation and therewith, at least in the first instance, a special task. The two million voters whom it sends to the ballot box, together with the young men and woman who stand behind them as non-voters, form the most numerous, most compact mass, the decisive 'shock force' of the international proletarian army. This mass already supplies over a fourth of the recorded votes: and as the by-elections to the Reichstag, the diet elections in individual states, the municipal council and industrial court elections demonstrate, it increases uninterruptedly. Its growth proceeds as spontaneously, as steadily, as irresistibly, and at the same time as tranquilly as a natural process. All government interventions have proved powerless against it. We can count even today on two and a half million voters. If it continues in this fashion, by the end of the century we shall conquer the greater part of the middle section of society, petty bourgeois and small peasants, and grow into the decisive power in the land, before which all other powers will have to bow, whether they like it or not. To keep this growth going without interruption until of itself it gets beyond the control of the ruling governmental system, not to fritter away this daily increasing shock force in advance guard fighting, but to keep it intact until the day of the decision, that is our main task. And there is only one means by which the steady rise of the Socialist fighting forces in Germany could be momentarily halted, and even thrown back for some time: a clash on a big scale with the military, a bloodbath like that of 1871 in Paris. In the long run that would also be overcome. To shoot out of the world a party which numbers millions—all the magazine rifles of Europe and America are not enough for this. But the normal development would be impeded, the shock force would, perhaps, not be available at the critical moment, the decisive struggle would be delayed, protracted, and attended by heavy sacrifices.

The irony of world history turns everything upside down. We, the 'revolutionaries', the 'rebels'—we are thriving far better on legal methods than on illegal methods and revolt. The parties of order, as they call themselves, are perishing under the legal conditions created by themselves. They cry despairingly with Odilon Barrot: *la légalité nous tue*, legality is the death of us; whereas we, under this legality, get firm muscles and rosy cheeks and look like eternal life. And if we are not so crazy as to let ourselves be driven into street fighting in order to please them then nothing else is finally left for them but themselves to break through this legality so fatal to them....

It is now, almost to the year, sixteen hundred years since a dangerous party of revolt made a great commotion in the Roman Empire. It undermined religion and all the foundations of the state; it flatly denied that Caesar's will was the supreme law; it was without a fatherland, international; it spread over all countries of the Empire from Gaul to Asia, and beyond the frontiers of the Empire. It had long carried on an underground agitation in secret; for a considerable time, however, it had felt itself strong enough to come out into the open. This party of revolt, who were known by the name of Christians, was also strongly represented in the army; whole legions were Christian. When they were ordered to attend the sacrificial ceremonies of the pagan established church, in order to do the honours there, the soldier rebels had the audacity to stick peculiar emblems—crosses—on their helmets in protest. Even the wonted barrack cruelties of their superior officers were fruitless. The Emperor Diocletian could not longer quietly look on while order, obedience and discipline in his army were being undermined. He intervened energetically, while there was still time. He passed an anti-Socialist, I should anti-Christian, law. The meetings of the rebels were forbidden, their meeting halls were closed or even pulled down, the Christian badges, crosses, etc., were, like the red handkerchiefs in Saxony, prohibited. Christians were declared incapable of holding offices in the state, they were not to be allowed even to become corporals. Since there were not available at that time judges so well trained in 'respect of persons' as Herr von Köller's anti-revolt bill assumes, the Christians were forbidden out of hand to seek justice before a court. This exceptional law was also without effect. The Christians tore it down from the walls with scorn; they are even supposed to have burnt the Emperor's palace in Nicomedia over his head. Then the latter revenged himself by the great persecution of Christians in the year 303, according to our chronology. It was the last of its kind. And it was so effective that seventeen years later the army consisted overwhelmingly of Christians, and the succeeding autocrat of the whole Roman Empire, Constantine, called the Great by the priests, proclaimed Christianity as the state religion.

Sources: F. Engels, Anti-Dühring (Moscow 1954), pp. 136 ff., 149 ff. K. Marx and F. Engels, Selected Correspondence (Moscow, 1962), pp. 488 ff. K. Marx and F. Engels, Selected Works (Moscow, 1962), ii. 124 ff., 135 ff.

FURTHER NOTES

WORKS

ENGELS, F., *Selected Works*, ed. W. Henderson (Baltimore, 1967).
MARX, K., AND F. ENGELS, *Collected Works* (New York, 1975–)
———— *Selected Works* (Moscow, 1962).

COMMENTARIES

CARLTON, G., *Friedrich Engels: The Shadow Prophet* (London, 1965).
CARVER, T., *Engels* (Oxford, 1982).
GOULDNER, A., *The Two Marxisms: Contradictions and Anomalies in the Development of Theory* (New York, 1980), ch. 9.
HENDERSON, W., *The Life of Frederick Engels*, 2 vols. (London, 1976).
HUNT, R., *The Political Ideas of Marx and Engels*, vol. i: *Marxism and Totalitarian Democracy* (Pittsburgh, 1974), ch. 4.
JORDAN, Z., *The Evolution of Dialectical Materialism* (New York, 1960).
KAPP, Y. *Eleanor Marx: The Crowded Years* (New York, 1976).
KOLAKOWSKI, L., *Main Currents of Marxism*, vol. i (New York, 1978).
LEVINE, N., *The Tragic Deception: Marx contra Engels* (Santa Barbara, 1975), chs. 9–14.
LICHTHEIM, G., *Marxism* (London 1964), pt. 5, ch. 3.
MAYER, G., *Friedrich Engels: A Biography* (New York, 1969).
MCLELLAN, D., *Engels* (Baltimore, 1978).
———— *Marxism after Marx* (Boston, 1981), ch. 1.
NOVA, F., *Friedrich Engels: His Contribution to Political Theory* (London, 1967).
PLAMENATZ, J., *German Marxism and Russian Communism* (London, 1954).

3

EDUARD BERNSTEIN

Peasants do not sink; middle class does not disappear; crises do not grow even larger; misery and serfdom do not increase.

Quoted in P. Gay, *The Dilemma of Democratic Socialism* (New York, 1962), p. 250.

EDUARD BERNSTEIN (1850–1932) *was one of the leading Marxist theoreticians at the end of the nineteenth century. He was particularly close to Engels, who made him his literary executor. Bernstein lived in London from 1880 to 1901, where he came under the influence of the Fabians. In the late 1890s he began to publish articles in which he sought to revise those elements in Marx's work that he considered to be incompatible with recent developments in economics and politics. In 1899 he published these ideas in a book entitled, in its English translation,* Evolutionary Socialism, *which remains the major work of classical revisionism and from which the extracts below are taken. In particular, Bernstein argued against the view that capitalism was faced with imminent collapse due to class polarization and deepening economic crisis: rather than revolution, he favoured a reforming Marxism in which the gradual extension of political and economic rights to the working class ensured a peaceful transition to Socialism. Bernstein's views were condemned by successive Congresses of the German Social Democratic Party but he continued to represent the party in the Reichstag.*

Evolutionary Socialism

(*a*) FROM THE PREFACE

IT has been maintained in a certain quarter that the practical deductions from my treatises would be the abandonment of the conquest of political power by the proletariat organized politically and economically. That is quite an arbitrary deduction, the accuracy of which I altogether deny

I set myself against the notion that we have to expect shortly a collapse of the bourgeois economy, and that social democracy should be induced by the prospect of such an imminent, great, social catastrophe to adapt its tactics to that assumption. That I maintain most emphatically.

The adherents of this theory of a catastrophe base it especially on the conclusions of the *Communist Manifesto*. This is a mistake in every respect.

The theory which the *Communist Manifesto* sets forth of the evolution of modern society was correct as far as it characterized the general tendencies of that evolution. But it was mistaken in several special deductions, above all in the estimate of the *time* the evolution would take. The last has been unreservedly acknowledged by Friedrich Engels, the joint author with Marx of the *Manifesto*, in his preface to the *Class War in France*. But it is evident that if social evolution takes a much greater period of time than was assumed, it must also take upon itself *forms* and lead to forms that were not foreseen and could not be foreseen then.

Social conditions have not developed to such an acute opposition of things and classes as is depicted in the *Manifesto*. It is not only useless, it is the greatest folly to attempt to conceal this from ourselves. The number of members of the possessing classes is today not smaller but larger. The enormous increase of social wealth is not accompanied by a decreasing number of large capitalists but by an increasing number of capitalists of all degrees. The middle classes change their character but they do not disappear from the social scale.

The concentration in productive industry is not being accomplished even today in all its departments with equal thoroughness and at an equal rate. In a great many branches of production it certainly justifies the forecasts of the socialist critic of society; but in other branches it lags even today behind them. The process of concentration in agriculture proceeds still more slowly. Trade statistics show an extraordinarily elaborated graduation of enterprises in regard to size. No rung of the ladder is disappearing from it. The significant changes in the inner structure of these enterprises and their interrelationship cannot do away with this fact.

In all advanced countries we see the privileges of the capitalist bourgeoisie yielding step by step to democratic organizations. Under the influence of this, and driven by the movement of the working classes, which is daily becoming stronger, a social reaction has set in against the exploiting tendencies of capital, a counteraction which, although it still proceeds timidly and feebly, yet does exist, and is always drawing more departments of economic life under its influence. Factory legislation, the democratization of local government, and the extension of its area of work, the freeing of trade unions and systems of co-operative trading from legal restrictions, the consideration of standard conditions of labour in the work undertaken by public authorities—all these characterize this phase of the evolution.

But the more the political organizations of modern nations are democratized the more the needs and opportunities of great political catastrophes are diminished. He who holds firmly to the catastrophic theory of evolution must, with all his power, withstand and hinder the evolution described above, which indeed, the logical defenders of that theory formerly did. But is the conquest of political power by the proletariat simply to be by a political catastrophe? Is it to be the appropriation and utilization of the power of the State by the proletariat exclusively against the whole non-proletarian world?

He who replies in the affirmative must be reminded of two things. In 1872 Marx and Engels announced in the preface to the new edition of the *Communist Manifesto* that the Paris Commune had exhibited a proof that 'the working classes cannot simply take possession of the ready-made State machine and set it in motion for their own aims'. And in 1895 Friedrich Engels stated in detail in the preface to *War of the Classes* that the time of political surprises, of the 'revolutions of small conscious minorities at the head of unconscious masses', was today at an end, that a collision on a large scale with the military would be the means of checking the steady growth of Social Democracy and of even throwing it back for a time—in short, that Social Democracy would flourish far better by lawful than by unlawful means and by violent revolution. And he points out in conformity with this opinion that the next task of the party should be 'to work for an uninterrupted increase of its votes' or to carry on a slow *propaganda of parliamentary activity*.

Thus Engels, who, nevertheless, as his numerical examples show, still somewhat overestimated the rate of process of the evolution! Shall we be told that he abandoned the conquest of political power by the working classes, because he wished to avoid the steady growth of Social Democracy secured by lawful means being interrupted by a political revolution?

If not, and if one subscribes to his conclusions, one cannot reasonably take any offence if it is declared that for a long time yet the task of Social Democracy is, instead of speculating on a great economic crash, 'to organize the working classes politically and develop them as a democracy and to fight for all reforms in the State which are adapted to raise the working classes and transform the State in the direction of democracy'.

That is what I have said in my impugned article and what I still maintain in its full import. As far as concerns the question propounded above it is equivalent to Engels' dictum, for democracy is, at any given time, as much government by the working classes as these are capable of practising according to their intellectual ripeness and the degree of social development they have attained. Engels, indeed, refers at the place just mentioned to the fact that the *Communist Manifesto* has 'proclaimed the conquest of the democracy as one of the first and important tasks of the fighting proletariat'.

In short, Engels is so thoroughly convinced that the tactics based on the presumption of a catastrophe have had their day, that he even considers a revision of them necessary in the Latin countries where tradition is much more favourable to them than in Germany. 'If the conditions of war between nations have altered,' he writes, 'no less have those for the war between classes.' Has this already been forgotten?

No one has questioned the necessity for the working classes to gain the control of government. The point at issue is between the theory of a social cataclysm and the question whether with the given social development in Germany and the present advanced state of its working classes in the towns and the country, a sudden catastrophe would be desirable in the interest of the Social Democracy. I have denied it and deny it again, because in my judgement a greater security for lasting success lies in a steady advance than in the possibilities offered by a catastrophic crash.

And as I am firmly convinced that important periods in the development of nations cannot be leapt over I lay the greatest value on the next tasks of Social Democracy, on the struggle for the political rights of the working man, on the political activity of working men in town and country for the interests of their class, as well as on the work of the industrial organization of the workers.

In this sense I wrote the sentence that the movement means everything for me and that what is *usually* called 'the final aim of socialism' is nothing; and in this sense I write it down again today. Even if the word 'usually' had not been shown that proposition was only to be understood conditionally, it was obvious that it *could* not express indifference concerning the final carrying out of Socialist principles, but only indifference—or, as it would be better expressed, carelessness—as to the form of the final arrangement of things. I have at no time had an excessive interest in the future, beyond general principles; I have not been able to read to the end any picture of the future. My thoughts and efforts are concerned with the duties of the present and the nearest future, and I only busy myself with the perspectives beyond so far as they give me a line of conduct for suitable action now.

The conquest of political power by the working classes, the expropriation of capitalists, are no ends in themselves but only means for the accomplishment of certain aims and endeavours. As such they are demands in the programme of Social Democracy and are not attacked by me. Nothing can be said beforehand as to the circumstances of their accomplishment; we can only fight for their realization. But the conquest of political power necessitates the possession of political *rights*; and the most important problem of tactics which German Social Democracy has at the present time to solve appears to me to be to devise the best ways for the extension of the political and economic rights of the German working classes.

(b) FROM THE TEXT

The trade unions are the democratic element in industry. Their tendency is to destroy the absolutism of capital, and to procure for the worker a direct influence in the management of an industry. It is only natural that the great differences of opinion should exist on the degree of influence to be desired. To a certain mode of thought it may appear a breach of principle to claim less for the union than an unconditional right of decision in the trade. The knowledge that such a right under present circumstances is just as Utopian as it would be contrary to the nature of a Socialist Community has led others to deny trade unions any lasting part in economic life, and to recognize them only temporarily as the lesser of various unavoidable evils. There are Socialists in whose eyes the union is only an object lesson to prove the uselessness of any other than political revolutionary action. As a matter of fact, the union today—and in the near future—has very important social tasks to fulfil for the trades, which, however, do not demand, nor are even consistent with, its omnipotence in any way....

The trade union, as mistress of a whole branch of production, the ideal of various older Socialists, would really be only a monopolist productive association, and as soon as it relied on its monopoly or worked upon it, it would be antagonistic to Socialism and democracy, let its inner constitution be what it may. Why it is contrary to Socialism needs no further explanation. Associations against the community are as little Socialism as is the oligarchic government of the State. But why should such a trade union not be in keeping with the principles of a democracy?

This question necessitates another. What is the principle of democracy?

The answer to this appears very simple. At first one would think it settled by the definition 'government by the people'. But even a little consideration tells us that by that only quite a superficial, purely formal definition is given, while nearly all who use the word democracy today understand by it more than a mere form of government. We shall come much nearer to the definition if we express ourselves negatively, and define democracy as an absence of class government, as the indication of a social condition where a political privilege belongs to no one class as opposed to the whole community. By that the explanation is already given as to why a monopolistic corporation is in principle anti-democratic. This negative definition has, besides, the advantage that it gives less room than the phrase 'government by the people' to the idea of the oppression of the individual by the majority which is absolutely repugnant to the modern mind. Today we find the oppression of the minority by the majority 'undemocratic', although it was originally held to be quite consistent with government by

the people. The idea of democracy includes, in the conception of the present
day, a notion of justice—an equality of rights for all members of the
community, and in that principle the rule of the majority, to which in every
concrete case the rule of the people extends, finds its limits. The more it is
adopted and governs the general consciousness, the more will democracy
be equal in meaning to the highest possible degree of freedom for all.

Democracy is in principle the suppression of class government, though
it is not yet the actual suppression of classes. They speak of the conservative
character of the democracy, and to a certain degree rightly. Absolutism, or
semi-absolutism, deceives its supporters as well as its opponents as to the
extent of their power. Therefore in countries where it obtains, or where its
traditions still exist, we have flitting plans, exaggerated language, zigzag
politics, fear of revolution, hope in oppression. In a democracy the parties,
and the classes standing behind them, soon learn to know the limits of their
power, and to undertake each time only as much as they can reasonably
hope to carry through under the existing circumstances. Even if they make
their demands rather higher than they seriously mean in order to give way
in the unavoidable compromise—and democracy is the high school of
compromise—they must still be moderate. The right to vote in a democracy
makes its members virtually partners in the community, and this virtual
partnership must in the end lead to real partnership. With a working class
undeveloped in numbers and culture the general right to vote may long
appear as the right to choose 'the butcher'; with the growing number and
knowledge of the workers it is changed, however, into the implement by
which to transform the representatives of the people from masters into real
servants of the people.

Universal franchise is, from two sides, the alternative to a violent revolution.
But universal suffrage is only a part of democracy, although a part which
in time must draw the other parts after it as the magnet attracts to itself
the scattered portions of iron. It certainly proceeds more slowly than many
would wish, but in spite of that it is at work. And Social Democracy cannot
further this work better than by taking its stand unreservedly on the
theory of democracy—on the ground of universal suffrage with all the
consequences resulting therefrom to its tactics.

In practice—that is, in its actions—it has in Germany always done so.
But in their explanations its literary advocates have often acted otherwise,
and still often do so today. Phrases which were composed in a time when
the political privilege of property ruled all over Europe, and which under
these circumstances were explanatory, and to a certain degree also justified,
but which today are only a dead weight, are treated with such reverence as

though the progress of the movement depended on them and not on the understanding of what can be done, and what should be done. Is there any sense, for example, in maintaining the phrase of the 'dictatorship of the proletariat' at a time when in all possible places representatives of Social Democracy have placed themselves practically in the arena of parliamentary work, have declared for the proportional representation of the people, and for direct legislation—all of which is inconsistent with a dictatorship.

The phrase is today so antiquated that it is only to be reconciled with reality by stripping the word dictatorship of its actual meaning and attaching to it some kind of weakened interpretation. The whole practical activity of Social Democracy is directed towards creating circumstances and conditions which shall render possible and secure a transition (free from convulsive outbursts) of the modern social order into a higher one. From the consciousness of being the pioneers of a higher civilization, its adherents are ever creating fresh inspiration and zeal. In this rests also, finally, the moral justification of the Socialist expropriation towards which they aspire. But the 'dictatorship of the classes' belongs to a lower civilization, and apart from the question of the expediency and practicability of the thing, it is only to be looked upon as a reversion, as political atavism. If the thought is aroused that the transition from a capitalist to a Socialist society must necessarily be accomplished by means of the development of forms of an age which did not know at all, or only in quite an imperfect form, the present methods of the initiating and carrying of laws, and which was without the organs fit for the purpose, reaction will set in.

I say expressly transition from a capitalist to a socialist society, and not from a 'civic society', as is so frequently the expression used today. This application of the word 'civic' is also much more an atavism, or in any case an ambiguous way of speaking, which must be considered an inconvenience in the phraseology of German Social Democracy, and which forms an excellent bridge for mistakes with friend and foe. The fault lies partly in the German language which has no special word for the idea of the citizen with equal civic rights separate from the idea of privileged citizens.

What is the struggle against, or the abolition of, a civic society? What does it mean specially in Germany, in whose greatest and leading state, Prussia, we are still constantly concerned with first getting rid of a great part of feudalism which stands in the path of civic development? No man thinks of destroying civic society as a civilized ordered system of society. On the contrary, Social Democracy does not wish to break up this society and make all its members proletarians together; it labours rather incessantly at raising the worker from the social position of a proletarian to that of a citizen, and thus to make citizenship universal. It does not want to set up a proletarian society instead of a civic society, but a Socialist order of society instead of a capitalist one. It would be well if one, instead of availing himself of the former ambiguous expression, kept to the latter quite clear declaration. Then one would be quite free of a good portion of other contradictions which opponents, not quite without reason, assert do exist between the phraseology and the practice of Social Democracy. A few Socialist newspapers find a pleasure today in forced anti-civic language, which at the most would be in place if we lived in a sectarian fashion as anchorites, but which is absurd in an age which declares it to be no offence to the Socialist sentiment to order one's private life throughout in a 'bourgeois fashion'.

Finally, it is to be recommended that some moderation should be kept in the declaration of war against 'liberalism'. It is true that the great liberal movement of modern times arose for the advantage of the capitalist bourgeoisie first of all, and the parties which assumed the names of liberals were, or became in due course, simple guardians of capitalism. Naturally, only opposition can reign between these parties and Social Democracy. But with respect to liberalism as a great historical movement, Socialism is its legitimate heir, not only in chronological sequence, but also in its spiritual qualities, as is shown moreover in every question of principle in which Social Democracy has had to take up an attitude.

Wherever an economic advance of the Socialist programme had to be carried out in a manner, or under circumstances, that appeared seriously to imperil the development of freedom, Social Democracy has never shunned taking up a position against it. The security of civil freedom has always seemed to it to stand higher than the fulfilment of some economic progress.

The aim of all Socialist measures, even of those which appear outwardly as coercive measures, is the development and the securing of a free personality. Their more exact examination always shows that the coercion included will raise the sum total of liberty in society, and will give more freedom over a more extended area than it takes away. The legal day of a maximum number of hours' work, for example, is actually a fixing of a minimum of freedom, a prohibition to sell freedom longer than for a certain number of hours daily, and, in principle, therefore, stands on the same ground as the prohibition agreed to by all liberals against selling oneself into personal slavery. It is thus no accident that the first country where a maximum hours' day was carried out was Switzerland, the most democratically progressive country in Europe, and democracy is only the political form of liberalism. Being in its origin a counter-movement to the oppression of nations under institutions imposed from without or having a justification only in tradition, liberalism first sought its realization as the principle of the sovereignty of the age and of the people, both of which principles formed the everlasting discussion of the philosophers of the rights of the state in the seventeenth and eighteenth centuries, until Rousseau set them up in his *Contrat Social* as the fundamental conditions of the legitimacy of every constitution, and the French Revolution proclaimed them—in the Democratic Constitution of 1793 permeated with Rousseau's spirit—as inalienable rights of men.

The Constitution of 1793 was the logical expression of the liberal ideas of the epoch, and a cursory glance over its contents shows how little it was, or is an obstacle to Socialism. Baboeuf, and the believers in absolute equality, saw in it an excellent starting-point for the realization of their Communistic strivings, and accordingly wrote 'The Restoration of the Constitution of 1793' at the head of their demands.

There is actually no really liberal thought which does not also belong to the elements of the ideas of Socialism. Even the principle of economic personal responsibility which belongs apparently so entirely to the Manchester School cannot, in my judgement, be denied in theory by Socialism nor be made inoperative under any conceivable circumstances. Without responsibility there is no freedom; we may think as we like theoretically about man's freedom of action, we must practically start from it as the foundation of the moral law, for only under this condition is social morality possible. And similarly, in our states which reckon with millions, a healthy social life is, in the age of traffic, impossible if the economic personal responsibility of all those capable of work is not assumed. The recognition of individual responsibility is the return of the individual to society for services rendered or offered him by society....

If democracy is not to excel centralized absolutism in the breeding of bureaucracies, it must be built up on an elaborately organized self-government with a corresponding economic, personal responsibility of all the units of administration as well as of the adult citizens of the State. Nothing is more injurious to its healthy development than enforced uniformity and a too abundant amount of protectionism or subventionism.

To create the organizations described—or, so far as they are already begun, to develop them further—is the indispensable preliminary to what we call Socialism of production. Without them the so-called social appropriation of the means of production would only result presumably in reckless devastation of productive forces, insane experimentalizing, and aimless violence, and the political sovereignty of the working class would, in fact, only be carried out in the form of a dictatorial revolutionary, central power, supported by the terrorist dictatorship of revolutionary clubs. As such it hovered before the Blanquists, and as such it is still represented in the *Communist Manifesto* and in the publications for which its authors were responsible at that time....

The future itself will reveal how far the municipalities and other self-governing bodies will discharge their duties under a complete democracy, and how far they will make use of these duties. But so much is clear; the more suddenly they come in possession of their freedom, the more experiments they will make in number and in violence and therefore be liable to greater mistakes, and the more experience the working-class democracy has had in the school of self-government, the more cautiously and practically will it proceed....

Meantime we are not yet so far on, and it is not my intention to unfold pictures of the future. I am not concerned with what will happen in the more distant future, but with what can and ought to happen in the present, for the present and the nearest future. And so the conclusion of this exposition is the very banal statement that the conquest of the democracy, the formation of political and social organs of the democracy, is the indispensable preliminary condition to the realization of Socialism....

My proposition, 'To me that which is generally called the ultimate aim of Socialism is nothing, but the movement is everything,' has often been conceived as a denial of every definite aim of the Socialist movement, and Mr George Plekhanov has even discovered that I have quoted this 'famous sentence' from the book *To Social Peace*, by Gerhard von Schulze-Gävernitz. There, indeed, a passage reads that it is certainly indispensable for revolutionary Socialism to take as its ultimate aim the nationalization of all the means of production, but not for practical political Socialism which places near aims in front of distant ones. Because an ultimate aim is here regarded as being dispensable for practical objects, and as I also have professed but little interest for ultimate aims, I am an 'indiscriminating follower' of Schulze-Gävernitz. One must confess that such demonstration bears witness to a striking wealth of thought.

When eight years ago I reviewed the Schulze-Gävernitz book in *Neue Zeit*, although my criticism was strongly influenced by assumptions which I now no longer hold, yet I put on one side as immaterial that opposition of ultimate aim and practical activity in reform, and admitted—without encountering a protest—that for England a further peaceful development, such as Schulze-Gävernitz places in prospect before her, was not improbable. I expressed the conviction that, with the continuance of free development, the English working classes would certainly increase their demands, but would desire nothing that could not be shown each time to be necessary and attainable beyond all doubt. That is at the bottom nothing else than what I say today. And if anyone wishes to bring up against me the advances in Social Democracy made since then in England, I answer that with this extension a development of the English Social Democracy has gone hand in hand from the Utopian, revolutionary sect, as Engels repeatedly represented it to be, to the party of political reform which we now know. No Socialist capable of thinking dreams today in England of an imminent victory for Socialism by means of a quick conquest of Parliament by a revolutionary proletariat. But they rely more and more on work in the municipalities and other self-governing bodies. The early contempt for the trade union movement has been given up; a closer sympathy has been won for it and, here and there also, for the co-operative movement.

And the ultimate aim? Well, that just remains an ultimate aim.

Source: E. Bernstein, *Evolutionary Socialism* (New York, 1961), pp. xxiii ff., 139 f., 141 ff., 145 ff., 155, 161, 163, 202 ff.

FURTHER NOTES

WORKS

BERNSTEIN, E., *Evolutionary Socialism: A Criticism and Affirmation*, introduced by Sydney Hook (New York, 1961).

COMMENTARIES

BAILEY, SYDNEY D., 'The Revision of Marxism', *The Review of Politics*, 16 (1954).

COLE, G. D. H., *A History of Socialist Thought*, vol. iii (New York, 1956).

COLLETTI, LUCIO, 'Bernstein and the Marxism of the Second International', in *From Rousseau to Lenin* (New York, 1972).

ELLIOTT, CHARLES F., '*Quis custodiet sacra?* Problems of Marxist Revisionism', *Journal of the History of Ideas*, 28 (1967).

GAY, PETER, *The Dilemma of Democratic Socialism: Edward Bernstein's Challenge to Marx* (New York 1962).

GNEUSS, CHRISTIAN, 'The Precursor: Edward Bernstein', in L. Labedz (ed.), *Revisionism: Essays in the History of Marxist Ideas* (London, 1962).

4

KARL KAUTSKY

For us . . . Socialism without democracy is unthinkable. We understand by Modern Socialism not merely social organization of production, but democratic organization of society as well.

The Dictatorship of the Proletariat (Ann Arbor, 1964), p. 6.

KARL KAUTSKY (1854–1938) *was the leading Marxist theorist during the two decades before the First World War. Born in Prague, he studied in Vienna and moved to London, where he spent several years collaborating with Engels. He wrote the theoretical section of the German Social Democratic Party's Erfurt Programme in 1891. Kautsky was the leading critic of Bernstein's revisionist ideas and his ascendancy was such that he was often referred to as the 'pope' of Marxism. Both Luxemburg and Lenin, however, accused Kautsky of being too 'centrist' in his politics. After the war Kautsky continued to write prolifically until he was forced to leave Germany by the Nazi seizure of power.*

Kautsky's version of Marxism was heavily influenced by Darwin in its determinist and evolutionary conception of history. In politics, Kautsky combined a commitment to working-class revolutionary action with a defence of parliamentary democracy. It was this latter that brought him into conflict with Lenin and the Bolsheviks. The extracts below are taken from Kautsky's short book The Dictatorship of the Proletariat, *written in 1918. Here he criticizes the Bolsheviks for instituting a party dictatorship, and insists on the importance of universal suffrage and parliamentary democracy, since the proletariat needs democracy not only before its conquest of power but also afterwards. Kautsky's criticisms earned him very sharp responses from Lenin and Trotsky, who branded him a 'renegade'. But his views undoubtedly represented the classical version of Marxian politics from Engels through to the Mensheviks.*

Socialist Democracy

THE modern State is a rigidly centralized organism, an organization comprising the greatest power within modern society, and influencing in the most effective way the fate of each individual, as is especially obvious in time of war.

The State is today what the family and community used to be for the individual. If communities were in their way democratically organized, the power of the State, on the contrary, including the bureaucracy and the army, looms over the people, even gaining such strength that at times it acquires an ascendancy over the classes which are socially and economically dominant, thus constituting itself an absolute government. Yet this latter condition is nowhere lasting. The absolute rule of bureaucracy leads to ossification and its absorption into endless time-wasting formulae, and that just at the time when industrial capitalism is developing, when the revolutionary methods of production which arise from it subject all economic and social conditions to constant change, and impart a quicker movement to industrial life, thus requiring the speediest political adjustments.

The absolute rule of bureaucracy, therefore, leads to arbitrariness and stultification, but a system of production like capitalism, in which each producer is dependent upon numerous others, needs for it prosperity the security and legality of social relations. The absolute State gets into conflict with the productive forces, and becomes a fetter on them. It is, then, urgently necessary for the executive to be subjected to public criticism, for free organizations of citizens to counterbalance the power of the State, for self-government in municipalities and provinces to be established, for the power of law-making to be taken from the bureaucracry, put under the control of a central assembly, freely chosen by the people, that is a Parliament. The control of the Government is the most important duty of Parliament, and in this it can be replaced by no other institution. It is conceivable, though hardly practicable, for the law-making power to be taken from the bureaucracy, and entrusted to various committees of experts, which would draft the laws and submit them to the people for their decision. The activities of the executive can only be supervised by another central body, and not by an unorganized and formless mass of people.

The attempts to overcome the absolute power of the State, as here described, are made by all classes in a modern State, with the exception of those which may share in its power, that is all except bureaucrats, court nobles, the State Church, as well as the great bankers who do a lucrative business with the State.

Before the united pressure of the other classes, which may include the landed gentry, the lower clergy, the industrial capitalists, the absolute regime must give way. In a greater or lesser degree it must concede freedom of the Press, of public meeting, or organization, and a Parliament. All the States of Europe have successfully passed through this development.

Every class will, however, endeavour to shape the new form of the State in a manner corresponding to its particular interests. This attempt is especially manifested in the struggle over the character of the Parliament, that is in the fight for the franchise. The watchword of the lower classes, of the people, is Universal Suffrage. Not only the wage-earner, but the small peasant and the lower middle classes, have an interest in the franchise.

Everywhere and under all circumstances these classes form the great majority of the population. Whether the proletariat is the predominant class among these depends on the extent of the economic development, although this factor does not determine whether the proletariat comprises the majority of the population. The exploiters are always a small minority of the population.

In the long run no modern State can withstand the pressure of these classes, and anything short of general suffrage in our society today would be an absurdity. In capitalist society, with its constantly changing conditions, the classes cannot be stereotyped in fixed grooves. All social conditions are in a state of flux. A franchise based on status is consequently excluded. A class which is not organized as such is a formless fluctuating mass, whose exact boundaries it is quite impossible to mark. A class is an economic entity, not a legal one. Class-membership is always changing. Many handworkers who, under the regime of small industry, think they are possessors, feel like proletarians under large industry, and are really proletarians even when for purposes of statistics they are included with the possessing classes and independent producers. There is also no franchise based on the census which would secure to the possessing classes a lasting monopoly of Parliament. It would be upset by every depreciation in money values. Finally, a franchise based on education would be even more futile, in view of the progress of culture among the masses. Thus various factors combine to render general suffrage the only solution in the society of today, and bring the question more and more to the front. Above all, it is the only rational solution from the standpoint of the proletariat as the lowest class of the population. The most effective weapon of the proletariat is its numerical strength. It cannot emancipate itself until it has become the largest class of the population, and until capitalist society is so far developed that the small peasants and the lower middle classes no longer overweight the proletariat.

The proletariat has also an interest in the fact that the suffrage should not only be universal and equal, but also non-discriminatory, so that men and women, or wage-earners and capitalists, do not vote in separate sections. Such a method would not only involve the danger that particular sections, who belong to the proletariat in reality, but are not wage-earners in form, would be separated from it, but it would also have the still worse result of narrowing the outlook of the proletariat. For its great historical mission consists in the fact that the collective interests of society fall into line with its permanent class interests, which are not always the same thing as special sectional interests. It is a symptom of the maturity of the proletariat when its class-consciousness is raised to the highest point by its grasp of large social relations and ends. This understanding is only made completely clear by scientific Socialism, not only by theoretical teaching, but by the habit of regarding things as a whole instead of looking at special interests which are furthered and extended by engaging in political action. Confining the outlook to trade interests narrows the mind, and this is one of the drawbacks to mere trade unionism. Herein lies the superiority of the organization of the Social Democratic Party, and also the superiority of a non-discriminatory, as compared with a franchise which divides the electors into categories.

In the struggle for the political rights referred to modern democracy arises, and the proletariat matures. At the same time a new factor appears, viz., the protection of minorities, the opposition in the State. Democracy signifies rule of majority, but not less the protection of minorities.

The absolute rule of bureaucracy strives to obtain for itself permanency. The forcible suppression of all opposition is its guiding principle. Almost everywhere it must do this to prevent its power being forcibly broken. It is otherwise with democracy, which means the rule of majorities. But majorities change. In a democracy no regime can be adapted to long duration.

Even the relative strength of classes is not a fixed quantity, at least in the capitalist era. But the strength of parties changes even quicker than the strength of classes, and it is parties which aspire to power in a democracy.

It must not here be forgotten, what so often happens, that the abstract simplification of theory, although necessary to a clear understanding of realities, is only true in the last resort, and between it and actualities there are many intervening factors. A class can rule, but not govern, for a class is a formless mass, while only an organization can govern. It is the political parties which govern in a democracy. A party is, however, not synonymous with a class, although it may, in the first place, represent a class interest. One and the same class interest can be represented in very different ways, by various tactical methods. According to their variety, the representatives of the same class interests are divided into different parties. Above all, the deciding factor is the position in relation to other classes and parties. Only seldom does a class dispose of so much power that it can govern the State by itself. If a class attains power, and finds that it cannot keep it by its own strength, it seeks for allies. If such allies are forthcoming, various opinions and standpoints prevail amongst the representatives of the dominant class interests.

In this way, during the eighteenth century Whigs and Tories represented the same landed interest, but one party endeavoured to further it by alliance with the middle classes of the towns at the expense of the Throne and its resources, while the other party conceived the monarchy to be its strongest support. Similarly today in England and also elsewhere, Liberals and Conservatives represent the same capitalist interests. But the one thinks they will be best served by an alliance with the landed class, and forcible suppression of the working classes, while the other fears dire consequences from this policy, and strives to conciliate the working classes by small concessions at the expense of the landed class.

As with the socially and economically ruling classes and their parties, so it is with the aspiring class and its parties.

Parties and classes are therefore not necessarily coterminous. A class can split up into various parties, and a party may consist of members of various classes. A class may still remain the rulers, while changes occur in the governing party, if the majority of the ruling class considers the methods of the existing governing party unsuitable, and that of its opponents to be more appropriate.

Government by parties in a democracy changes more rapidly than the rule of classes. Under these circumstances, no party is certain of retaining power, and must always count on the possibility of being in the minority, but by virtue of the nature of the State no party need remain in a minority for ever.

These conditions account for the growing practice of protecting minorities in a democracy. The deeper the roots which a democracy has struck, and the longer it has lasted and influenced political customs, the more effective is the minority, and the more successfully it can oppose the pretensions of any party which seems to remain in power at all costs.

What significance the protection of minorities has for the early stages of the Socialist Party, which everywhere started as a small minority, and how much it has helped the proletariat to mature, is clear. In the ranks of the Socialist Party the protection of minorities is very important. Every new doctrine, be it of a theoretical or a tactical nature, is represented in the first place by minorities. If these are forcibly suppressed, instead of being discussed, the majority is spared much trouble and inconvenience. Much unnecessary labour might be saved—a doctrine does not mean progress because it is new and championed by a minority. Most of what arises as new thought has already been discussed long before, and recognised as untenable, either by practice or by refutation.

Ignorance is always bringing out old wares as if they were something new. Other new ideas may be original, but put in a perverted shape. Although only a few of the new ideas and doctrines may spell real progress, yet progress is only possible through new ideas, which at the outset are put forward by minorities. The suppression of the new ideas of minorities in the Party would only cause harm to the proletarian class struggle, and an obstacle to the development of the proletariat. The world is always bringing us against new problems, which are not to be solved by the existing methods.

Tedious as it may be to sift the wheat from the chaff, this is an unavoidable task if our movement is not to stagnate, and is to rise to the height of the tasks before it. And what is needful for a party is also needful for the State. Protection of minorities is an indispensable condition for democratic development, and no less important than the rule of the majority.

Another characteristic of democracy is here brought in view, which is the form it gives to the class struggle.

In 1893 and in 1900 I have already discussed this matter, and give below some quotations from my writings:

Freedom of combination and of the Press and universal suffrage (under circumstances, even conscription) are not only weapons which are secured to the proletariat in the modern State by the revolutionary struggle of the middle class, but these institutions throw on the relative strength of parties and classes, and on the mental energy which vitalizes them, a light which is absent in the time of Absolutism. At that time the ruling, as well as the revolutionary, classes were fighting in the dark. As every expression of opposition was rendered impossible, neither the government nor the revolutionists were aware of their strength. Each of the two sides was thus exposed to the danger of overestimating its strength, so long as it refrained from measuring itself in a struggle with the opponent, and of underestimating its strength the moment it suffered a single defeat, and then threw its arms away.

This is really one of the chief reasons why, in the revolutionary period of the middle class, so many institutions collapsed at one blow, and so many governments were overthrown at a single stroke, and it also explains all the vicissitudes of revolution and counter-revolution.

It is quite different today, at least in countries which possess some measure of democratic government. These democratic institutions have been called the safety valve of society. It is quite false to say that the proletariat in a democracy ceases to be revolutionary, that it is contented with giving public expression to its indignation and its sufferings, and renounces the idea of social and political revolution. Democracy cannot remove the class antagonisms of capitalist society, nor prevent the overthrow of that society, which is their inevitable outcome. But if it cannot prevent the revolution, it can avoid many reckless and premature attempts at revolution, and render many revolutionary movements unnecessary. It gives a clear indication of the relative strength of classes and parties; it does not do away with their antagonism, nor does it avoid the ultimate outcome of their struggle, but it serves to prevent the rising classes from attempting tasks to which they are not equal, and it also restrains the ruling classes from refusing concessions when they no longer have the strength to maintain such refusal. The direction of evolution is not thereby altered, but the pace is made more even and steady. The coming to the front of the proletariat in a State with some measure of democratic government will not be marked by such a striking victory as attended the middle classes in their revolutionary period, nor will it be exposed to a violent overthrow.

Since the rise of the modern Social Democratic working-class movement in the sixties, the European proletariat has only suffered one great defeat, in the Paris Commune of 1871. At the time France was still suffering from the consequences of the Empire, which had withheld real democratic institutions from the people, the French proletariat had only attained to the slightest degree of class-consciousness, and the revolt was provoked.

The proletarian-democratic method of conducting the struggle may seem to be a slower affair than the revolutionary period of the middle class; it is certainly less dramatic and striking, but it also exacts a smaller measure of sacrifice. This may be quite indifferent to the finely endowed literary people who find in Socialism an interesting pastime, but not to those who really carry on the fight.

This so-called peaceful method of the class struggle, which is confined to non-militant methods, parliamentarism, strikes, demonstrations, the Press, and similar means of pressure, will retain its importance in every country according to the effectiveness of the democratic institutions which prevail there, the degree of political and economic enlightenment, and the self-mastery of the people.

On these grounds, I anticipate that the social revolution of the proletariat will assume quite other forms than that of the middle class, and that it will be possible to carry it out by peaceful economic, legal, and moral means, instead of by physical force, in all places where democracy has been established.

The above is my opinion today.

Of course, every institution has its bad side, and disadvantages can be discovered in democracy.

Where the proletariat is without rights, it can develop no mass organization, and normally cannot promote mass action; there it is only possible for a handful of reckless fighters to offer lasting opposition to the governing regime. But this élite is daily confronted with the necessity of bringing the entire system to an end. Undistracted by the small demands of daily politics, the mind is concentrated on the largest problems, and learns constantly to keep in view the entire political and social relations.

Only a small section of the proletariat takes part in the fight, but it cherishes keen theoretical interest, and is inspired by the great aims.

Quite differently does democracy affect the proletariat, when it has only a few hours a day at its disposal under present-day conditions. Democracy develops mass organizations involving immense administrative work: it calls on the citizen to discuss and solve numerous questions of the day, often of the most trivial kind. The whole of the free time of the proletariat is more and more taken up with petty details and its attention occupied by passing events. The mind is contracted within a narrow circle. Ignorance and even contempt of theory, opportunism in place of broad principles, tend to get the upper hand. Marx and Engels praised the theoretical mind of the German working class, in contrast with the workers of Western Europe and America. They would today find the same theoretical interest among the Russian workers, in comparison with the Germans.

Nevertheless, everywhere the class-conscious proletariat and their representatives fight for the realization of democracy, and many of them have shed their life's blood for it.

They know that without democracy nothing can be done. The stimulating results of the struggle with a despotism are confined to a handful, and do not touch the masses. On the other hand, the degenerating influence of democracy on the proletariat need not be exaggerated. Often is it the consequence of the lack of leisure from which the proletariat suffers, not of democracy itself.

It were indeed extraordinary if the possession of freedom necessarily made men more narrow and trivial than its absence. The more democracy tends to shorten the working day, the greater the sum of leisure at the disposal of the proletariat, the more it is enabled to combine devotion to large problems with attention to necessary detail. And the impulse thereto is not lacking. For whatever democracy may be able to accomplish it cannot resolve the antagonisms inherent in a capitalist system of production, so long as it refrains from altering this system. On the contrary, the antagonisms in capitalist society become more acute and tend to provoke bigger conflicts, in this way forcing great problems on the attention of the proletariat, and taking its mind off routine and detail work.

Under democracy this moral elevation is no longer confined to a handful, but is shared in by the whole of the people, who are at the same time gradually accustomed to self-government by the daily performance of routine work.

Again, under democracy, the proletariat does not always think and talk of revolution, as under despotism. It may for years, and even decades, be immersed in detail work, but everywhere situations must arise which will kindle in it revolutionary thought and aspirations.

When the people are roused to action under a democracy, there is less danger than under despotism that they have been prematurely provoked, or will waste their energy in futile efforts. When victory is achieved, it will not be lost, but successfully maintained. And that is better in the end that the mere nervous excitement of a fresh revolutionary drama.

The pernicious features of the method of dictatorship here discussed must now be contrasted with more favourable aspects. It furnishes a striking object lesson, and even if it cannot last it is able to accomplish many things to the advantage of the proletariat, which cannot be lost.

Let us look closely at the object lesson. This argument obviously rests on the following consideration: Under democracy, by virtue of which the majority of the people rule, Socialism can only be brought about when a majority in its favour, is gained. A long and tedious way. We reach our goal far quicker if an energetic minority which knows its aims, seizes hold of the power of the State, and uses it for passing Socialist measures. Its success would at once compel conviction, and the majority, which hitherto had opposed, would quickly rally to Socialism.

This sounds very plausible, and sounded so in the mouth of old Weitling. It has only the one defect that it assumes that which has to be proved. The opponents of the method of dictatorship contest the assumption that Socialist production can be brought about by a minority without the co-operation of the great mass of the people. If the attempt fails, it certainly is an object lesson, but in the wrong sense, not by attracting, but by frightening.

People who are influenced by such an object lesson, and not by examining and verifying social relations, thoughtless worshippers of mere success, would, in the case of the attempt failing, not inquire from what causes it did not succeed. They would not seek for the explanation in the unfavourable or unripe conditions, but in Socialism itself, and would conclude that Socialism is realizable under no circumstances.

It is apparent that the object lesson has a very dangerous side.

How has it been represented to us?

We may popularly express the essentials of Socialism in the words: Freedom and bread for all. That is what the masses expect from it, and why they rally to it. Freedom is not less important than bread. Even well-to-do and rich classes have fought for their freedom, and not seldom have made the biggest sacrifices for their convictions in blood and treasure. The need for freedom, for self-determination, is as natural as the need for food.

Hitherto Social Democracy did represent to the masses of the people the object lesson of being the most tireless champion of the freedom of all who were oppressed, not merely the wage-earner, but also of women, persecuted religions and races, the Jews, Negroes, and Chinese. By this object lesson it has won adherents quite outside the circle of wage-earners.

Now, so soon as Social Democracy attains to power, this object lesson is to be replaced by one of an opposite character. The first step consists in the suspension of universal suffrage and of liberty of the Press, the disfranchisement of large masses of the people, for this must always take place if dictatorship is substituted for democracy. In order to break the political influence of the upper ten thousand, it is not necessary to exclude them from the franchise. They exercise this influence not by their personal votes. As regards small shopkeepers, home workers, peasants who are well off and in moderate condition, the greater part of the intellectuals, so soon as the dictatorship deprives them of their rights, they are changed at once into enemies of Socialism by this kind of object lesson, so far as they are not inimical from the beginning. Thus all those who adhere to Socialism on the ground that it fights for the freedom of all would become enemies of the proletarian dictatorship.

This method will win nobody who is not already a Socialist. It can only increase the enemies of Socialism.

But we saw that Socialism promised not only freedom, but also bread. This ought to reconcile those whom the Communist dictatorship robbed of freedom.

They are not the best of the masses who are consoled in their loss of freedom with bread and pleasure. But without doubt material well-being will lead many to Communism who regard it sceptically, or who are by it deprived of their rights. Only this prosperity must really come, and that quickly, not as a promise for the future, if the object lesson is to be effective.

How is this prosperity to be attained? The necessity for dictatorship presupposes that a minority of the population have possessed themselves of the power of the State. A minority composed of those who possess nothing. The greatest weapon of the proletariat is, however, its numbers, and in normal times it can only progress on these lines, conquering the political power only when it forms the majority. As a minority it can only achieve power by the combination of extraordinary circumstances, by a catastrophe which causes the collapse of a regime, and leaves the State helpless and impoverished.

Under such circumstances, Socialism, that is general well-being within modern civilization, would only be possible through a powerful development of the productive forces which capitalism brings into existence, and with the aid of the enormous riches which it creates and concentrates in the hands of the capitalist class. A State which, by a foolish policy or by unsuccessful war has dissipated these riches, is by its nature condemned to be an unfavourable starting-point for the rapid diffusion of prosperity in all classes.

If, as the heir of the bankrupt State, not a democratic but a dictatorial regime enters into power, it even renders the position worse, as civil war is its necessary consequence. What might still be left in the shape of material resources is wasted by anarchy.

In fine, the uninterrupted progress of production is essential for the prosperity of all. The destruction of capitalism is not Socialism. Where capitalist production cannot be transformed at once into Socialist production, it must go on as before, otherwise the process of production will be interrupted, and that hardship for the masses will ensue which the modern proletariat so much fears in the shape of general unemployment.

In those places where, under the new conditions, capitalist production has been rendered impossible, Socialist production will only be able to replace it if the proletariat has acquired experience in self-government, in trade unions, and on town councils, and has participated in the making of laws and the control of government, and if numerous intellectuals are prepared to assist with their services the new methods.

In a country which is so little developed economically that the proletariat only forms a minority, such maturity of the proletariat is not to be expected.

It may therefore be taken for granted that in all places where the proletariat can only maintain itself in power by a dictatorship, instead of by democracy, the difficulties with which Socialism is confronted are so great that it would seem to be out of the question that dictatorship could rapidly bring about prosperity for all, and in this manner reconcile to the reign of force the masses of the people who are thereby deprived of political rights.

As a matter of fact, we see that the Soviet Republic, after nine months of existence, instead of diffusing general prosperity, is obliged to explain how the general poverty arises.

We have lying before us: 'Theses respecting the Socialist Revolution and the tasks of the proletariat during its dictatorship in Russia', which emanates from the Bolshevist side. A passage deals with 'the difficulties of the position'.

Paragraph 28 reads as follows:

28. The proletariat has carried out positive organic work under the greatest difficulties. The internal difficulties are: The wearing out and enormous exhaustion of the social resources and even their dissolution in consequence of the war, the policy of the capitalist class before the October revolution (their calculated policy of disorganization, in order, after the 'anarchy', to create a middle-class dictatorship), the general sabotage of the middle class and the intellectuals after the October revolution; the permanent counter-revolutionary revolt of the ex-officers, generals, and middle classes, with arms or without; *lack of technical skill and experience on*

the part of the working class itself [italicized in original], lack of organizing experience; the existence of large masses of the small middle class which are an unorganized class, par excellence, etc.

This is all very true. But it does not indicate anything else than that the conditions are not ripe. And does it not strikingly show that an object lesson on the lines of Socialism is, under these conditions in present-day Russia, not to be thought of? It is a famous object lesson which makes it necessary for theoretical arguments to be set out why that which is to be shown is not possible at the moment. Will it convert those who have hitherto opposed Socialism, and who are only to be convinced by its practical success?

Of course, a new regime will come up against unexpected difficulties. It is wrong to lay the blame for them on this regime, as a matter of course, and to be discouraged by them without closer examination of the circumstances. But if one is to persevere, in spite of these difficulties, then it is necessary to win beforehand a strong conviction of the justice and necessity of this regime. Only then will confusion be avoided. Success worshippers are always uncertain Cantonists.

So we are driven back upon democracy, which obliges us to strive to enlighten and convince the masses by intensive propaganda before we can reach the point of bringing Socialism about. We must here again repudiate the method of dictatorship, which substitutes compulsory object lessons for conviction.

This is not to say that object lessons may avail nothing in the realization of Socialism. On the contrary, they can and will play a great part in this, but not through the medium of dictatorship.

The various States of the world are at very different stages of economic and political development. The more a State is capitalistic on the one side and democratic on the other, the nearer it is to Socialism. The more its capitalist industry is developed, the higher is its productive power, the greater its riches, the more socially organized its labour, the more numerous its proletariat; and the more democratic a State is, the better trained and organized is its proletariat. Democracy may sometimes repress its revolutionary thought, but it is the indispensable means for the proletariat to attain that ripeness which it needs for the conquest of political power, and the bringing about of the social revolution. In no country is a conflict between the proletariat and the ruling classes absent, but the more a country is progressive in capitalism and democracy, the greater is the prospect of the proletariat, in such a conflict, not merely of gaining a passing victory, but also of maintaining it.

Where a proletariat, under such conditions, gains control of the State, it will discover sufficient material and intellectual resources to permit it at once to give the economic development a Socialist direction, and immediately to increase the general well-being.

This will then furnish a genuine object lesson to countries which are economically and politically backward. The mass of their proletariat will now unanimously demand measures on the same lines, and also all other sections of the poorer classes, as well as numerous intellectuals, will demand that the State should take the same road to general prosperity. Thus, by the example of the progressive countries, the cause of Socialism will become irresistible in countries which today are not so advanced as to allow their proletariat of its own strength to conquer the power of the State, and put Socialism into operation.

And we need not place this period in the distant future. In a number of industrial States the material and moral prerequisites for Socialism appear already to exist in sufficient measure. The question of the political dominion of the proletariat is merely a question of power alone, above all of the determination of the proletariat to engage in resolute class struggle. But Russia is not one of these leading industrial States. What is being enacted there now is, in fact, the last of middle class, and not the first of Socialist revolutions. This shows itself ever more distinctly. Its present Revolution could only assume a Socialist character if it coincided with Socialist revolutions in Western Europe.

That, by an object lesson of this kind in the more highly developed nations, the pace of social development may be accelerated, was already recognized by Marx in the preface to the first edition of 'Capital':

One nation can and should learn from others. And even when a society has got upon the right track for the discovery of the natural laws of its movement—it can neither clear by bold leaps, nor remove by legal enactments the obstacles offered by the successive phases of its normal development. But it can shorten and lessen the birth-pangs.

In spite of their numerous calls on Marx, our Bolshevist friends seem to have quite forgotten this passage, for the dictatorship of the proletariat, which they preach and practise, is nothing but a grandiose attempt to clear by bold leaps or remove by legal enactments the obstacles offered by the successive phases of normal development. They think that it is the least painful method for the delivery of Socialism, for 'shortening and lessening its birth-pangs'. But if we are to continue in metaphor, then their practice reminds us more of a pregnant woman who performs the most foolish exercises in order to shorten the period of gestation, which makes her impatient, and thereby causes a premature birth.

The result of such proceedings is, as a rule, a child incapable of life.

Marx speaks here of the object lesson which one nation may afford another. Socialism is, however, concerned with yet another kind of object lesson, viz., that which a highly developed industry may furnish to an industry which is backward.

To be sure, capitalist competition everywhere tends to displace old-fashioned industrial methods, but under capitalist conditions this is so painful a process that those threatened by its operation strive to avert it by all means. The Socialist method of production would therefore find in existence a number of processes which are technically obsolete; for example, in agriculture, where large-scale production has made little progress, and in places is even receding.

Socialist production can only develop on the basis of the large industry. Socialist agriculture would have to consist solely in the socialization of what large-scale production already exists. If good results are thereby obtained, which is to be expected, provided the social labour of freely organized men is substituted for wage labour (which only produces very inadequate results in agriculture), the conditions of the workers in the large Socialist industry will be seen to be more favourable than those of the small peasants, and it may then be anticipated with certainty that the latter will voluntarily pass over to the new productive methods, when society furnishes them with the necessary means. But not before. In agriculture the way for Socialism is not prepared by capitalism in any adequate measure. And it is quite hopeless to try to convince peasant proprietors of the theoretical super-iority of Socialism. Only the object lesson of the socialization of peasant agriculture can help. This, however, presupposes a certain extension of large-cale agriculture. The object lesson will be the quicker and more effective ac-cording to the degree of development of large-scale industry in the country.

The policy of the lower middle class democrats, which has been taken up by Social Democrats of the David school, and in some respects made more extreme, that is, the destruction of any large-scale agriculture and its partition into small-scale industry, is sharply opposed to Socialism as applied to agriculture, and therefore to Socialism as applied to society generally.

The most striking feature of the present Russian revolution is its working out on the lines of Eduard David. He, and not Lenin, has given the revolution its peculiar direction in this respect. That is the socialist instruction which it imparts. It testifies, in fact, to its middle-class character.

We have seen that the method of dictatorship does not promise good results for the proletariat, either from the standpoint of theory or from that of the special Russian conditions; nevertheless, it is understandable only in the light of these conditions.

The fight against tsarism was for a long time a fight against a system of government which had ceased to be based on the conditions prevailing, but was only maintained by naked force, and only by force was to be overthrown. This fact would easily lead to a cult of force even among the revolutionaries, and to overestimating what could be done by the powers over them, which did not repose on the economic conditions, but on special circumstances. Accordingly, the struggle against tsarism was carried on secretly, and the method of conspiracy created the manners and the habits proper to dictatorship, and not to democracy.

The operation of these factors was however, crossed by another consequence of the struggle against absolutism. We have already referred to the fact that, in contradistinction to democracy, which awakens an interest for wider relations and greater objects side by side with its constant pre-occupations with momentary ends, absolutism arouses theoretical interest. There is today, however, only one revolutionary theory of society, that of Karl Marx.

This became the theory of Russian Socialism. Now what this theory teaches is that our desires and capabilities are limited by the material conditions, and it shows how powerless is the strongest will which would rise superior to them. It conflicted sharply with the cult of mere force, and caused the Social Democrats to recognize that definite boundaries were set to their participation in the coming revolution, which, owing to the economic backwardness of Russia, could only be a middle-class one.

Then the second revolution came, and suddenly brought a measure of power to the Socialists which surprised them, for this revolution led to the complete demobilization of the army, which was the strongest support of property and middle-class order. And at the same time as the physical support collapsed, the moral support of this order went to pieces, neither the Church nor the intellectuals being able to maintain their pretensions. The rule devolved on the lower classes in the State, the workers and peasants, but the peasants do not form a class which is able itself to govern. They willingly permitted themselves to be led by a Proletarian Party, which promised them immediate peace, at whatever price, and immediate satisfaction of their land hunger. The masses of the proletariat rallied to the same party, which promised them peace and bread.

Thus the Bolshevist Party gained the strength which enabled it to seize political power. Did this not mean that at length the prerequisite was obtained which Marx and Engels had postulated for the coming of Socialism, viz., the conquest of political power by the proletariat? In truth, economic theory discountenanced the idea that Socialist production was realizable at once under the social conditions of Russia, and not less unfavourable to it was the practical confirmation of this theory, that the new regime in no way signified the sole rule of the proletariat, but the rule of a coalition of proletarian and peasant elements, which left each section free to behave as it liked on its own territory. The proleteriat put nothing in the way of the peasants as regards the land, and the peasants put no obstacle in the way of the proletariat as regards the factories. None the less, a Socialist Party had become the ruler in a great State, for the first time in the world's history. Certainly a colossal and, for the fighting proletariat, a glorious event.

But for what can a Socialist Party use its power except to bring about Socialism? It must at once proceed to do so, and, without thought or regard, clear out of the way all obstacles which confront it. If democracy thereby comes in conflict with the new regime, which, in spite of the great popularity which it so quickly won, cannot dispose of a majority of the votes in the Empire, then so much the worse for democracy. Then it must be replaced by dictatorship, which is all the easier to accomplish, as the people's freedom is quite a new thing in Russia, and as yet has struck no deep roots amongst the masses of the people. It was now the task of dictatorship to bring about Socialism. This object lesson must not only suffice for the elements in its own country which are still in opposition, but must also compel the proletariat of other capitalist countries to imitation, and provoke them to revolution.

This was assuredly a train of thought of outstanding boldness and fascinating glamour for every proletarian and every Socialist. What we have struggled for during half a century, what we have so often thought ourselves to be near, what has always again evaded us, is at length going to be accomplished. No wonder that the proletarians of all countries have hailed Bolshevism. The reality of proletarian rule weighs heavier in the scale than theoretical considerations. And that consciousness of victory is still more strengthened by mutual ignorance of the conditions of the neighbour. It is only possible for a few to study foreign countries, and the majority believe that in foreign countries it is at bottom the same as with us, and when this is not believed, very fantastic ideas about foreigners are entertained.

Consequently, we have the convenient conception that everywhere the same imperialism prevails, and also the conviction of the Russian Socialists that the political revolution is as near to the peoples of Western Europe as it is in Russia, and, on the other hand, the belief that the conditions necessary for Socialism exist in Russia as they do in Western Europe.

What happened, once the army had been dissolved and the Assembly had been proscribed, was only the consequence of the step that had been taken.

All this is very understandable, if not exactly encouraging. On the other hand, it is not so conceivable why our Bolshevist comrades do not explain their measures on the ground of the peculiar situation in Russia, and justify them in the light of the pressure of the special circumstances, which, according to their notions, left no choice but dictatorship or abdication. They went beyond this by formulating quite a new theory, on which they based their measures, and for which they claimed universal application.

For us the explanation of this is to be found in one of their characteristics, for which we should have great sympathy, viz., their great interest in theory.

The Bolshevists are Marxists, and have inspired the proletarian sections coming under their influence with great enthusiasm for Marxism. Their dictatorship, however, is in contradiction to the Marxist teaching that no people can overcome the obstacles offered by the successive phases of their development by a jump, or by legal enactment. How is it that they find a Marxist foundation for their proceedings?

They remembered opportunely the expression, 'the dictatorship of the proletariat', which Marx used in a letter written in 1875. In so doing he had, indeed, only intended to describe a political *condition*, and not a *form of government*. Now this expression is hastily employed to designate the latter, especially as manifested in the rule of the Soviets.

Now if Marx had somewhere said that under certain circumstances things might come to a dictatorship of the proletariat, he has described this condition as one unavoidable for the transition to Socialism. In fact, as he declared, almost at the same time, that in countries like England and America a peaceful transition to Socialism was possible, which would only be on the basis of democracy and not of dictatorship, he has also shown that he did not mean by dictatorship the suspension of democracy. Yet this does not disconcert the champions of dictatorship. As Marx once stated that the dictatorship of the proletariat might be unavoidable, so they announce that the Soviet Constitution, and the disfranchising of its opponents, was recognized by Marx himself as the form of government corresponding to the nature of the proletariat, and indissolubly bound up with its rule. As such it must last as long as the rule of the proletariat itself, and until Socialism is generally accomplished and all class distinctions have disappeared.

In this sense dictatorship does not appear to be a transitory emergency measure, which, so soon as calmer times have set in, will again give place to democracy, but as a condition for the long duration of which we must adapt ourselves.

This interpretation is confirmed by Theses 9 and 10 respecting the Social Revolution, which state:

(9) Hitherto, the necessity of the Dictatorship of the Proletariat was taught, without enquiring as to the form it would take. The Russian Socialist Revolution has discovered this form. It is the form of the Soviet Republic as the type of the permanent Dictatorship of the Proletariat and (in Russia) of the poorer classes of peasant. It is therefore necessary to make the following remarks. We are speaking now, not of a passing phenomenon, in the narrower sense of the word, but of a particular form of the State during the whole historical epoch. What needs now to be done is to organize a new form of the State, and this is not to be confused with special measures directed against the middle class which are only functions of a special State organization appropriate to the colossal tasks and struggle.

(10) The proletarian dictatorship accordingly consists, so to speak, in a permanent state of war against the middle class. It is also quite clear that all those who cry out about the violence of the Communists completely forget what dictatorship really is. The Revolution itself is an act of naked force. The word dictatorship signifies in all languages nothing less than government by force. The class meaning of force is here important, for it furnishes the historical justification of revolutionary force. It is also quite obvious that the more difficult the situation of the Revolution becomes, the sharper the dictatorship must be.

From the above it is also apparent that dictatorship as a form of government is not only to be a permanent thing, but will also arise in all countries.

If in Russia now the newly acquired general freedom is put an end to again, this must also happen after the victory of the proletariat in countries where the people's freedom is already deeply rooted, where it has existed for half a century and longer, and where the people have won it and maintained it in frequent bloody revolutions. The new theory asserts this in all earnestness. And stranger still it finds support not only amongst the workers of Russia, who still remember the yoke of the old tsardom, and now rejoice to be able to turn the handle for once, even as apprentices when they become journeymen rejoice when they may give the apprentices who come after them the drubbing they used to receive themselves—no, the new theory finds support even in old democracies like Switzerland.

Yet something stranger still and even less understandable is to come.

A complete democracy is to be found nowhere, and everywhere we have to strive after modifications and improvements. Even in Switzerland there is an agitation for the extension of the legislative powers of the people, for proportional representation, and for woman suffrage. In America the power and mode of selection of the highest judges need to be very severely restricted. Far greater are the demands that should be put forward by us in the great bureaucratic and militarist States in the interests of democracy.

And in the midst of these struggles, the most extreme fighters raise their heads, and say to the opponents: That which we demand for the protection of minorities, the opposition, we only want so long as we ourselves are the opposition, and in the minority. As soon as we have become the majority, and gained the power of government, our first act will be to abolish as far as you are concerned all that we formerly demanded for ourselves, viz., franchise, freedom of Press and of organization, etc.

The Theses respecting the Socialist revolution are quite unequivocal on this point:

(17) The former demands for a democratic republic, and general freedom (that is freedom for the middle classes as well) were quite correct in the epoch that is now passed, the epoch of preparation and gathering of strength. The worker needed freedom for his Press, while the middle-class Press was noxious to him, but he could not at this time put forward a demand for the suppression of the middle-class Press. Consequently, the proletariat demanded general freedom, even freedom for reactionary assemblies, for black labour organizations.

(18) Now we are in the period of the direct attack on capital, the direct overthrow and destruction of the imperialist robber State, and the direct suppression of the middle class. It is therefore absolutely clear that in the present epoch the principle of defending general freedom (that is also for the counter-revolutionary middle class) is not only superfluous, but directly dangerous.

(19) This also holds good for the Press, and the leading organizations of the social traitors. The latter have been unmasked as the active elements of the counter-revolution. They even attack with weapons the proletarian Government. Supported by former officers and the money bags of the defeated finance capital, they appear on the scene as the most energetic organizations for various conspiracies. The proletariat dictatorship is their deadly enemy. Therefore, they must be dealt with in a corresponding manner.

(20) As regards the working class and the poor peasants, these possess the fullest freedom.

Do they really possess the fullest freedom?

The 'social traitors' are proletarians and Socialists, too, but they offer opposition, and are therefore to be deprived of rights like the middle-class opposition. Would we not display the liveliest anger and fight with all our strength in any case where a middle-class government endeavoured to employ similar measures against its opposition?

Certainly we should have to do so but our efforts would only have a laughable result if the middle-class government could refer to Socialist precepts like the foregoing, and a practice corresponding with them.

How often have we reproached the Liberals that they are different in government from what they are in opposition, and that then they abandon all their democratic pretensions. Now the Liberals are at least sufficiently prudent to refrain from the formal abandonment of any of their democratic demands. They act according to the maxim; one does this, but does not say so.

The authors of the Theses are undeniably more honourable; whether they are wiser may be doubted. What would be thought of the wisdom of the German Social Democrats, if they openly announced that the democracy for which they fight today would be abandoned the day after victory. That they have perverted their democratic principles to their opposites, or that they have no democratic principles at all; that democracy is merely a ladder for them, up which to climb to governmental omnipotence, a ladder they will no longer need, and will push away, as soon as they have reached the top, that, in a word, they are revolutionary opportunists.

Even for the Russian revolutionaries it is a short-sighted policy of expediency, if they adopt the method of dictatorship, in order to gain power, not to save the jeopardized democracy, but in order to maintain themselves in spite of it. This is quite obvious.

On the other hand, it is less obvious why some German Social Democrats who are not yet in power, who furthermore only at the moment represent a weak opposition, accept this theory. Instead of seeing something which should be generally condemned in the method of dictatorship, and the disfranchising of large sections of the people, which at the most is only defensible as a product of the exceptional conditions prevailing in Russia, they go out of their way to praise this method as a condition which the German Social Democracy should also strive to realize.

This assertion is not only thoroughly false, it is in the highest degree destructive. If generally accepted, it would paralyse the propagandist strength of our party to the utmost, for, with the exception of a small handful of sectarian fanatics, the entire German, as also the whole proletariat of the world, is attached to the principle of general democracy. The proletariat would angrily repudiate every thought of beginning its rule with a new privileged class, and a new disfranchised class. It would repudiate every suggestion of coupling its demand for general rights for the whole people with a mental reservation, and in reality only strive for privileges for itself. And not less would it repudiate the comic insinuation of solemnly declaring now that its demand for democracy is a mere deceit.

Dictatorship as a form of government in Russia is as understandable as the former anarchism of Bakunin. But to understand it does not mean that we should recognize it; we must reject the former as decisively as the latter. The dictatorship does not reveal itself as a resource of a Socialist Party to secure itself in the sovereignty which has been gained in opposition to the majority of the people, but only as means of grappling with tasks which are beyond its strength, and the solution of which exhausts and wears it; in doing which it only too easily compromises the ideas of Socialism itself, the progress of which it impedes rather than assists.

Happily, the failure of the dictatorship is not synonynous with a collapse of the Revolution. It would be so only if the Bolshevist dictatorship was the mere prelude to a middle-class dictatorship. The essential achievements of the Revolution will be saved, if dictatorship is opportunely replaced by democracy.

Source: K. Kautsky, *The Dictatorship of the Proletariat* (Ann Arbor, 1964), pp. 25 ff., 88 ff., 135 ff.

FURTHER NOTES

WORKS

KAUTSKY, K., *The Road to Power* (Chicago, 1971).
—— *The Class Struggle* (New York, 1971).
—— *The Dictatorship of the Proletariat* (Ann Arbor, 1964).
—— *Terrorism and Communism* (London, 1920).
—— *Selected Political Writings*, ed. P. Goode (London, 1984).

COMMENTARIES

GEARY, D., *Karl Kautsky* (Manchester, 1987).
KOLAKOWSKI, L., *Main Currents of Marxism* (Oxford, 1978), vol. ii, ch. 2.
LICHTHEIM, G., *Marxism* (London, 1964), pt. 5, section 5.
McLELLAN, D., *Marxism After Marx* (New York, 1980), ch. 2.
SALVADORI, M., *Karl Kautsky and the Socialist Revolution* (London 1979).
STEENSON, G., *Karl Kautsky 1854–1938: Marxism in the Classical Era* (Pittsburgh, 1978).

5

ROSA LUXEMBURG

> The masses are the decisive factor; they are the rock upon which the
> final victory of the revolution is erected.
>
> *Selected Political Writings*, ed. R. Looker (London, 1972), p. 306.

ROSA LUXEMBURG (1871–1919) *was born in Poland but spent most of her
very active political life in Germany where she rapidly became prominent in
the left wing of the Marxist movement. The first extract below, from* Social
Reform or Revolution?, *written in 1900, contains the best Marxist reply
to Bernstein's revisionism. Here Luxemburg claims that trade-union and
parliamentary activity may reform, but can never abolish, capitalist
relations of production: such an abolition would only be achieved by a
revolution involving the conquest of political power by the working class.
This theme is continued in the second extract, from* Mass Strike, Party and
Trade Unions, *written in 1906, in which, inspired by the 1905 revolution
in Russia, she advocated the mass strike as the most important weapon to
promote spontaneous creative working-class involvement in revolutionary
activity—an advocacy which led to her break with Kautsky, who favoured
a more cautious parliamentary approach. The last two extracts indicate
Luxemburg's disagreements with Lenin. In the first, from* Organizational
Questions of Russian Social Democracy, *written in 1904, Luxemburg criticizes Lenin's conception of a highly centralized vanguard party, which, she
felt, would prevent the broad mass of the workers from giving themselves
political expression. In the second, from* The Russian Revolution, *written
from prison in 1919, Luxemburg, after making clear her fundamental
support for the Bolshevik revolution of 1917, takes Lenin and Trotsky to
task for their tendency to substitute dictatorship for democracy. Shortly
afterwards Luxemburg was brutally murdered by right-wing soldiers during
a failed uprising in Berlin.*

1.   Social Reform or Revolution?

IN the first chapter we aimed to show that Bernstein's theory lifted the
programme of the Socialist movement off its material base and tried to
place it on an idealist base. How does this theory fare when translated into
practice?

Upon the first comparison, the party practice resulting from Bernstein's theory does not seem to differ from the practice followed by the Social Democracy up to now. Formerly, the activity of the Social Democratic Party consisted of trade-union work, of agitation for social reforms, and the democratization of existing political institutions. The difference is not in the *what* but in the *how*.

At present, the trade-union struggle and parliamentary practice are considered to be the means of guiding and educating the proletariat in preparation for the task of taking over power. From the revisionist standpoint, this conquest of power is at the same time impossible and useless. And therefore, trade-union and parliamentary activity are to be carried on by the party only for their immediate results, that is, for the purpose of bettering the present situation of the workers, for the gradual reduction of capitalist exploitation, for the extension of social control.

So that if we do not consider momentarily the immediate amelioration of the workers' condition—an objective common to our party programme as well as to revisionism—the difference between the two outlooks is, in brief, the following. According to the present conception of the party, trade-union and parliamentary activity are important for the Socialist movement because such activity prepares the proletariat, that is to say, creates the *subjective* factor of the Socialist transformation, for the task of realizing Socialism. But according to Bernstein, trade unions and parliamentary activity gradually reduce capitalist exploitation itself. They remove from capitalist society its capitalist character. They realize *objectively* the desired social change.

Examining the matter closely, we see that the two conceptions are diametrically opposed. Viewing the situation from the current standpoint of our party, we say that, as a result of its trade-union and parliamentary struggles, the proletariat becomes convinced of the impossibility of accomplishing a fundamental social change through such activity and arrives at the understanding that the conquest of power is unavoidable. Bernstein's theory, however, begins by declaring that this conquest is impossible. It concludes by affirming that Socialism can only be introduced as a result of the trade-union struggle and parliamentary activity. For as seen by Bernstein, trade-union and parliamentary action has a Socialist character because it exercises a progressively socializing influence on capitalist economy.

We tried to show that this influence is purely imaginary. The relations between capitalist property and the capitalist State develop in entirely opposite directions, so that the daily practical activity of the present Social Democracy loses, in the last analysis, all connection with work for Socialism.

From the viewpoint of a movement for Socialism, the trade-union struggle and our parliamentary practice are vastly important in so far as they make Socialistic the *awareness*, the consciousness, of the proletariat and help to organize it as a class. But once they are considered as instruments of the direct socialization of capitalist economy, they lose not only their usual effectiveness but cease being means of preparing the working class for the conquest of power. Eduard Bernstein the Konrad Schmidt suffer from a complete misunderstanding when they console themselves with the belief that even though the programme of the party is reduced to work for social reforms and ordinary trade-union work, the final objective of the labour movement is not thereby discarded, for each forward step reaches beyond the given immediate aim and the Socialist goal is implied as a tendency in the supposed advance.

That is certainly true about the present procedure of the German Social Democracy. It is true whenever a firm and conscious effort for the conquest of political power impregnates the trade-union struggle and the work for social reforms. But if this effort is separated from the movement itself and social reforms are made an end in themselves, then such activity not only does not lead to the final goal of Socialism but moves in a precisely opposite direction.

Konrad Schmidt simply falls back on the idea that an apparently mechanical movement, once started, cannot stop by itself, because 'one's appetite grows with eating', and the working class will not supposedly content itself with reforms till the final Socialist transformation is realized.

Now the last mentioned condition is quite real. Its effectiveness is guaranteed by the very insufficiency of capitalist reforms. But the conclusion drawn from it could only be true if it were possible to construct an unbroken chain of augmented reforms leading from the capitalism of today to Socialism. This is, of course, sheer fantasy. In accordance with the nature of things as they are the chain breaks quickly, and the paths that the supposed forward movement can take from that point on are many and varied.

What will be the immediate result should our party change its general procedure to suit a viewpoint that wants to emphasize the practical results of our struggle, that is, social reforms? As soon as 'immediate results' become the principal aim of our activity, the clear-cut, irreconcilable point of view, which has meaning only in so far as it proposes to win power, will be found more and more inconvenient. The direct consequence of this will be the adoption by the party of a 'policy of compensation', a policy of political trading, and an attitude of diffident, diplomatic conciliation. But this attitude cannot be continued for a long time. Since the social reforms

can only offer an empty promise, the logical consequence of such a programme must necessarily be disillusionment.

*It is not true that Socialism will arise automatically from the daily struggle
of the working class. Socialism will be the consequence of* (1) *the growing
contradictions of capitalist economy and* (2) *the comprehension by the
working class of the unavoidability of the suppression of these contradictions through a social transformation.* When, in the manner of revisionism, the first condition is denied and the second rejected, the labour
movement finds itself reduced to a simple co-operative and reformist movement. We move here in a straight line toward the total abandonment of the
class viewpoint.

This consequence also becomes evident when we investigate the general
character of revisionism. It is obvious that revisionism does not wish to
concede that its standpoint is that of the capitalist apologist. It does not
join the bourgeois economists in denying the existence of the contradictions
of capitalism. But, on the other hand, what precisely constitutes the fundamental point of revisionism and distinguishes it from the attitude taken by
the Social Democracy up to now, is that it does not base its theory on the
belief that the contradictions of capitalism will be suppressed as a result of
the logical inner development of the present economic system.

We may say that the theory of revisionism occupies an intermediate place
between two extremes. Revisionism does not expect to see the contradictions of capitalism mature. It does not propose to suppress these
contradictions through a revolutionary transformation. It wants to lessen,
to attenuate, the capitalist contradictions. So that the antagonism existing
between production and exchange is to be mollified by the cessation of
crises and the formation of capitalist combines. The antagonism between
Capital and Labour is to be adjusted by bettering the situation of the
workers and by the conservation of the middle classes. And the contradiction between the class State and society is to be liquidated through
increased State control and the progress of democracy.

It is true that the present procedure of the Social Democracy does not
consist in waiting for the antagonisms of capitalism to develop and in
passing on, only then, to the task of suppressing them. On the contrary,
the essence of revolutionary procedure is to be guided by the direction of
this development, once it is ascertained, and inferring from this direction
what consequences are necessary for the political struggle. Thus the Social
Democracy has combatted tariff wars and militarism without waiting for
their reactionary character to become fully evident. Bernstein's procedure
is not guided by a consideration of the development of capitalism, by the
prospect of the aggravation of its contradictions. It is guided by the prospect

of the attenuation of these contradictions. He shows this when he speaks of the 'adaptation' of capitalist economy.

Now when can such a conception be correct? If it is true that capitalism will continue to develop in the direction it takes at present, then its contradictions must necessarily become sharper and more aggravated instead of disappearing. The possibility of the attenuation of the contradictions of capitalism presupposes that the capitalist mode of production itself will stop its progress. In short, the general condition of Bernstein's theory is the cessation of capitalist development.

This way, however, his theory condemns itself in a twofold manner.

In the first place, it manifests its *Utopian* character in its stand on the establishment of Socialism. For it is clear that a defective capitalist development cannot lead to a Socialist transformation.

In the second place, Bernstein's theory reveals its *reactionary* character when it is referred to the rapid capitalist development that is taking place at present. Given the development of real capitalism, how can we explain, or rather state, Bernstein's position?

We have demonstrated in the first chapter the baselessness of the economic conditions on which Bernstein builds his analysis of existing social relationships. We have seen that neither the credit system nor cartels can be said to be 'means of adaptation' of capitalist economy. We have seen that not even the temporary cessation of crises nor the survival of the middle class can be regarded as symptoms of capitalist adaptation. But even though we should fail to take into account the erroneous character of all these details of Bernstein's theory we cannot help but be stopped short by one feature common to all of them. Bernstein's theory does not seize these manifestations of contemporary economic life as they appear in their organic relationship with the whole of capitalist development, with the complete economic mechanism of capitalism. His theory pulls these details out of their living economic context. It treats them as the *disjecta membra* (separate parts) of a lifeless machine.

Consider, for example, his conception of the adaptive effect of credit. If we recognize credit as a higher natural stage of the process of exchange and, therefore, of the contradictions inherent in capitalist exchange, we cannot at the same time see it as a mechanical means of adaptation existing outside of the process of exchange. It would be just as impossible to consider money, merchandise, capital as 'means of adaptation' of capitalism.

However, credit, like money, commodities, and capital, is an organic link of capitalist economy at a certain stage of its development. Like them, it is an indispensable gear in the mechanism of capitalist economy and, at the same time, an instrument of destruction, since it aggravates the internal contradictions of capitalism.

The same thing is true about cartels and the new, perfected means of communication.

The same mechanical view is presented by Bernstein's attempt to describe the promise of the cessation of crises as a symptom of the 'adaptation' of capitalist economy. For him, crises are simply derangements of the economic mechanism. With their cessation, he thinks, the mechanism could function well. But the fact is that crises are not 'derangements' without which capitalist economy could not develop at all. For if crises constitute the only method possible in capitalism—and therefore the normal method—of solving periodically the conflict existing between the unlimited extension of production and the narrow limits of the world market, then crises are an organic manifestation inseparable from capitalist economy.

Revisionism is nothing else than a theoretic generalization made from the angle of the isolated capitalist. Where does this viewpoint belong theoretically if not in vulgar bourgeois economics?

All the errors of this school rest precisely on the conception that mistakes the phenomena of competition, as seen from the angle of the isolated capitalist, for the phenomena of the whole of capitalist economy. Just as Bernstein considers credit to be a means of 'adaptation', so vulgar economy considers money to be a judicious means of 'adaptation' to the needs of exchange. Vulgar economy, too, tries to find the antidote against the ills of capitalism, that is at the belief in the possibility of patching up the sores of capitalism. It ends up by subscribing to a programme of reaction. It ends up in Utopia.

The theory of revisionism can therefore be defined in the following way. It is a theory of standing still in the Socialist movement, built, with the aid of vulgar economy, on a theory of a capitalist standstill. . . .

The fate of democracy is bound up, we have seen, with the fate of the labour movement. But does the development of democracy render superfluous or impossible a proletarian revolution, that is, the conquest of the political power by the workers?

Bernstein settles the question by weighing minutely the good and bad sides of social reform and social revolution. He does it almost in the same manner in which cinnamon or pepper is weighed out in a consumers' co-operative store. He sees the legislative course of historic development as the action of 'intelligence', while the revolutionary course of historic development is for him the action of 'feeling'. Reformist activity, he recognizes as a slow method of historic progress, revolution as a rapid method of progress. In legislation he sees a methodic force; in revolution, a spontaneous force.

We have known for a long time that the petty-bourgeois reformer finds 'good' and 'bad' sides in everything. He nibbles a bit at all grasses. But the real course of events is that affected by such combination. The carefully gathered little pile of the 'good sides' of all things possible collapses at the first fillip of history. Historically, legislative reform and the revolutionary method function in accordance with influences that are much more profound than the consideration of the advantages or inconveniences of the method or another.

In the history of bourgeois society, legislative reform served to strengthen progressively the rising class till the latter was sufficiently strong to seize political power, to suppress the existing juridical system, and to construct itself a new one. Bernstein, thundering against the conquest of political power as a theory of Blanquist violence, has the misfortune of labelling as a Blanquist error that which has always been the pivot and the motive force of human history. From the first appearance of class societies having the class struggle as the essential content of their history, the conquest of political power has been the aim of all rising classes. Here is the starting-point and end of every historic period. This can be seen in the long struggle of the Latin peasantry against the financiers and nobility of ancient Rome, in the struggle of the medieval nobility against the bishops, and in the struggle of the artisans against the nobles, in the cities of the Middle Ages. In modern times, we see it in the struggle of the bourgeoisie against feudalism.

Legislative reform and revolution are not different methods of historic development that can be picked out at pleasure from the counter of history, just as one chooses hot or cold sausages. Legislative reform and revolution are different *factors* in the development of class society. They condition and complement each other, and are at the same time reciprocally exclusive, as are the North and South Poles, the bourgeoisie and the proletariat.

Every legal constitution is the *product* of a revolution. In the history of classes, revolution is the act of political creation, while legislation is the political expression of the life of a society that has already come into being. Work for reform does not contain its own force, independent from revolution. During every historic period, work for reforms is carried on only in the direction given to it by the impetus of the last revolution, and continues as long as the impulsion of the last revolution continues to make itself felt. Or, to put it more concretely, in each historic period work for reforms is carried on only in the framework of the social form created by the last revolution. Here is the kernel of the problem.

It is contrary to history to represent work for reforms as a long-drawn-out revolution and revolution as a condensed series of reforms. A social transformation and a legislative reform do not differ according to their duration but according to their content. The secret of historic change through the utilization of political power resides precisely in the transformation of simple quantitative modification into a new quality, or, to speak more concretely, in the passage of a historic period from one given form of society to another.

That is why people who pronounce themselves in favour of the method of legislative reform *in place of and in contradistinction to* the conquest of political power and social revolution, do not really choose a more tranquil, calmer, and slower road to the *same* goal, but a *different* goal. Instead of taking a stand for the establishment of a new society they take a stand for surface modifications of the old society. If we follow the political conceptions of revisionism, we arrive at the same conclusion that is reached when we follow the economic theories of revisionism. Our programme becomes not the realization of *Socialism*, but the reform of *capitalism*; not the suppression of the system of wage labour, but the diminution of exploitation, that is, the suppression of the abuses of capitalism instead of the suppression of capitalism itself.

Does the reciprocal role of legislative reform and revolution apply only to the class struggles of the past? Is it possible that now, as a result of the development of the bourgeois juridical system, the function of moving society from one historic phase to another belongs to legislative reform, and that the conquest of State power by the proletariat has really become 'an empty phrase', as Bernstein puts it?

The very opposite is true. What distinguishes bourgeois society from other class societies—from ancient society and from the social order of the Middle Ages? Precisely the fact that class domination does not rest on 'acquired rights' but on *real economic relations*—the fact that wage labour is not a juridical relation, but purely an economic relation. In our juridical system there is not a single legal formula for the class domination of today. The few remaining traces of such formulae of class domination are (as that concerning servants), survivals of feudal society.

How can wage slavery be suppressed the 'legislative way', if wage slavery is not expressed in laws? Bernstein, who would do away with capitalism by means of legislative reform, finds himself in the same situation as Uspensky's Russian policeman who tells: 'Quickly I seized the rascal by the collar! But what do I see? The confounded fellow has no collar!' And that is precisely Bernstein's difficulty.

'All previous societies were based on an antagonism between an oppressing class and an oppressed class' (*Communist Manifesto*). But in the preceding phases of modern society, this antagonism was expressed in distinctly determined juridical relations and could, especially because of that, accord, to a certain extent, a place to new relations within the framework of the old. 'In the midst of serfdom, the serf raised himself to the rank of a member of the town community' (*Communist Manifesto*). How was that made possible? It was made possible by the progressive suppression of all feudal privileges in the environs of the city: the corvée, the right to special dress, the inheritance tax, the lord's claim to the best cattle, the personal levy, marriage under duress, the right to succession, etc., which all together constituted serfdom.

In the same way, the small bourgeoisie of the Middle Ages succeeded in raising itself, while it was still under the yoke of feudal absolutism, to the rank of bourgeoisie (*Communist Manifesto*). By what means? By means of the formal partial suppression or complete loosening of the corporative bonds, by the progressive transformation of the fiscal administration and of the army.

Consequently, when we consider the question from the abstract viewpoint, not from the historic viewpoint, we can *imagine* (in view of the former class relations) a legal passage, according to the reformist method, from feudal society to bourgeois society. But what do we see in reality? In reality, we see that legal reforms not only did not obviate the seizure of political power by the bourgeoisie, but have, on the contrary, prepared for it and led to it. A formal social-political transformation was indispensable for the abolition of slavery as well as for the complete suppression of feudalism.

But the situation is entirely different now. Now law obliges the proletariat to submit itself to the yoke of capitalism. Poverty, the lack of means of production, obliges the proletariat to submit itself to the yoke of capitalism. And no law in the world can give to the proletariat the means of production while it remains in the framework of bourgeois society, for not laws but economic development have torn the means of production from the producers' possession.

And neither is the exploitation inside the system of wage labour based on laws. The level of wages is not fixed by legislation, but by economic factors. The phenomenon of capitalist exploitation does not rest on a legal disposition, but on the purely economic fact that labour power plays in this exploitation the role of a merchandise possessing, among other characteristics, the agreeable quality of producing value—*more* than the value it consumes in the form of the labourer's means of subsistence. In short, the fundamental relations of the domination of the capitalist class cannot be transformed by means of legislative reforms, on the basis of capitalist society, because these relations have not been introduced by bourgeois laws, nor have they received the form of such laws. Apparently Bernstein is not aware of this, for he speaks of 'Socialist reforms'. On the other hand, he seems to express implicit recognition of this when he writes, on page 10 of his book, that 'the economic motive acts freely today, while formerly it was masked by all kinds of relations of domination, by all sorts of ideology'.

It is one of the peculiarities of the capitalist order that within it all the elements of the future society first assume, in their development, a form not approaching Socialism but, on the contrary, a form moving more and more away from Socialism. Production takes on a progressively increasing social character. But under what form is the social character of capitalist production expressed? It is expressed in the form of the large enterprise, in the form of the shareholding concern, the cartel, within which the capitalist antagonisms, capitalist exploitation, the oppression of labour-power, are augmented to the extreme.

In the army, capitalist development leads to the extension of obligatory military service, to the reduction of the time of service, and, consequently, to a material approach to a popular militia. But all of this takes place under the form of modern militarism, in which the domination of the people by the militarist State and the class character of the State manifest themselves most clearly.

In the field of political relations, the development of democracy brings— in the measure that it finds a favourable soil—the participation of all popular strata in political life and, consequently, some sort of 'people's State'. But this participation takes the form of bourgeois parliamentarism, in which class antagonisms and class domination are not done away with, but are, on the contrary, displayed in the open. Exactly because capitalist development moves through these contradictions, it is necessary to extract the kernel of Socialist society from its capitalist shell. Exactly for this reason must the proletariat seize political power and suppress completely the capitalist system.

Of course, Bernstein draws other conclusions. If the development of democracy leads to the aggravation and not to the lessening of capitalist antagonisms, 'the Social Democracy', he answers us, 'in order not to render its task more difficult, must by all means try to stop social reforms and the extension of democratic institutions' (page 71). Indeed, that would be the right thing to do if the Social Democracy found to its taste, in the petty-bourgeois manner, the futile task of picking for itself all the good sides of history and rejecting the bad sides of history. However, in that case, it should at the same time 'try to stop' capitalism in general, for there is no doubt that the latter is the rascal placing all these obstacles in the way of Socialism. But capitalism furnishes besides the *obstacles* also the only *possibilities* of realizing the socialist programme. The same can be said about democracy.

If democracy has become superfluous or annoying to the bourgeoisie, it is on the contrary necessary and indispensable to the working class. It is necessary to the working class because it creates the political forms (autonomous administration, electoral rights, etc.) which will serve the proletariat as fulcrums in its task of transforming bourgeois society. Democracy is indispensable to the working class, because only through the exercise of its democratic rights, in the struggle for democracy, can the proletariat become aware of its class interests and its historic task.

In a word, democracy is indispensable not because it renders superfluous the conquest of political power by the proletariat, but because it renders this conquest of power both *necessary* and *possible*. When Engels, in his preface to the *Class Struggles in France*, revised the tactics of the modern labour movement and urged the legal struggle as opposed to the barricades, he did not have in mind—this comes out of every line of the preface—the question of a definite conquest of political power, but the contemporary daily struggle. He did not have in mind the attitude that the proletariat must take toward the capitalist State at the time of its seizure of power, but the attitude of the proletariat while in the bounds of the capitalist State. Engels was giving directions to the proletariat *oppressed*, and not the proletariat victorious.

On the other hand, Marx's well-known sentence on the agrarian question in England (Bernstein leans on it heavily), in which he says: 'We shall probably succeed easier by buying the estates of the landlords,' does not refer to the stand of the proletariat *before, but after its victory*. For there evidently can be a question of buying the property of the old dominant class only when the workers are in power. The possibility envisaged by Marx is that of the *pacific exercise of the dictatorship of the proletariat* and not the replacement of the dictatorship with capitalist social reforms. There was no doubt for Marx and Engels about the necessity of having the proletariat conquer political power. It is left to Bernstein to consider the poultry-yard of bourgeois parliamentarism as the organ by means of which we are to realize the most formidable social transformation of history, *the passage from capitalist society to Socialism*.

Bernstein introduces his theory by warning the proletariat against the danger of acquiring power too early. That is, according to Bernstein, the proletariat ought to leave the bourgeois society in its present condition and itself suffer a frightful defeat. If the proletariat came to power, it could draw from Bernstein's theory the following 'practical' conclusion: to go to sleep. His theory condemns the proletariat, at the most decisive moments of the struggle, to inactivity, to a passive betrayal of its own cause.

Our programme would be a miserable scrap of paper if it could not serve us in *all* eventualities, at *all* moments of the struggle, and if it did not serve us by its *application* and not by its non-application. If our programme contains the formula of the historic development of society from capitalism to Socialism, it must also formulate, in all its characteristic fundamentals, all the transitory phases of this development, and it should, consequently, be able to indicate to the proletariat what ought to be its corresponding action at every moment on the road toward Socialism. There can be no time for the proletariat when it will be obliged to abandon its programme or be abandoned by it.

Practically, this is manifested in the fact that there can be no time when the proletariat, placed in power by the force of events, is not in the condition, or is not morally obliged, to take certain measures for the realization of its programme, that is, take transitory measures in the direction of Socialism. Behind the belief that the Socialist programme can collapse completely at any point of the dictatorship of the proletariat lurks the other belief that the *Socialist programme is, generally and at all times, unrealizable.*

And what if the transitory measures are premature? The question hides a great number of mistaken ideas concerning the real course of a social transformation.

In the first place, the seizure of political power by the proletariat, that is to say by a large popular class, is not produced artificially. It presupposes (with the exception of such cases as the Paris Commune, when power was not obtained by the proletariat after a conscious struggle for its goal, but fell into its hands, like a good thing abandoned by everybody else) a definite degree of maturity of economic and political relations. Here we have the essential difference between *coups d'état* along Blanqui's conception, which are accomplished by an 'active minority', and burst out like pistol shot, always inopportunely, and the conquest of political power by a great conscious popular mass, which can only be the product of the decomposition of bourgeois society and therefore bears in itself the economic and political legitimation of its opportune appearance.

If, therefore, considered from the angle of political effect, the conquest of political power by the working class cannot materialize itself 'too early', then from the angle of conservation of power, the premature revolution, the thought of which keeps Bernstein awake, menaces us like a sword of Damocles. Against that neither prayers nor supplication, neither scares nor any amount of anguish, are of any avail. And this for two very simple reasons.

In the first place, it is impossible to imagine that a transformation as formidable as the passage from capitalist society to Socialist society can be realized in one happy act. To consider that as possible is again to lend colour to conceptions that are clearly Blanquist. The Socialist transformation supposes a long and stubborn struggle, in the course of which, it is quite probable, the proletariat will be repulsed more than once, so that the first time, from the viewpoint of the final outcome of the struggle, it will have necessarily come to power 'too early'.

In the second place, it will be impossible to avoid the 'premature' conquest of State power by the proletariat precisely because these 'premature' attacks of the proletariat constitute a factor, and indeed a very important factor, creating the political conditions of the final victory. In the course of the political crisis accompanying its seizure of power, in the course of the long and stubborn struggles, the proletariat will acquire the degree of political maturity permitting it to obtain in time a definitive victory of the revolution. Thus these 'premature' attacks of the proletariat against the State power are in themselves important historic factors helping to provoke and determine the *point* of the definite victory. Considered from this viewpoint, the idea of a 'premature' conquest of political power by the labouring class appears to be a political absurdity derived from a mechanical conception of the development of society, and positing for the victory of the class struggle a point fixed *outside* and *independent of* the class struggle.

Since the proletariat is not in the position to seize political power in any other way than 'prematurely', since the proletariat is absolutely obliged to seize power once or several times 'too early' before it can maintain itself in power for good, the objection to the 'premature' conquest of power is at bottom nothing more than a *general opposition to the aspiration of the proletariat to possess itself of State power.* Just as all roads lead to Rome, so too, do we logically arrive at the conclusion that the revisionist proposal to slight the final aim of the Socialist movement is really a recommendation to renounce the socialist movement itself.

2.   The Mass Strike

The first revision of the question of the mass strike which results from the experience of Russia relates to the general conception of the problem. Till the present time the zealous advocates of an 'attempt with the mass strike' in Germany of the stamp of Bernstein, Eisner, etc., and also the strongest opponents of such an attempt as represented in the trade-union camp by, for example, Bomelburg, stand, when all is said and done, on the same conception, and that the anarchist one. The apparent polar opposites do not mutually exclude each other but, as always, condition and, at the same time, supplement each other. For the anarchist mode of thought is direct speculation on the 'great Kladderadatsch', on the social revolution merely as an external and inessential characteristic. According to it, what is essential is the whole abstract, unhistorical view of the mass strike and of all the conditions of the proletarian struggle generally. For the anarchist there exist only two things as material suppositions of his 'revolutionary' specu-lations—first imagination, and second goodwill and courage to rescue humanity from the existing capitalist vale of tears. This fanciful mode of reasoning sixty years ago gave the result that the mass strike was the shortest, surest, and easiest means of springing into the better social future. The same mode of reasoning recently gave the result that the trade-union struggle was the only real 'direct action of the masses' and also the only real revolutionary struggle—which, as is well known, is the latest notion of the French and Italian 'Syndicalists'. The fatal thing for anarchism has always been that the methods of struggle improvised in the air were not only a reckoning without their host, that is, they were purely Utopian, but that they, while not reckoning in the least with the despised evil reality, un-expectedly became in this evil reality, practical helps to the reaction, where previously they had only been, for the most part, revolutionary speculations.

On the same ground of abstract, unhistorical methods of observation stand those today who would, in the manner of a board of directors, put the mass strike in Germany on the calendar on an appointed day, and those who, like the participants in the trade-union congress at Cologne, would by a prohibition of 'propaganda' eliminate the problem of the mass strike from the face of the earth. Both tendencies proceed on the common purely anarchistic assumption that the mass strike is a purely technical means of struggle which can be 'decided' at pleasure and strictly according to conscience, or 'forbidden'—a kind of pocket-knife which can be kept in the pocket clasped 'ready for any emergency', and according to decision, can be unclasped and used. The opponents of the mass strike, do indeed claim for themselves the merit of taking into consideration the historical groundwork and the material conditions of the present situation in Germany in opposition to the 'revolutionary romanticists' who hover in the air, and do not at any point reckon with the hard realities and their possibilities and impossibilities. 'Facts and figures; figures and facts!' they cry, like Mr Gradgrind in Dicken's *Hard Times*. What the trade-union opponent of the mass strike understands by the 'historical basis' and 'material conditions' is two things—on the one hand the weakness of the proletariat, and on the other hand, the strength of Prussian-German militarism. The inadequate organization of the workers and the imposing Prussian bayonet—these are the facts and figures upon which these trade-union leaders base their practical policy in the given case. Now while it is quite true that the trade-union cash box and the Prussian bayonet are material and very historical phenomena, the conception based upon them is not historical materialism in Marx's sense but a policemanlike materialism in the sense of Puttkammer. The representatives of the capitalist police state reckon much, and, indeed, exclusively, with the occasional real power of the organized proletariat as well as with the material might of the bayonet, and from the comparative example of these two rows of figures the comforting conclusion is always drawn that the revolutionary labour movement is produced by individual demagogues and agitators; and that therefore there is in the prisons and bayonets an adequate means of subduing the unpleasant 'passing phenomena'.

The class-conscious German workers have at last grasped the humour of the policemanlike theory that the whole modern Labour movement is an artificial arbitrary product of a handful of conscienceless 'demagogues and agitators'.

It is exactly the same conception, however, that finds expression when two or three worthy comrades unite in a voluntary column of night-watchmen in order to warn the German working class against the dangerous agitation of a few 'revolutionary romanticists' and their 'propaganda of the mass strike'; or, when, on the other side, a noisy indignation campaign is engineered by those who, by means of 'confidential' agreements between the executive of the party and the general commission of the trade unions, believe they can prevent the outbreak of the mass strike in Germany. If it depended on the inflammatory 'propaganda' of revolutionary romanticists or on confidential or public decisions or the party direction, then we should not even yet have had in Russia a single serious mass strike. In no country in the world—as I pointed out in March 1905 in the *Sachische Arbeiterzeitung*—was the mass strike so little 'propagated' or even 'discussed' as in Russia. And the isolated examples of decisions and agreements of the Russian party executive which really sought to proclaim the mass strike of their own accord—as, for example, the last attempt in August of this year after the dissolution of the Duma—are almost valueless. If, therefore, the Russian revolution teaches us anything, it teaches above all that the mass strike is not artificially 'made', not 'decided' at random, not 'propagated', but that it is an historical phenomenon which, at a given moment, results from social conditions with historical inevitability. It is not therefore by abstract speculations on the possibility or impossibility, the utility or the injuriousness, of the mass strike, but only by an examination of those factors and social conditions out of which the mass strike grows in the present phase of the class struggle—in other words, it is not by *subjective criticism* of the mass strike from the standpoint of what is desirable, but only by *objective investigation* of the sources of the mass strike from the standpoint of what is historically inevitable, that the problem can be grasped or even discussed.

In the unreal sphere of abstract logical analysis it can be shown with exactly the same force on either side that the mass strike is absolutely impossible and sure to be defeated, and that it is possible and that its triumph cannot be questioned. And therefore the value of the evidence produced on each side is exactly the same—and that is nil. Therefore the fear of the 'propagation' of the mass strike, which has even led to formal anathemas against the persons alleged to be guilty of this crime, is solely the product of the droll confusion of persons. It is just as impossible to 'propagate' the mass strike as an abstract means of struggle as it is to propagate the 'revolution'. 'Revolution', like 'mass strike', signifies nothing but an external form of the class struggle, which can have sense and meaning only in connection with definite political situations.

If anyone were to undertake to make the mass strike generally as a form of proletarian action the object of methodical agitation, and to go house-to-house canvassing with this 'idea' in order gradually to win the working class to it, it would be as idle and profitless and absurd an occupation as it would be to seek to make the idea of the revolution or of the fight at the barricades the object of a special agitation. The mass strike has now become the centre of the lively interest of the German and the international working class because it is a new form of struggle, and as such is the sure symptom of a thoroughgoing internal revolution in the relations of the classes and in the conditions of the class struggle. It is a testimony to the sound revolutionary instinct and to the quick intelligence of the mass of the German proletariat that, in spite of the obstinate resistance of their trade-union leaders, they are applying themselves to this new problem with such keen interest. But it does not meet the case, in the presence of this interest and of this fine, intellectual thirst and desire for revolutionary deeds on the part of the workers, to treat them to abstract mental gymnastics on the possibility or impossibility of the mass strike; they should be enlightened on the development of the Russian revolution, the international significance of that revolution, the sharpening of class antagonisms in Western Europe, the wider political perspectives of the class struggle in Germany, and the role and the tasks of the masses in the coming struggles. Only in this form will the discussion on the mass strike lead to the widening of the intellectual horizon of the proletariat, to the sharpening of their way of thinking, and to the steeling of their energy. . . .

3. Lenin's Centralism

In general, it is rigorous, despotic centralism that is preferred by opportunist intellectuals at a time when the revolutionary elements among the workers still lack cohesion and the movement is groping its way, as is the case now in Russia. In a later phase, under a parliamentary regime and in connection with a strong labour party, the opportunist tendencies of the intellectuals express themselves in an inclination toward 'decentralization'.

If we assume the viewpoint claimed as his own by Lenin and we fear the influence of intellectuals in the proletarian movement, we can conceive of no greater danger to the Russian party than Lenin's plan of organization. *Nothing will more surely enslave a young labour movement to an intellectual élite hungry for power than this bureaucratic strait-jacket, which will immobilize the movement and turn it into an automaton manipulated by a Central Committee.* On the other hand, there is no more effective guarantee against opportunist intrigue and personal ambition than the independent revolutionary action of the proletariat, as a result of which the workers acquire the sense of political responsibility and self-reliance.

What is today only a phantom haunting Lenin's imagination may become reality tomorrow.

Let us not forget that the revolution soon to break out in Russia will be a bourgeois and not a proletarian revolution. This modifies radically all the conditions of Socialist struggle. The Russian intellectuals, too, will rapidly become imbued with bourgeois ideology. The Social Democracy is at present the only guide of the Russian proletariat. But on the day after the revolution, we shall see the bourgeoisie, and above all the bourgeois intellectuals, seek to use the masses as a stepping-stone to their domination.

The game of the bourgeois demagogues will be made easier if at the present stage, the spontaneous action, initiative, and political sense of the advanced sections of the working class are hindered in their development and restricted by the protectorate of an authoritarian Central Committee.

More important is the fundamental falseness of the idea underlying the plan of unqualified centralism—the idea that the road to opportunism can be barred by means of clauses in a party constitution.

Impressed by recent happenings in the Socialist parties of France, Italy, and Germany, the Russian Social Democrats tend to regard opportunism as an alien ingredient, brought into the labour movement by representatives of bourgeois democracy. If that were so, no penalties provided by a party constitution could stop this intrusion. The afflux of non-proletarian recruits to the party of the proletariat is the effect of profound social causes, such as the economic collapse of the petty bourgeoisie, the bankruptcy of bourgeois liberalism, and the degeneration of bourgeois democracy. It is naïve to hope to stop this current by means of a formula written down in a constitution.

A manual of regulations may master the life of a small sect or a private circle. A historic current, however, will pass through the mesh of the most subtly worded statutory paragraph. It is furthermore untrue that to repel the elements pushed toward the Socialist movement by the decomposition of bourgeois society means to defend the interests of the working class. The Social Democracy has always contended that it represents not only the class interests of the proletariat but also the progressive aspirations of the whole of contemporary society. It represents the interests of all who are oppressed by bourgeois domination. This must not be understood merely in the sense that all these interests are ideally contained in the Socialist programme. Historic evolution translates the given proposition into reality. In its capacity as a political party, the Social Democracy becomes the haven of all discontented elements in our society and thus of the entire people, as contrasted to the tiny minority of the capitalist masters.

But Socialists must always know how to subordinate the anguish, rancour, and hope of this motley aggregation to the supreme goal of the working class. The Social Democracy must enclose the tumult of the non-proletarian protestants against existing society within the bounds of the revolutionary action of the proletariat. It must assimilate the elements that come to it.

This is only possible if the Social Democracy already contains a strong, politically educated proletarian nucleus class conscious enough to be able, as up to now in Germany, to pull along in its tow the declassed and petty-bourgeois elements that join the party. In that case, greater strictness in the application of the principle of centralization and more severe discipline, specifically formulated in party by-laws, may be an effective safeguard against the opportunist danger. That is how the revolutionary Socialist movement in France defended itself against the Jaurèsist confusion. A modification of the constitution of the German Social Democracy in that direction would be a very timely measure.

But even here we should not think of the party constitution as a weapon that is, somehow, self-sufficient. It can be at most a coercive instrument enforcing the will of the proletarian majority in the party. If this majority is lacking, then the most dire sanctions on paper will be of no avail.

However, the influx of bourgeois elements into the party is far from being the only cause of the opportunist trends that are now raising their heads in the Social Democracy. Another cause is the very nature of Socialist activity and the contradictions inherent in it.

The international movement of the proletariat toward its complete emancipation is a process peculiar in the following respect. For the first time in the history of civilization, the people are expressing their will consciously and in opposition to all ruling classes. But this will can only be satisfied beyond the limits of the existing system.

On the one hand, we have the mass; on the other, its historic goal, located outside of existing society. On one hand, we have the day-to-day struggle; on the other, the social revolution. Such are the terms of the dialectical contradiction through which the Socialist movement makes its way.

It follows that this movement can best advance by tacking betwixt and between the two dangers by which it is constantly being threatened. One is the loss of its mass character; the other, the abandonment of its goal. One is the danger of sinking back to the condition of a sect; the other, the danger of becoming a movement of bourgeois social reform.

That is why it is illusory, and contrary to historic experience, to hope to fix, once for always, the direction of the revolutionary Socialist struggle with the aid of formal means, which are expected to secure the labour movement against all possibilities of opportunist digression.

Marxist theory offers us a reliable instrument enabling us to recognize and combat typical manifestations of opportunism. But the socialist movement is a mass movement. Its perils are not the product of the insidious machinations of individuals and groups. They arise out of unavoidable social conditions. We cannot secure ourselves in advance against all possibilities of opportunist deviation. Such dangers can be overcome only by the movement itself—certainly with the aid of Marxist theory, but only after the dangers in question have taken tangible form in practice.

Looked at from this angle, opportunism appears to be a product and an inevitable phase of the historic development of the labour movement.

The Russian Social Democracy arose a short while ago. The political conditions under which the proletarian movement is developing in Russia are quite abnormal. In that country, opportunism is to a large extent a by-product of the groping and experimentation of Socialist activity seeking to advance over a terrain that resembles no other in Europe.

In view of this, we find most astonishing the claim that it is possible to avoid any possibility of opportunism in the Russian movement by writing down certain words, instead of others, in the party constitution. *Such an attempt to exorcise opportunism by means of a scrap of paper may turn out to be extremely harmful—not to opportunism but to the socialist movement.*

Stop the natural pulsation of a living organism, and you weaken it, and you diminish its resistance and combative spirit—in this instance, not only against opportunism but also (and that is certainly of great importance) against the existing social order. The proposed means turn against the end they are supposed to serve.

In Lenin's over-anxious desire to establish the guardianship of an omniscient and omnipotent Central Committee in order to protect so promising and vigorous a labour movement against any misstep, we recognize the symptoms of the same subjectivism that has already played more than one trick on socialist thinking in Russia.

It is amusing to note the strange somersaults that the respectable human 'ego' has had to perform in recent Russian history. Knocked to the ground, almost reduced to dust, by Russian absolutism, the 'ego' takes revenge by turning to revolutionary activity. In the shape of a committee of conspirators, in the name of a non-existent Will of the People, it seats itself on a kind of throne and proclaims it is all-powerful. But the 'object' proves to be the stronger. The knout is triumphant, for tsarist might seems to be the 'legitimate' expression of history.

In time we see appear on the scene an even more 'legitimate' child of history—the Russian labour movement. For the first time, bases for the formation of a real 'people's will' are laid in Russian soil.

But here is the 'ego' of the Russian revolutionary again! Pirouetting on its head, it once more proclaims itself to be the all-powerful director of history—this time with the title of His Excellency the Central Committee of the Social Democratic Party of Russia.

The nimble acrobat fails to perceive that the only 'subject' which merits today the role of director is the collective 'ego' of the working class. The working class demands the right to make its mistakes and learn in the dialectic of history.

Let us speak plainly. Historically, the errors committed by a truly revolutionary movement are infinitely more fruitful than the infallibility of the cleverest Central Committee.

4. The Russian Revolution

Lenin says: the bourgeois State is an instrument of oppresson of the working class; the Socialist state, of the bourgeoisie. To a certain extent, he says, it is only the capitalist State stood on its head. This simplified view misses the most essential thing: bourgeois class rule has no need of the political training and education of the entire mass of the people, at least not beyond certain narrow limits. But for the proletarian dictatorship that is the life element, the very air without which it is not able to exist.

'Thanks to the open and direct struggle for governmental power,' writes Trotsky, 'the labouring masses accumulate in the shortest time a considerable amount of political experience and advance quickly from one stage to another of their development.'

Here Trotsky refutes himself and his own friends. Just because this is so, they have blocked up the fountain of political experience and the source of this rising development by their suppression of public life! Or else we would have to assume that experience and development were necessary up to the seizure of power by the Bolsheviks, and then, having reached their highest peak, became superfluous thereafter. (Lenin's speech: Russia is won for Socialism!!!)

In reality, the opposite is true! It is the very giant tasks which the Bolsheviks have undertaken with courage and determination that demand the most intensive political training of the masses and the accumulation of experience.

Freedom only for the supporters of the government, only for the members of one party—however numerous they may be—is no freedom at all. Freedom is always and exclusively freedom for the one who thinks differently. Not because of any fanatical concept of 'justice' but because all that is instructive, wholesome, and purifying in political freedom depends on this essential characteristic, and its effectiveness vanishes when 'freedom' becomes a special privilege.

The Bolsheviks themselves will not want, with hand on heart, to deny that, step by step, they have to feel out the ground, try out, experiment, test now one way now another, and that a good many of their measures do not represent priceless pearls of wisdom. Thus it must and will be with all of us when we get to the same point—even if the same difficult circumstances may not prevail everywhere.

The tacit assumption underlying the Lenin–Trotsky theory of the dictatorship is this: that the Socialist transformation is something for which a ready-made formula lies completed in the pocket of the revolutionary party, which needs only to be carried out energetically in practice. This is, unfortunately—or perhaps fortunately—not the case. Far from being a sum of ready-made prescriptions which have only to be applied, the practical realization of socialism as an economic, social, and juridical system is something which lies completely hidden in the mists of the future. What we possess in our programme is nothing but a few main signposts which indicate the general direction in which to look for the necessary measures, and the indications are mainly negative in character at that. Thus we know more or less what we must eliminate at the outset in order to free the road for a Socialist economy. But when it comes to the nature of the thousand concrete, practical measures, large and small, necessary to introduce Socialist principles into economy, law, and all social relationships, there is no key in any socialist Party programme or textbook. That is not a shortcoming but rather the very thing that makes scientific Socialism superior to the Utopian varieties. The Socialist system of society should only be, and can only be, a historical product, born out of the school of its own experiences, born in the course of its realization, as a result of the developments of living history, which—just like organic nature of which, in the last analysis, it forms a part—has the fine habit of always producing along with any real social need the means to its satisfaction, along with the task simultaneously the solution. However, if such is the case, then it is clear that Socialism by its very nature cannot be decreed or introduced by ukase. It has as its prerequisite a number of measures of force—against property, etc. The negative, the tearing down, can be decreed; the building up, the positive, cannot. New territory. A thousand problems. Only experience is capable of correcting and opening new ways. Only unobstructed, effervescing life falls into a thousand new forms and improvisations, brings to light creative force, itself corrects all mistaken attempts. The public life of countries with limited freedom is so poverty-stricken, so miserable, so rigid, so unfruitful, precisely because, through the exclusion of democracy, it cuts off the living sources of all spiritual riches and progress. (Proof: the year 1905 and the months from February to October 1917.) There it was political in character; the same thing applies to economic and social life also. The whole mass of the people must take part in it. Otherwise, socialism will be decreed from behind a few official desks by a dozen intellectuals.

Public control is indispensably necessary. Otherwise the exchange of experiences remains only with the closed circle of the officials of the new regime. Corruption becomes inevitable. (Lenin's words, Bulletin No. 29.) Socialism in life demands a complete spiritual transformation in the masses degraded by centuries of bourgeois class rule. Social instincts in place of egotistical ones, mass initiative in place of inertia, idealism which conquers all suffering, etc., etc. No one knows this better, describes it more penetratingly, repeats it more stubbornly than Lenin. But he is completely mistaken in the means he employs. Decree, dictatorial force of the factory overseer, Draconic penalties, rule by terror—all these things are but palliatives. The only way to a rebirth is the school of public life itself, the most unlimited, the broadest democracy and public opinion. It is rule by terror which demoralizes.

When all this is eliminated, what really remains? In place of the representative bodies created by general, popular elections, Lenin and Trotsky have laid down the soviets as the only true representation of the labouring masses. But with the repression of political life in the land as a whole, life in the soviets must also become more and more crippled. Without general elections, without unrestricted freedom of press and assembly, without a free struggle of opinion, life dies out in every public institution, becomes a mere semblance of life in which only the bureaucracy remains as the active element. Public life gradually falls asleep, a few dozen party leaders of inexhaustible energy and boundless experience direct and rule. Among them, in reality only a dozen, outstanding heads do the leading and an élite of the working class is invited from time to time to meetings where they are to applaud the speeches of the leaders, and to approve proposed resolutions unanimously—at bottom, then, a clique affair—a dictatorship, to be sure, not the dictatorship of the proletariat, however, but only the dictatorship in the bourgeois sense, in the sense of the rule of the Jacobins (the postponement of the Soviet Congress from three-month periods to six-month period!) Yes, we can go even further: such conditions must inevitably cause a brutalization of public life: attempted assassinations, shooting of hostages, etc. (Lenin's speech on discipline and corruption.) . . .

The basic error of the Lenin–Trotsky theory is that they too, just like Kautsky, oppose dictatorship to democracy. 'Dictatorship *or* democracy' is the way the question is put by Bolsheviks and Kautsky alike. The latter naturally decides in favour of 'democracy', that is, of bourgeois democracy, precisely because he opposes it to the alternative of the Socialist revolution. Lenin and Trotsky, on the other hand, decide in favour of dictatorship in contradistinction to democracy, and thereby, in favour of the dictatorship of a handful of persons, that is, in favour of dictatorship on the bourgeois model. They are two opposite poles, both alike being far removed from a genuine Socialist policy. The proletariat, when it seizes power, can never follow the good advice of Kautsky, given on the pretext of the 'unripeness of the country', the advice being to renounce the Socialist revolution and devote itself to democracy. It cannot follow this advice without betraying thereby itself, the International, and the revolution. It should and must at once undertake Socialist measures in the most energetic, unyielding, and unhesitant fashion, in other words, exercise a dictatorship, but a dictatorship of the *class*, not of a party or of a clique—dictatorship of the class, that means in the broadest public form on the basis of the most active, unlimited participation of the mass of the people, of unlimited democracy.

'As Marxists', writes Trotsky, 'we have never been idol worshippers of formal democracy.' Surely, we have never been idol worshippers of formal democracy. Nor have we ever been idol worshippers of Socialism or Marxism either. Does it follow from this that we may also throw Socialism on the scrap-heap, à la Cunow, Lensch, and Parvus, if it becomes uncomfortable for us? Trotsky and Lenin are the living refutation of this answer.

'We have never been idol-worshippers of formal democracy.' All that that really means is: We have always distinguished the social kernel from the political form of *bourgeois* democracy; we have always revealed the hard kernel of social inequality and lack of freedom hidden under the sweet shell of formal equality and freedom—not in order to reject the latter but to spur the working class into not being satisfied with the shell, but rather, by conquering political power, to create a Socialist democracy to replace bourgeois democracy—not to eliminate democracy altogether.

But Socialist democracy is not something which begins only in the promised land after the foundations of Socialist economy are created; it does not come as some sort of Christmas present for the worthy people who, in the interim, have loyally supported a handful of Socialist dictators. Socialist democracy begins simultaneously with the beginnings of the destruction of class rule and of the construction of Socialism. It begins at the very moment of the seizure of power by the Socialist Party. It is the same thing as the dictatorship of the proletariat.

Yes, dictatorship! But this dictatorship consists in the *manner of applying democracy*, not in its *elimination*, in energetic, resolute attacks upon the well-entrenched rights and economic relationships of bourgeois society, without which a Socialist transformation cannot be accomplished. But this dictatorship must be the work of the *class* and not of a little leading minority in the name of the class—that is, it must proceed step by step out of the active participation of the masses; it must be under their direct influence, subjected to the control of complete public activity; it must arise out of the growing political training of the mass of the people.

Doubtless the Bolsheviks would have proceeded in this very way were it not that they suffered under the frightful compulsion of the World War, the German occupation, and all the abnormal difficulties connected therewith, things which were inevitably bound to distort any Socialist policy, however imbued it might be with the best intentions and the finest principles.

A crude proof of this is provided by the use of terror to so wide an extent by the Soviet government, especially in the most recent period just before the collapse of German imperialism, and just after the attempt on the life of the German ambassador. The commonplace to the effect that revolutions are not pink teas is in itself pretty inadequate.

Everything that happens in Russia is comprehensible and represents an

inevitable chain of causes and effects, the starting-point and end term of
which are: the failure of the German proletariat and the occupation of
Russia by German imperialism. It would be demanding something super-
human from Lenin and his comrades if we should expect of them that under
such circumstances they should conjure forth the finest democracy, the
most exemplary dictatorship of the proletariat, and a flourishing Socialist
economy. By their determined revolutionary stand, their exemplary strength
in action, and their unbreakable loyalty to international Socialism, they
have contributed whatever could possibly be contributed under such devil-
ishly hard conditions. The danger begins only when they make a virtue of
necessity and want to freeze into a complete theoretical system all the tactics
forced upon them by these fatal circumstances, and want to recommend
them to the international proletariat as a model of Socialist tactics. When
they get in their own light in this way, and hide their genuine, unquestionable historical service under the bushel of false steps forced upon them
by necessity, they render a poor service to international socialism for the
sake of which they have fought and suffered; for they want to place in its
storehouse as new discoveries all the distortions prescribed in Russia by
necessity and compulsion—in the last analysis only by-products of the
bankruptcy of international Socialism in the present World War.

Let the German government Socialists cry that the rule of the Bolsheviks
in Russia is a distorted expression of the dictatorship of the proletariat. If
it was or is such, that is only because it is a product of the behaviour of
the German proletariat, in itself a distorted expression of the Socialist
class struggle. All of us are subject to the laws of history, and it is only
internationally that the Socialist order of society can be realized. The
Bolsheviks have shown that they are capable of everything that a genuine
revolutionary party can contribute within the limits of the historical pos-
sibilites. They are not supposed to perform miracles. For a model and
faultless proletarian revolution in an isolated land, exhausted by world
war, strangled by imperialism, betrayed by the international proletariat,
would be a miracle.

What is in order is to distinguish the essential from the non-essential, the
kernel from the accidental excrescences in the policies of the Bolsheviks. In
the present period, when we face decisive final struggles in all the world,
the most important problem of Socialism was and is the burning question
of our time. It is not a matter of this or that secondary question of tactics,
but of the capacity for action of the proletariat, the strength to act, the will
to power of Socialism as such. In this, Lenin and Trotsky and their friends
were the *first*, those who went ahead as an example to the proletariat of

the world; they are still the *only ones* up to now who can cry with Huten: 'I have dared!'

This is the essential and *enduring* in Bolshevik policy. In *this* sense theirs is the immortal historical service of having marched at the head of the international proletariat with the conquest of political power and the practical placing of the problem of the realization of Socialism, and of having advanced mightily the settlement of the score between capital and labour in the entire world. In Russia the problem could only be posed. It could not be solved in Russia. And in *this* sense, the future everywhere belongs to 'Bolshevism'.

Sources: *Social Reform or Revolution* (New York, 1937), pp. 35 ff., 58 ff. *The Mass Strike* (Detroit, 1925), pp. 15 ff. *The Russian Revolution* and *Leninism or Marxism?* (Ann Arbor, 1961), pp. 101 ff., 68 ff., 76 ff.

FURTHER NOTES

WORKS

DAVIS, HORACE B. (ed.), *The National Question: Selected Writings by Rosa Luxemburg* (New York, 1976).

HOWARD, D. (ed.), *Selected Political Writings of Rosa Luxemburg* (London and New York, 1971).

LOOKER, R. (ed.), *Rosa Luxemburg: Selected Political Writings* (London, 1972).

LUXEMBURG, ROSA *The Accumulation of Capital*, Introduction by J. Robinson (London, 1971).

—— *The Russian Revolution* and *Leninism or Marxism?*, Introduction by B. D. Wolfe (Ann Arbor, 1961).

WATERS, M.-A. (ed.), *Rosa Luxemburg Speaks* (New York, 1970).

COMMENTARIES

BASSO, L., *Rosa Luxemburg: A Reappraisal* (London, 1975).

CLIFF, TONY, *Rosa Luxemburg* (London, 1968).

ETTINGER, E., *Rosa Luxemburg: A Life* (New York, 1987).

FROLICH, PAUL, *Rosa Luxemburg: Her Life and Work* (London, 1940).

GERAS, N., *The Legacy of Rosa Luxemburg* (London, 1976).

NETTL, J. P., *Rosa Luxemburg*, 2 vols. (Oxford, 1966).

6

GEORGY PLEKHANOV

Darwin succeeded in solving the problem of how there originate veg-
etable and animal species in the struggle for existence. Marx solved the
problem of how there arise different species of social organization in
the struggle of men for their existence ... Marxism is Darwinism in its
application to social science.

The Development of the Monist View of History (London, 1947),
p. 24.

GEORGY PLEKHANOV (1856–1918) *was not only the first Russian completely to embrace Marxism: he was also the undisputed intellectual leader
of the Russian Marxists from the early 1880s until the turn of the century.
As such, he was the educator of a whole generation of revolutionaries—
including, most importantly, Lenin, whom he taught to see Russia as an
already capitalist country where it was incumbent on the proletariat, led by
its intelligentsia, to spearhead a movement to overthrow tsarist autocracy.
Plekhanov set great store by doctrinal orthodoxy and spent a lot of time
elaborating it. In the first extract below, from an essay written in 1898, he
tries to account for the role of personality in history—obviously a potential
difficulty for Marxism. In the second extract, from* Fundamental Problems
of Marxism, *published in 1908, Plekhanov elaborates on the materialist
conception of history, especially in two areas that considerably interested
him: the influence of geography on human evolution, and the explanation
of art.*

1.   The Role of Personality in History

WE know now that individuals often exercise considerable influence upon
the fate of society, but this influence is determined by the internal structure
of that society and by its relation to other societies. But this is not all that
has to be said about the role of the individual in history. We must approach
this question from still another side.

Sainte-Beuve thought that had there been a sufficient number of petty
and dark causes of the kind that he had mentioned, the outcome of the
French Revolution would have been the opposite of what we know it to
have been. This is a great mistake. No matter how intricately the petty,

psychological, and physiological causes may have been interwoven, under no circumstances would they have eliminated the great social needs that gave rise to the French Revolution; and as long as these needs remained unsatisfied the revolutionary movement in France would have continued. To make the outcome of this movement the opposite of what it was, the needs that gave rise to it would have had to be the opposite of what they were; and this, of course, no combination of petty causes would ever be able to bring about.

The causes of the French Revolution lay in the character of *social relations*; and the petty causes assumed by Sainte-Beuve could lie only in the *personal qualities of individuals*. The final cause of social relationships lies in the state of the productive forces. This depends on the qualities of individuals only in the sense, perhaps, that these individuals possess more or less talent for making technical improvements, discoveries, and inventions. Sainte-Beuve did not have these qualities in mind. No other qualities, however, enable individuals directly to influence the state of productive forces, and, hence, the social relations which they determine, i.e., *economic relations*. No matter what the qualities of the given individual may be, they cannot eliminate the given economic relations if the latter conform to the given state of productive forces. But the personal qualities of individuals make them more or less fit to satisfy those social needs which arise out of the given economic relations, or to counteract such satisfaction.

The urgent social need of France at the end of the eighteenth century was the substitution for the obsolete political institutions of new institutions that would conform more to her economic system. The most prominent and useful public men of that time were those who were more capable than others of helping to satistfy this most urgent need.

We will assume that Mirabeau, Robespierre, and Napoleon were men of that type. What would have happened had premature death not removed Mirabeau from the political stage? The constitutional monarchist party would have retained its considerable power for a longer period; its resistance to the republicans would, therefore, have been more energetic. But that is all. No Mirabeau could, at that time, have averted the triumph of the republicans. Mirabeau's power rested entirely on the sympathy and confidence of the people; but the people wanted a republic, as the Court irritated them by its obstinate defence of the old order. As soon as the people had become convinced that Mirabeau did not sympathize with their republican strivings they would have ceased to sympathize with him; and then the great orator would have lost nearly all influence, and in all probability would have fallen a victim to the very movement that he vainly would have tried to check.

Approximately the same thing may be said about Robespierre. Let us assume that he was an absolutely indispensable force. If the accidental fall of a brick had killed him, say, in January 1793, his place, of course, would have been taken by somebody else, and although this person might have been inferior to him in every respect, nevertheless, events would have taken *the same course* as they did when Robespierre was alive. For example, even under these circumstances the Gironde would probably not have escaped defeat; but it is possible that Robespierre's party would have lost power somewhat earlier and we would now be speaking, not of the *Thermidor* reaction, but of the *Floréal, Prairial*, or *Messidor* reaction. Perhaps some will say that, with his inexorable Terror, Robespierre did not delay but hastened the downfall of his party. We will not stop to examine this supposition here; we will accept it as if it were quite sound. In that case we must assume that Rosespierre's party would have fallen not in *Thermidor*, but in *Fructidor, Vendemiaire*, or *Brumaire*. In short, it may have fallen sooner or perhaps later, but it certainly would have fallen, because the section of the people which supported Robespierre's party was totally unprepared to hold power for a prolonged period. At all events, results 'opposite' to those which arose from Robespierre's energetic action are out of the question.

Nor could they have arisen even if Bonaparte had been struck down by a bullet, let us say, at the Battle of Arcole. What he did in the Italian and other campaigns other generals would have done. Probably they would not have displayed the same talent as he did, and would not have achieved such brilliant victories; nevertheless the French Republic would have emerged victorious from the wars it waged at that time because its soldiers were incomparably the best in Europe.

As for the eighteenth of *Brumaire* and its influence on the internal life of France, here too, *in essence* the general course and outcome of events would probably have been the same as they were under Napoleon. The Republic, mortally wounded by the events of the ninth of *Thermidor*, was slowly dying. The *Directoire* was unable to restore order which the bourgeoisie, having rid itself of the rule of the aristocracy, now desired most of all. To restore order a '*good sword*', as Siéyès expressed it, was needed. At first it was thought that General Jourdan would serve in this virtuous role, but when he was killed at Novi, the names of Moreau, MacDonald, and Bernadotte were mentioned. Bonaparte was only mentioned later; and had he been killed, like Jourdan, he would not have been mentioned at all, and some other 'sword' would have been put forward.

It goes without saying that the man whom events had elevated to the position of dictator tirelessly must have been aspiring to power himself, energetically pushing aside and crushing ruthlessly all who stood in his way. Bonaparte was a man of iron energy and was remorseless in the pursuit of his goal. But in those days there were not a few energetic, talented, and ambitious egoists besides him. The place Bonaparte succeeded in occupying probably would not have remained vacant. Let us assume that the other general who had secured this place would have been more peaceful than Napoleon, that he would not have roused the whole of Europe against himself, and, therefore, would have died in the Tuileries and not on the island of St Helena. In that case, the Bourbons would not have returned to France at all; for them such a result would certainly have been the 'opposite' of what it was. In its relation to the internal life of France as a whole, however, this result would have differed little from the actual result. After the 'good sword' had restored order and had consolidated the power of the bourgeoisie, the latter would have soon tired of its barrack-room habits and despotism. A liberal movement would have arisen, similar to the one that arose after the Restoration; the fight would have gradually flared up, and, as good swords' are not distinguished for their yielding nature, the virtuous Louis Philippe perhaps would have ascended the throne of his beloved kinsmen, not in 1839, but in 1820, or in 1825.

All such changes in the course of events to some extent might have influenced the subsequent political and, through it, the economic life of Europe. Nevertheless, under no circumstances would the final outcome of the revolutionary movement have been the 'opposite' of what it was. Owing to the specific qualities of their minds and characters, influential individuals can change the *individual features of events and some of their particular consequences*, but they cannot change their general *trend*, which is determined by other forces....

But let us return to our subject. A great man is great not because his personal qualities give individual features to great historical events, but because he possesses qualities which make him most capable of serving the great social needs of his time, needs which arose as a result of general and particular causes. In his well-known book on heroes and hero-worship, Carlyle calls great men *beginners*. This is a very apt description. A great man is a beginner precisely because he sees *further* than others and desires things *more strongly* than others. He solves the scientific problems brought up by the preceding process of intellectual development of society; he points to the new social needs created by the preceding development of social relationships; he takes the initiative in satisfying these needs. He is a hero. But he is a hero not in the sense that he can stop or change the natural course of things, but in the sense that his activities are the conscious and free expression of this inevitable and unconscious course. Herein lies all his significance; herein lies his whole power. But this significance is colossal, and the power is terrible.

2.   The Marxist Conception of Historical Determinism

Marxism is a complete theoretical system. To put the matter in a nutshell, Marxism is contemporary materialism, the highest stage of development of that philosophy, that view of the universe, whose foundations were laid down in ancient Greece by Democritus, and in part by his predecessors, the Ionian thinkers. In fact, what is called hylozoism is nothing more than a naïve materialism. The chief credit for the elaboration of contemporary materialism undoubtedly belongs to Karl Marx and his friend Friedrich Engels. The historical and economic aspects of this philosophy, that is to say of historical materialism, together with the system of views concerning the problems, the method, and the categories of political economy and concerning the economic evolution of society (in especial, of capitalist society)—for these we are almost wholly indebted to Marx and Engels. What their predecessors achieved in this domain can only be regarded as preparatory. Before Marx and Engels got to work, materials, often abundant and valuable, had been collected; but these materials had not been systematized, had not been analysed from a general outlook, and, consequently, had not been appraised or utilized as they should have been. On the other hand, what the followers of Marx and Engels in Europe and America have achieved in this domain has been nothing beyond the more or less successful study of special problems—some of them, it is true, of great importance. That is why, in many cases, when people speak of Marxism they refer only to the two before-mentioned aspects of the materialist conception of the world. This is true, not only of the general public, which is not yet competent to understand philosophical doctrines, but also of whose who (both in Russia and elsewhere in the civilized world) regard themselves as faithful disciples of Marx and Engels. These two aspects are believed to be something altogether independent of 'philosophical materialism', and are even regarded as something contrasted therewith. Since, however, these two aspects, when forcibly detached from the totality of kindred conceptions which form their theoretical foundation, cannot hang in the void, the persons who have thus detached them naturally consider it necessary to 'provide a new basis for Marxism' by founding it (quite arbitrarily, and as a rule under the influence of philosophical currents that prevail among bourgeois savants) upon this or that philosophy....

Many people confound dialectic with the theory of evolution. Dialectic is, in fact a theory of evolution. But it differs profoundly from the vulgar theory of evolution, which is based substantially upon the principle that neither in nature nor in history do sudden changes occur, and that all changes taking place in the world occur gradually. Hegel had already shown that, understood in such a sense, the theory of evolution was inconsistent and absurd....

Marx and Engels whole-heartedly adopted this dialectical view of Hegel's as to the inevitability of jumps in the process of evolution. Engels developed it in a detailed fashion in his controversy with Dühring, and here he 'set it on its feet', that is to say, he placed it on a materialist foundation.

Thus he showed that the passage from one form of energy to another could only occur suddenly. He sought in modern chemistry a confirmation of the dialectical theory of the transformation of quantity into quality. Speaking generally, he considered that the laws of dialectical thought were confirmed by the dialectical properties of being. Here, as usual, being conditions thought....

Hegel in his *Philosophy of History*, had already drawn attention to the importance of the 'geographical basis of universal history'. But since, in his view, the ultimate cause of all evolution was the idea, and since he referred only in passing and reluctantly, as concerns matters of secondary import-ance, to the materialist explanation of phenomena, it was impossible that the thoroughly sound idea he entertained as to the significance of geo-graphical environment should bear fruit. The supremely important con-clusions of this theory could only be drawn by the materialist Marx. The properties of the geographical environment determine the character, not only of the natural products with which man satisfies his wants, but also of the objects which man himself produces in order to satisfy these wants. Where there were no metals, aboriginal tribes could not, unaided, get beyond the limits of what is termed the Stone Age. In like manner, if primitive fishers and primitive hunters were to pass on to the stage of cattle-breeding and that of agriculture, suitable geographical conditions were requisite, a suitable fauna and flora. Lewis Morgan has shown that the remarkable difference between the social evolution of the New World and that of the Old, is to be explained by the lack in the New World of animals capable of being domesticated, and by the differences between the flora of the New World and the Old.

Writing of the redskins of North America, Waitz says: 'They have no domesticated animals. This is extremely important, for it is the principal reason why they have remained at such a low stage of development.' Schweinfurth relates that in Africa, when a region is overcrowded, part of the population emigrates and thereupon changes its mode of life in accordance with the geographical environment: 'Tribes which have hitherto been engaged in agriculture will take to hunting, and tribes which have lived by cattle-breeding will turn to agriculture.' Schweinfurth also points out that the inhabitants of a region which, like much of Central Africa, is rich in iron, naturally take to smelting and to making iron weapons and tools.

Nor is this all. Already in the lowest stages of social evolution, tribes enter into relation one with another, mutually exchanging some of their products. The result is an enlargement of the boundaries of their geographical environment, and that in its turn has an effect upon the evolution of the productive forces in each of the tribes, quickening this evolution. It will readily be understood that the ease with which such relations become established and developed depends also upon the characteristics of the geographical environment. Hegel said that seas and rivers bring men closer together whereas mountains keep them apart. Though this is true, the seas only bring men closer together when the development of the forces of production has already attained a tolerably high level. As long as that level is low, the sea (as Ratzel rightly points out) is a great hindrance to intercourse between the races which it separates. However this may be, it is certain that the more variable the properties of the geographical environment, the more favourable they are to the development of the forces of production. Marx writes: 'It is not the absolute fertility of the soil but the multifariousness of its natural products which constitutes the natural foundation of the social division of labour, and, by changing the natural conditions of his environment, spurs man on to multiply his own needs, capacities, means of labour, and methods of labour.' Using almost the same terms as Marx, Ratzel says: 'The most important thing is, not that there is a greater facility in procuring food, but that certain inclinations, certain wants are awakened in man.'

Thus the peculiarities of the geographical environment determine the evolution of the forces of production, and this, in its turn, determines the development of economic forces and, therefore, the development of all the other social relations. Marx explains the matter in the following terms:

The social relations which the producers enter into one with another, the conditions of their reciprocal activities, and their participation in the totality of production, differ in accordance with differences in the character of the forces of production. The invention of a new weapon, the firearm, necessarily modified the whole internal organization of an army, the relations within whose scope individuals form an army—the relations which make of it an organized whole. Necessarily, too, this invention modified the relations between different armies.

To make this explanation more convincing, I will give another instance. The Masai, in eastern equatorial Africa, do not take prisoners, give no quarter, the reason being that, as Ratzel points out, they are pastoral people, so that the technical possibility of making use of slave labour has not arisen among them. But the Wakamba, who, though neighbours of the Masai, are agriculturists, have a use for slave labour, and they, therefore, give quarter, take prisoners, and make slaves of them. The appearance of slavery as an institution thus presupposes that the social forces have reached a degree of development at which the exploitation of the labour of prisoners has become possible. But slavery is a relation of production whose appearance indicates the beginnings of class division in a society which has hitherto known no other divisions than those of sex and age. When slavery is in full bloom, it puts its stamp on the whole economy of the society in which it exists, and thereby upon all the other social relations—especially upon the political regime. However diversified the States of classical antiquity were in the matter of political regime, they all had this characteristic in common, that every one of them was a political organization concerned to express and to defend the interests of freemen alone.

We now know that the development of the forces of production (which for its part, in the last resort, determines the development of all social relations) itself primarily depends upon the pecularities of the geographical environment. But as soon as specific social relations have come into existence, they, in their turn, exercise a marked influence upon the development of the forces of production. Hence that which primarily was a consequence, becomes in its turn a cause; between the evolution of the forces of production, on the one hand, and the social regime, on the other, there occurs a play of action and reaction which assumes, at various epochs, the most divergent forms.

Nor must we fail to note that the state of the forces of production determines, not only the internal relations existing within a given society, but also the external relations between this society and other societies. To each phase of the development of the forces of production there corresponds a determinate character of armaments, of the military art, and, finally, of international law—or, to speak more accurately, of intersocial or intertribal law [custom]. Hunting tribes are not in a position to constitute large political organizations, were it only for the reason that, owing to the low level attained by the forces of production, primitive hunters are compelled to undertake scattered efforts, each one for himself, in the search for food, and therefore can only form tiny social groups. But the more widely these social groups are dispersed, each fighting for its own band, the more inevitable it is that there should occur sanguinary combats for the settlement of disputes which, in a civilized society could easily be settled by a magistrate. . . .

The theory of the influence which the geographical environment exerts upon the historical evolution of mankind has often been reduced to a simple recognition of the direct influence of climate upon man in society. It has been supposed that, under the influence of one climate, a race becomes passionately devoted to liberty; that under the influence of another climate, another race will become inclined to endure patiently the rule of a more or less despotic sovereign; that under the influence of a third climate, yet another race will become superstitious, and will therefore accept the sway of a priesthood. Such, for example, were the views held by Buckle. According to Marx, the geographical environment acts on man through the instrumentality of the relations of production which arise in a given area upon the foundation of the given forces of production, whose primary condition of development consists in the properties of this very environment. Modern ethnology inclines more and more towards such an outlook, and consequently ascribes less and less importance to race as a factor in the history of civilization. 'The possession of a certain amount of civilization', writes Ratzel, 'has nothing whatever to do with race in itself.'

Still, there can be no doubt that, as soon as a certain level of civilization has been attained, this civilization exercises an influence upon the bodily and mental qualities of the race.

The influence of the geographical environment upon man in society is a variable one. The evolution of the forces of production determined by the properties of the geographical environment increases the power of man over nature, and thereby brings into being a new relation between man and the geographical environment. The English today react upon the geographical environment of their island in a way which differs from that in which the British tribes of Caesar's day reacted upon the same environment. This answers the objection that the character of the population of a given country can be essentially transformed although the geographical conditions remain unaltered. . . .

In primitive society, where class divisions do not yet exist, productive activity has a direct influence upon the conception of the universe and upon aesthetic taste. Decorative art draws its motifs from technique; and dancing, which is perhaps the most important of the arts in such a society, is usually content to mimic one of the processes of production. This is especially obvious in hunting tribes at the lowest level of economic development accessible to our observation. That is why I have directed my readers' attention mainly to these tribes when discussing the way in which primitive man's mental condition is dependent upon his economic activity. When we are concerned with a society divided into classes, the direct influence of economic activity upon ideology is far less obvious. We can easily understand why this should be so. If, for instance, the dance performed by Australian Blackfellows is a reproduction of the activities of the same tribesmen when engaged in collecting roots, we know where we are; but a dance on the part of fine ladies in eighteenth-century France could not, of course, be a representation of their productive activities, seeing that they had no such activities and preferred to devote themselves to 'the science of love'. If we want to understand a dance performed by Australian indigens, it suffices that we should know what part is played by the women of the tribe in collecting the roots of wild plants. But a knowledge of the economic life of France in the eighteenth century will not explain to us the origin of the minuet. In the latter case we have to do with a dance which is an expression of the psychology of a non-productive class. Most of the habits and customs of what is called good society depend upon the same sort of psychology. Here, then, the economic factor yields place to the psychological factor. We must not forget, however, that the appearance of non-productive classes in a society is itself the outcome of the economic development of that society. This means that the economic factor remains predominant, even when its activity is overlaid by that of other factors. In truth, then only does its significance become conspicuous, for then it determines the possibility of the working of other factors and the limits of their influence.

If we wish to summarize the views of Marx and Engels on the relation between the famous 'foundation' and the no less famous 'superstructure', we shall get something like this.

1. The state of the forces of production;
2. Economic relations conditioned by these forces;
3. The socio-political regime erected upon a given economic foundation;
4. The psychology of man in society, determined in part directly by economic conditions, and in part by the whole socio-political regime erected upon the economic foundation;
5. Various ideologies reflecting this psychology.

The formula is sufficiently comprehensive to embrace all the forms of historical development. At the same time it is utterly alien to the eclecticism which cannot get beyond the idea of a reciprocal action between the various social forces and does not realize that such reciprocal action between forces cannot solve the problem of their origin. Our formula is a monistic formula, and this monistic formula is impregnated with materialism. In his

Philosophy of Spirit Hegel said 'Spirit [Mind] is the only motive principle of history.' We cannot take any other view if we accept that idealist standpoint according to which being is determined by thought. Marxian materialism, shows how the history of thought is determined by the history of being. But idealism did not prevent Hegel from recognizing that economic conditions were causes 'which become effective through the instrumentality of the development of the spirit'. In the same way, materialism did not prevent Marx from recognizing that in history 'spirit' acted as a force whose direction, at any given moment, was determined by the development of economic conditions.

Sources: G. Plekhanov, *Fundamental Problems of Marxism* (London, 1937), pp. 1 f., 27 f., 32 ff., 36, 40 ff., 60 f., 72 f. *Fundamental Problems of Marxism* (London, 1969), pp. 64 ff., 176.

FURTHER NOTES

WORKS

PLEKHANOV, G. V., *In Defense of Materialism: The Development of the Monist View of History* (London, 1947).
—— *The Materialist Conception of History* (London, 1940).
—— *The Role of the Individual in History* (London, 1940).

COMMENTARIES

ASCHER, A., *Pavel Axelrod and the Development of Menshevism* (Cambridge, Mass., 1972).
BARON, SAMUEL H., 'Between Marx and Lenin: George Plekhanov', in L. Labedz (ed.), *Revisionism: Essays on the History of Marxist Ideas* (London, 1962).
—— 'Plekhanov and the Revolution of 1905', in J. S. Curtiss (ed.), *Essays in Russian and Soviet History* (Leiden, 1963).
—— *Plekhanov: The Father of Russian Marxism* (London, 1963).
HAIMSON, L., *The Russian Marxists and the Origins of Bolshevism* (Cambridge, Mass., 1955).

7

VLADIMIR LENIN

A Marxist is solely someone who *extends* the recognition of the class struggle to the recognition of the *dictatorship of the proletariat*. This is what constitutes the most profound distinction between the Marxist and the ordinary petty (as well as big) bourgeois. This is the touchstone on which the *real* understanding and recognition of Marxism is to be tested.

> *State and Revolution*, in *Selected Works* (MOSCOW, 1960–1), ii. 328.

VLADIMIR ILICH ULYANOV (1870–1924), *who adopted the pseudonym of Lenin, was the most influential Marxist theorist and political leader of the twentieth century. Born into a fairly comfortable middle-class family (his father was a school inspector), Lenin became a Marxist while still a student. In 1900 he joined forces with Plekhanov and soon emerged as the driving force behind the Bolshevik faction of the Russian Marxists. Lenin displayed tireless energy in keeping the Bolshevik Party together during the lean years between the revolutions of 1905 and 1917: following the October revolution these energies were devoted to coping with the crises of civil war, famine, and foreign intervention. Each of the extracts below illustrates an important facet of the originality of Lenin's contribution to the Marxist tradition. In the first,* What Is To Be Done? *(1902), Lenin outlines his view that the workers, if left to their own devices, can achieve no more than 'trade-union consciousness' and consequently need to be led by a vanguard party of professional revolutionaries. In the second, from* Two Tactics *(1905), he explains why he thinks that the revolution imminent in Russia, though bourgeois in its political and economic form, will be led by the proletariat. The next two extracts show the shift in Lenin's perspective caused by the outbreak of the First World War. In the third, from* Imperialism: The Highest Stage of Capitalism *(1916), Lenin argues that the imperialist tendencies of monopoly finance capital have laid the basis for an immediate move to Socialism. The nature of this Socialism is explored in the fourth extract, from* The State and Revolution *(1917), where Lenin expounds his confidence in the libertarian and participatory nature of the future state. In the fifth extract, from* Left-wing Communism *(1920), Lenin opposes those ultra-left Marxists who rejected the idea of participating in bourgeois trade unions or parliaments. The final extract, from* Better Fewer but Better *(1923) demonstrates Lenin's concern at the increasingly bureau-cratic and high-handed nature of the emergent Soviet state.*

1. The Proletarian Party

SINCE there can be no talk of an independent ideology formulated by the working masses themselves in the process of their movement, the *only* choice is—either bourgeois or Socialist ideology. There is no middle course (for mankind has not created a 'third' ideology, and, moreover, in a society torn by class antagonisms there can never be a non-class or an above-class ideology). Hence, to belittle the Socialist ideology *in any way, to turn aside from it in the slightest degree,* means to strengthen bourgeois ideology. There is much talk of spontaneity. But the *spontaneous* development of the working-class movement leads to its subordination to bourgeois ideology, *to its development along the lines of the Credo programme;* for the spontaneous working-class movement is trade-unionism, is *Nur-Gew-erkschaftlerei,* and trade-unionism means the ideological enslavement of the workers by the bourgeoisie. Hence, our task, the task of Social Demo-cracy, is *to combat spontaneity, to divert* the working-class movement from this spontaneous, trade-unionist striving to come under the wing of the bourgeoisie, and to bring it under the wing of revolutionary Social Democracy....

We have seen that the conduct of the broadest political agitation and, consequently, of all-sided political exposures is an absolutely necessary and a *paramount* task of our activity, if this activity is to be truly Social Democratic. However, we arrived at this conclusion *solely* on the grounds of the pressing needs of the working class for political knowledge and political training. But such a presentation of the question is too narrow, for it ignores the general democratic tasks of Social Democracy, in particular of present-day Russian Social Democracy. In order to explain the point more concretely we shall approach the subject from an aspect that is 'nearest' to the economist, namely, from the practical aspect. 'Everyone agrees' that it is necessary to develop the political consciousness of the working class. The question is, *how* that is to be done and what is required to do it. The economic struggle merely 'impels' the workers to realize the government's attitude towards the working class. Consequently, *however much we may try* to 'lend the economic struggle itself a political character', we *shall never be able* to develop the political consciousness of the workers (to the level of Social Democratic political consciousness) by keeping within the framework of the economic struggle, for *that framework is too narrow....*

Class political consciousness can be brought to the workers *only from without*, that is, only from outside the economic struggle, from outside the sphere of relations between workers and employers. The sphere from which alone it is possible to obtain this knowledge is the sphere of relationships of *all* classes and strata to the state and the government, the sphere of the interrelations between *all* classes. For that reason, the reply to the question as to what must be done to bring political knowledge to the workers cannot be merely the answer with which, in the majority of cases, the practical workers, especially those inclined towards economism, mostly content themselves, namely: 'To go among the workers'. To bring political knowledge to the *workers* the Social Democrats must *go among all classes of the population*; they must dispatch units of their army *in all directions*.

We deliberately select this blunt formula, we deliberately express ourselves in this sharply simplified manner, not because we desire to indulge in paradoxes, but in order to 'impel' the economists to a realization of their tasks which they unpardonably ignore, to suggest to them strongly the difference between trade-unionist and Social Democratic politics, which they refuse to understand. We therefore beg the reader not to get wrought up, but to hear us patiently to the end.

Let us take the type of Social Democratic study circle that has become most widespread in the past few years and examine its work. It has 'contacts with the workers' and rests content with this, issuing leaflets in which abuses in the factories, the government's partiality towards the capitalists, and the tyranny of the police are strongly condemned. At workers' meetings the discussions never, or rarely ever, go beyond the limits of these subjects. Extremely rare are the lectures and discussions held on the history of the revolutionary movement, on questions of the economic evolution of Russia and of Europe, on the position of the various classes in modern society, etc. As to systematically acquiring and extending contact with other classes of society, no one even dreams of that. In fact, the ideal leader, as the majority of the members of such circles picture him, is something far more in the nature of a trade-union secretary than a Socialist political leader. For the secretary of any, say English, trade union always helps the workers to carry on the economic struggle, he helps them to expose factory abuses, explains the injustice of the laws and of measures that hamper the freedom to strike and to picket (i.e., to warn all and sundry that a strike is proceeding at a certain factory), explains the partiality of arbitration court judges who belong to the bourgeois classes, etc., etc. In a word, every trade-union secretary conducts and helps to conduct 'the economic struggle against the employers and the government'. It cannot be too strongly maintained that *this is still not* Social Democracy, that the Social Democrat's ideal should not be the trade-union secretary, but *the tribune of the people*, who is able to react to every manifestation of tyranny and oppression, no matter where it appears, no matter what stratum or class of the people it affects; who is able to generalize all these manifestations and produce a single picture of police violence and capitalist exploitation; who is able to take advantage of every event, however small, in order to set forth *before all* his socialist convictions and his democratic demands, in order to clarify for *all* and everyone the world-historic significance of the struggle for the emancipation of the proletariat....

As I have stated repeatedly, by 'wise men', in connection with organization, I mean *professional revolutionaries*, irrespective of whether they have developed from among students or working men. I assert: (1) that no revolutionary movement can endure without a stable organization of leaders maintaining continuity; (2) that the broader the popular mass drawn spontaneously into the struggle, which forms the basis of the movement and participates in it, the more urgent the need for such an organization, and the more solid this organization must be (for it is much easier for all sorts of demagogues to side-track the more backward sections of the masses); (3) that such an organization must consist chiefly of people professionally engaged in revolutionary activity; (4) that in an autocratic state, the more we *confine* the membership of such an organization to people who are professionally engaged in revolutionary activity and who have been professionally trained in the art of combating the political police, the more difficult will it be to unearth the organization; and (5) the *greater* will be the number of people from the working class and from the other social classes who will be able to join the movement and perform active work in it.

I invite our economists, terrorists, and 'economists-terrorists' to confute these propositions. At the moment, I shall deal only with the last two points. The question as to whether it is easier to wipe out 'a dozen wise men' or 'a hundred fools' reduces itself to the question, above considered, whether it is possible to have a mass *organization* when the maintenance of strict secrecy is essential. We can never give a mass organization that degree of secrecy without which there can be no question of persistent and continuous struggle against the government. To concentrate all secret functions in the hands of as small a number of professional revolutionaries as possible does not mean that the latter will 'do the thinking for all' and that the rank and file will not take an active part in the *movement*. On the contrary, the membership will promote increasing numbers of the professional revolutionaries from its ranks; for it will know that it is not enough for a few students and for a few working men waging the economic struggle to gather in order to form a 'committee', but that it takes years to train oneself to be a professional revolutionary; and the rank and file will 'think', not only of amateurish methods, but of such training. Centralization of the secret functions of the *organization* by no means implies centralization of all the functions of the *movement*. Active participation of the widest masses in the illegal press will not diminish because a 'dozen' professional revolutionaries centralize the secret functions connected with this work; on the contrary, it will *increase* tenfold. In this way, and in this way alone, shall we ensure that reading the illegal press, writing for it, and to some extent even distributing it, will *almost cease to be secret work*, for the police will soon come to realize the folly and impossibility of judicial and administrative red-tape procedure over every copy of a publication that is being distributed in the thousands. This holds not only for the press, but for every function of the movement, even for demonstrations. The active and widespread participation of the masses will not suffer; on the contrary, it will benefit by the fact that a 'dozen' experienced revolutionaries, trained professionally no less than the police, will centralize all the secret aspects of the work—the drawing up of leaflets, the working out of approximate plans; and the appointing of bodies of leaders for each urban district, for each factory district, and for each educational institution, etc. (I know that exception will be taken to my 'undemocratic' views, but I shall reply below fully to this anything but intelligent objection.) Centralization of the most secret functions in an organization of revolutionaries will not diminish, but rather increase, the extent and enhance the quality of the activity of a large number of other organizations that are intended for a broad public and are therefore as loose and as non-secret as possible, such as workers' trade unions; workers' self-education circles and circles for reading illegal literature; and Socialist, as well as democratic, circles among *all* other sections of the population; etc., etc. We must have such circles, trade unions, and organizations everywhere in *as large a number as possible* and with the widest variety of functions; but it would be absurd and harmful *to confound* them with the organization of *revolutionaries*, to efface the border-line between them, to make still more hazy the all too faint recognition of the fact that in order to 'serve' the mass movement we must have people who will devote themselves exclusively to Social Democratic activities, and that such people must *train* themselves patiently and steadfastly to be professional revolutionaries. . . .

2.   Revolution: Bourgeois or Proletarian?

The idea of seeking salvation for the working class in anything save the further development of capitalism is *reactionary*. In countries like Russia the working class suffers not so much from capitalism as from the insufficient development of capitalism. The working class is, therefore, *most certainly interested* in the broadest, freest, and most rapid development of capitalism. The removal of all the remnants of the old order which hamper the broad, free, and rapid development of capitalism is of absolute *advantage* to the working class. The bourgeois revolution is precisely an upheaval that most resolutely sweeps away survivals of the past, survivals of the serf-owning system (which include not only the autocracy but the monarchy as well), and most fully guarantees the broadest, freest, and most rapid development of capitalism.

That is why a *bourgeois* revolution is *in the highest degree advantageous to the proletariat*. A bourgeois revolution is *absolutely* necessary in the interests of the proletariat. The more complete, determined, and consistent the bourgeois revolution, the more assured will the proletariat's struggle be against the bourgeoisie and for socialism. Only those who are ignorant of the ABC of scientific Socialism can regard this conclusion as new, strange, or paradoxical. And from this conclusion, among other things, follows the thesis that *in a certain sense* a bourgeois revolution is *more advantageous* to the proletariat than to the bourgeoisie. . . .

On the other hand, it is more advantageous to the working class for the necessary changes in the direction of bourgeois democracy to take place by way of revolution and not by way of reform, because the way of reform is one of delay, procrastination, the painfully slow decomposition of the putrid parts of the national organism. It is the proletariat and the peasantry that suffer first of all and most of all from the putrefaction. The revolutionary path is one of rapid amputation, which is the least painful to the proletariat, the path of the immediate removal of what is putrescent, the path of least compliance with and consideration for the monarchy and the abominable, vile, rotten, and noxious institutions that go with it. . . .

We must be perfectly certain in our minds as to what real social forces are opposed to 'tsarism' (which is a real force perfectly intelligible to all) and are capable of gaining a 'decisive victory' over it. The big bourgeoisie, the landlords, the factory owners, the 'society' which follows the *Osvobozhdeniye* lead, cannot be such a force. We see that they do not even want a decisive victory. We know that owing to their class position they are incapable of waging a decisive struggle against tsarism; they are too heavily fettered by private property, by capital, and land to enter into a decisive struggle. They stand in too great need of tsarism, with its bureaucratic, police, and military forces for use against the proletariat and the peasantry, to want it to be destroyed. No, the only force capable of gaining 'a decisive victory over tsarism' is the *people*, i.e., the proletariat and the peasantry, if we take the main, big forces, and distribute the rural and urban petty bourgeoisie (also part of 'the people') between the two. 'The revolution's decisive victory over tsarism' means the establishment of the *revolutionary-democratic dictatorship of the proletariat and the peasantry....*

Vperyod stated quite definitely wherein lies the real 'possibility of retaining power'—namely, in the revolutionary-democratic dictatorship of the proletariat and the peasantry; in their joint mass strength, which is capable of outweighing all the forces of counter-revolution; in the inevitable concurrence of their interests in *democratic* reforms. Here, too, the resolution of the Conference gives us nothing positive; it merely evades the issue. Surely, the possibility of retaining power in Russia must be determined by the composition of the social forces in Russia herself, by the circumstances of the democratic revolution now taking place in our country. A victory of the proletariat in Europe (it is still quite a far cry from bringing the revolution into Europe to the victory of the proletariat) will give rise to a desperate counter-revolutionary struggle on the part of the Russian bourgeoisie—yet the resolution of the new-Iskrists does not say a word about this counter-revolutionary force whose significance was appraised in the resolution of the RSDLP's Third Congress. If, in our fight for a republic and democracy, we could not rely upon the peasantry as well as upon the proletariat, the prospect of our 'retaining power' would be hopeless. But if it is not hopeless, if the 'revolution's decisive victory over tsarism' opens up such a possibility, then we must indicate it, call actively for its transformation into reality, and issue practical slogans not only *for the contingency* of the revolution being brought into Europe, but also *for the purpose* of taking it there. The reference made by tail-ist Social-Democrats to the 'limited historical scope of the Russian revolution' merely serves to cover up their limited understanding of the aims of this democratic revolution, and of the proletariat's leading role in it!

One of the objections raised to the slogan of 'the revolutionary-democratic dictatorship of the proletariat and the peasantry' is that dictatorship presupposes a 'single will' (*Iskra*, No. 95), and that there can be no single will of the proletariat and the petty bourgeoisie. This objection is unsound, for it is based on an abstract, 'metaphysical' interpretation of the term 'single will'. There may be a single will in one respect and not in another. The absence of unity on questions of Socialism and in the struggle for Socialism does not preclude singleness of will on questions of democracy and in the struggle for a republic. To forget this would be tantamount to forgetting the logical and historical difference between a democratic revolution and a Socialist revolution. To forget this would be tantamount to forgetting the character of the democratic revolution as one of *the whole people*: if it is 'of the whole people', that means that there is 'singleness of will' precisely in so far as this revolution meets the needs and requirements of the whole people. Beyond the bounds of democratism there can be no question of the proletariat and the peasant bourgeoisie having a single will. Class struggle between them is inevitable, but it is in a democratic republic that this struggle will be the most thoroughgoing and widespread struggle of the people *for Socialism*. Like everything else in the world, the revolutionary-democratic dictatorship of the proletariat and the peasantry has a past and a future. Its past is autocracy, serfdom, monarchy, and privilege. In the struggle against this past, in the struggle against counter-revolution, a 'single will' of the proletariat and the peasantry is possible, for here there is unity of interests.

Its future is the struggle against private property, the struggle of the wage-worker against the employer, the struggle for Socialism. Here singleness of will is impossible. Here the path before us lies not from autocracy to a republic, but from a petty-bourgeois democratic republic to Socialism.

Of course, in actual historical circumstances, the elements of the past become interwoven with those of the future; the two paths cross. Wage labour with its struggle against private property exists under the autocracy as well; it arises even under serfdom. But this does not in the least prevent us from logically and historically distinguishing between the major stages of development. We all contrapose bourgeois revolution and Socialist revolution; we all insist on the absolute necessity of strictly distinguishing between them; however, can it be denied that in the course of history individual, particular elements of the two revolutions become interwoven? Has the period of democratic revolutions in Europe not been familiar with a number of Socialist movements and attempts to establish Socialism? And will not the future Socialist revolution in Europe still have to complete a great deal left undone in the field of democratism?

A Social Democrat must never for a moment forget that the proletariat will inevitably have to wage a class struggle for Socialism even against the most democratic and republican bourgeoisie and petty bourgeoisie. This is beyond doubt. Hence, the absolute necessity of a separate, independent, strictly class party of Social Democracy. Hence, the temporary nature of our tactics of 'striking a joint blow' with the bourgeoisie and the duty of keeping a strict watch 'over our ally, as over an enemy', etc. All this also leaves no room for doubt. However, it would be ridiculous and reactionary to deduce from this that we must forget, ignore, or neglect tasks which, although transient and temporary, are vital at the present time. The struggle against the autocracy is a temporary and transient task for Socialists, but to ignore or neglect this task in any way amounts to betrayal of Socialism and service to reaction. The revolutionary-democratic dictatorship of the proletariat and the peasantry is unquestionably only a transient, temporary Socialist aim, but to ignore this aim in the period of a democratic revolution would be downright reactionary.

3. Imperialism

We must now try to sum up, put together, what has been said above on the subject of imperialism. Imperialism emerged as the development and direct continuation of the fundamental characteristics of capitalism in general. But capitalism only became capitalist imperialism at a definite and very high stage of its development, when certain of its fundamental characteristics began to change into their opposites, when the features of the epoch of transition from capitalism to a higher social and economic system had taken shape and revealed themselves all along the line. Economically, the main thing in this process is the displacement of capitalist free competition by capitalist monopoly. Free competition is the fundamental characteristic of capitalism, and of commodity production generally; monopoly is the exact opposite of free competition, but we have seen the latter being transformed into monopoly before our eyes, creating large-scale industry and forcing out small industry, replacing large-scale by still larger-scale industry, and carrying concentration of production and capital to the point where out of it has grown and is growing monopoly: cartels, syndicates, and trusts, and, merging with them, the capital of a dozen or so banks, which manipulate thousands of millions. At the same time the monopolies, which have grown out of free competition, do not eliminate the latter, but exist over it and alongside of it, and thereby give rise to a number of very acute, intense antagonisms, frictions, and conflicts. Monopoly is the transition from capitalism to a higher system.

If it were necessary to give the briefest possible definition of imperialism we should have to say that imperialism is the monopoly stage of capitalism. Such a definition would include what is most important, for, on the one hand, finance capital is the bank capital of a few very big monopolist banks, merged with the capital of the monopolist combines of industrialists; and, on the other hand, the division of the world is the transition from a colonial policy which has extended without hindrance to territories unseized by any capitalist power, to a colonial policy of monopolistic possession of the territory of the world which has been completely divided up.

But very brief definitions, although convenient, for they sum up the main points, are nevertheless inadequate, since very important features of the phenomenon that has to be defined have to be especially deduced. And so, without forgetting the conditional and relative value of all definitions in general, which can never embrace all the concatenations of a phenomenon in its complete development, we must give a definition of imperialism that will include the following five of its basic features: (1) the concentration of production and capital has developed to such a high stage that it has created monopolies which play a decisive role in economic life; (2) the merging of bank capital with industrial capital, and the creation, on the basis of this 'finance capital', of a financial oligarchy; (3) the export of capital as distinguished from the export of commodities acquires exceptional importance; (4) the formation of international monopolist capitalist combines which share the world among themselves, and (5) the territorial division of the whole world among the biggest capitalist powers is completed. Imperialism is capitalism in that stage of development in which the dominance of monopolies and finance capital has established itself; in which the export of capital has acquired pronounced importance; in which the division of the world among the international trusts has begun, in which the division of all territories of the globe among the biggest capitalist powers has been completed.

We shall see later that imperialism can and must be defined differently if we bear in mind, not only the basic, purely economic concepts—to which the above definition is limited—but also the historical place of this stage of capitalism in relation to capitalism in general, or the relation between imperialism and the two main trends in the working-class movement. The point to be noted just now is that imperialism, as interpreted above, undoubtedly represents a special stage in the development of capitalism. To enable the reader to obtain the most well-grounded idea of imperialism, we deliberately tried to quote as largely as possible *bourgeois* economists who are obliged to admit the particularly incontrovertible facts concerning the latest stage of capitalist economy. With the same object in view, we have quoted detailed statistics which enable one to see to what degree bank capital, etc., has grown, in what precisely the transformation of quantity into quality, of developed capitalism into imperialism, was expressed. Needless to say, of course, all boundaries in nature and in society are conditional and changeable, and it would be absurd to argue, for example, about the particular year or decade in which imperialism 'definitely' became established.

In the matter of defining imperialism, however, we have to enter into controversy, primarily, with K. Kautsky, the principal Marxian theoretician of the epoch of the so-called Second International—that is, of the twenty-five years between 1889 and 1914. The fundamental ideas expressed in our definition of imperialism were very resolutely attacked by Kautsky in 1915, and even in November 1914, when he said that imperialism must not be regarded as a 'phase' or stage of economy, but as a policy, a definite policy 'preferred' by finance capital; that imperialism must not be 'identified' with 'present-day capitalism'; that if imperialism is to be understood to mean 'all the phenomena of present-day capitalism'—cartels, protection, the domination of the financiers, and colonial policy—then the question as to whether imperialism is necessary to capitalism becomes reduced to the 'flattest tautology', because, in that case, 'imperialism is naturally a vital necessity for capitalism', and so on. The best way to present Kautsky's idea is to quote his own definition of imperialism, which is diametrically opposed to the substance of the ideas which we have set forth (for the objections coming from the camp of the German Marxists, who have been advocating similar ideas for many years already, have been long known to Kautsky as the objections of a defnite trend in Marxism).

Kautsky's definition is as follows:

Imperialism is a product of highly developed industrial capitalism. It consists in the striving of every industrial capitalist nation to bring under its control or to annex larger and larger areas of *agrarian* [Kautsky's italics] territory, irrespective of what nations inhabit those regions.

This definition is utterly worthless because it one-sidedly, i.e., arbitrarily, singles out only the national question (although the latter is extremely important in itself as well as in its relation to imperialism), arbitrarily and *inaccurately* connects this question *only* with industrial capital in the countries which annex other nations, and in an equally arbitrary and inaccurate manner pushes into the forefront the annexation of agrarian regions.

Imperialism is a striving for annexations—this is what the *political* part of Kautsky's definition amounts to. It is correct, but very incomplete, for politically imperialism is, in general, a striving towards violence and reaction. For the moment, however, we are interested in the *economic* aspect of the question, which Kautsky *himself* introduced into *his* definition. The inaccuracies in Kautsky's definition are glaring. The characteristic feature of imperialism is *not* industrial *but* finance capital. It is not an accident that in France it was precisely the extraordinarily rapid development of *finance* capital, and the weakening of industrial capital, that, from the eighties onwards, gave rise to the extreme intensification of annexationist (colonial) policy. The characteristic feature of imperialism is precisely that it strives to annex *not only* agrarian territories, but even most highly industrialized regions (German appetite for Belgium; French appetite for Lorraine), because, (1) the fact that the world is already divided up obliges those contemplating a *redivision* to reach out for *every kind* of territory, and (2) an essential feature of imperialism is the rivalry between several Great Powers in the striving for hegemony, i.e., for the conquest of territory, not so much directly for themselves as to weaken the adversary and undermine *his* hegemony. (Belgium is particularly important for Germany as a base for operations against England; England needs Baghdad as a base for operations against Germany, etc.)

Kautsky refers especially—and repeatedly—to Englishmen who, he alleges, have given a purely political meaning to the word 'imperialism' in the sense that he, Kautsky, understands it. We take up the work by the Englishman Hobson, *Imperialism* which appeared in 1902, and there we read:

> The new imperialism differs from the older, first, in substituting for the ambition of a single growing empire the theory and the practice of competing empires, each motived by similar lusts of political aggrandizement and commercial gain; secondly, in the dominance of financial or investing over mercantile interests.

We see that Kautsky is absolutely wrong in referring to Englishmen generally (unless he meant the vulgar English imperialists, or the avowed apologists for imperialism). We see that Kautsky, while claiming that he continues to advocate Marxism, as a matter of fact takes a step backward compared with the *social-liberal* Hobson, who *more correctly* takes into account two 'historically concrete' (Kautsky's definition is a mockery of historical concreteness!) features of modern imperialism: (1) the competition between *several* imperialisms, and (2) the predominance of the financier over the merchant. If it is chiefly a question of the annexation of agrarian countries by industrial countries, then the role of the merchant is put in the forefront.

Kautsky's definition is not only wrong and un-Marxian. It serves as a basis for a whole system of views which signify a rupture with Marxian theory and Marxian practice all along the line. We shall refer to this later. The argument about words which Kautsky raises as to whether the latest stage of capitalism should be called 'imperialism' or 'the stage of finance capital' is absolutely frivolous. Call it what you will, it makes no difference. The essence of the matter is that Kautsky detaches the politics of imperialism from its economics, speaks of annexations as being a policy 'preferred' by finance capital, and opposes to it another bourgeois policy which, he alleges, is possible on this very same basis of finance capital. It follows, then, that monopolies in economics are compatible with non-monopolistic, non-violent, non-annexationist methods in politics. It follows, then, that the territorial division of the world, which was completed precisely during the epoch of finance capital, and which constitutes the basis of the present peculiar forms of rivalry between the biggest capitalist states, is compatible with a non-imperialist policy. The result is a slurring-over and a blunting of the most profound contradictions of the latest stage of capitalism, instead of an exposure of their depth; the result is bourgeois reformism instead of Marxism.

Kautsky enters into controversy with the German apologist of imperialism and annexations, Cunow, who clumsily and cynically argues that imperialism is present-day capitalism; the development of capitalism is inevitable and progressive; therefore imperialism is progressive; therefore, we should grovel before it and glorify it! This is something like the caricature of the Russian Marxists which the Narodniks drew in 1894–95. They argued: if the Marxists believe that capitalism is inevitable in Russia, that it is progressive, then they ought to open a tavern and begin to implant capitalism! Kautsky's reply to Cunow is as follows: imperialism is not present-day capitalism; it is only one of the forms of the policy of present-day capitalism. This policy we can and should fight, fight imperialism, annexations, etc.

The reply seems quite plausible, but in effect it is a more subtle and more disguised (and therefore more dangerous) advocacy of conciliation with imperialism, because a 'fight' against the policy of the trusts and banks that does not affect the basis of the economics of the trusts and banks is nothing more than bourgeois reformism and pacifism, the benevolent and innocent expression of pious wishes. Evasion of existing contradictions, forgetting the most important of them, instead of revealing their full depth—such is Kautsky's theory, which has nothing in common with Marxism. Naturally, such a 'theory' can only serve the purpose of advocating unity with the Cunows!

'From the purely economic point of view', writes Kautsky, 'it is not impossible that capitalism will yet go through a new phase, that of the extension of the policy of the cartels to foreign policy, the phase of ultra-imperialism,' i.e., of a super-imperialism, of a union of the imperialisms of the whole world and not struggles among them, a phase when wars shall cease under capitalism, a phase of 'the joint exploitation of the world by internationally united finance capital'.

We shall have to deal with this 'theory of ultra-imperialism' later on in order to show in detail how definitely and utterly it breaks with Marxism.

At present, in keeping with the general plan of the present work, we must examine the exact economic data on this question. 'From the purely economic point of view', is 'ultra-imperialism' possible, or is it ultra-nonsense?

If by purely economic point of view a 'pure' abstraction is meant, then all that can be said reduces itself to the following proposition: development is proceeding towards monopolies, hence, towards a single world monopoly, towards a single world trust. This is indisputable, but it is also as completely meaningless as is the statement that 'development is proceeding' towards the manufacture of foodstuffs in laboratories. In this sense the 'theory' of ultra-imperialism is no less absurd than a 'theory of ultra-agriculture' would be.

If, however, we are discussing the 'purely economic' conditions of the epoch of finance capital as a historically concrete epoch which opened at the beginning of the twentieth century, then the best reply that one can make to the lifeless abstractions of 'ultra-imperialism' (which serve exclusively a most reactionary aim: that of diverting attention from the depth of *existing* antagonisms) is to contrast them with the concrete economic realities of present-day world economy. Kautsky's utterly meaningless talk about ultra-imperialism encourages, among other things, that profoundly mistaken idea which only brings grist to the mill of the apologists of imperialism, viz., that the rule of finance capital *lessens* the unevenness and contradictions inherent in world economy, whereas in reality it *increases* them....

There are two areas where capitalism is little developed: Russia and Eastern Asia. In the former, the density of population is extremely low, in the latter it is extremely high; in the former political concentration is high, in the latter it does not exist. The partition of China is only just beginning, and the struggle for it between Japan, the USA, etc., is continually gaining in intensity.

Compare this reality—the vast diversity of economic and political conditions, the extreme disparity in the rate of development of the various countries, etc., and the violent struggles among the imperialist states—with Kautsky's silly little fable about 'peaceful' ultra-imperialism. Is this not the reactionary attempt of a frightened Philistine to hide from stern reality? Are not the international cartels, which Kautsky imagines are the embryos of 'ultra-imperialism' (in the same way as one 'can' describe the manufacture of tabloids in a laboratory as ultra-agriculture in embryo), an example of the division *and the redivision* of the world, the transition from peaceful division to non-peaceful division and vice versa? Is not American and other finance capital, which divided the whole world peacefully with Germany's participation in, for example, the international rail syndicate, or in the

international mercantile shipping trust, now engaged in *redividing* the
world on the basis of a new relation of forces, which is being changed by
methods altogether more peaceful?

Finance capital and the trusts do not diminish but increase the differences
in the rate of growth of the various parts of the world economy. Once the
relation of forces is changed, what other solution of the contradictions can
be found *under capitalism* than that of force?...

It must be observed that in Great Britain the tendency of imperialism to
split the workers, to strengthen opportunism among them, and to cause
temporary decay in the working-class movement, revealed itself much
earlier than the end of the nineteenth and the beginning of the twentieth
centuries; for two important distinguishing features of imperialism were
already observed in Great Britain in the middle of the nineteenth century,
viz., vast colonial possessions and a monopolist position in the world
market....

This clearly shows the causes and effects. The causes are: (1) exploitation
of the whole world by this country; (2) its monopolist position in the world
market; (3) its colonial monopoly. The effects are: (1) a section of the British
proletariat becomes bourgeois; (2) a section of the proletariat allows itself
to be led by men bought by, or at least paid by, the bourgeoisie....

The distinctive feature of the present situation is the prevalence of such
economic and political conditions as could not but increase the irreconcilability between opportunism and the general and vital interests of the
working-class movement: imperialism has grown from an embryo into the
predominant system; capitalist monopolies occupy first place in economics
and politics; the division of the world has been completed on the other
hand, instead of the undivided monopoly of Great Britain we see a few
imperialist powers contending for the right to share in this monopoly, and
this struggle is characteristic of the whole period of the beginning of the
twentieth century. Opportunism cannot now be completely triumphant in
the working-class movement of one country for decades as it was in England
in the second half of the nineteenth century; but in a number of countries
it has grown ripe, overripe, and rotten and has become completely merged
with bourgeois policy in the form of 'social-chauvinism'....

We have seen that in its economic essence imperialism is monopoly
capitalism. This in itself determines its place in history, for monopoly
that grows out of the soil of free competition, and precisely out of free
competition, is the transition from the capitalist system to a higher social-
economic order. We must take special note of the four principal types of
monopoly, or principal manifestations of monopoly capitalism, which are
characteristic of the epoch we are examining.

Firstly, monopoly arose out of a very high stage of development of the concentration of production. This refers to the monopolist capitalist combines, cartels, syndicates, and trusts. We have seen the important part these play in present-day economic life. At the beginning of the twentieth century, monopolies had acquired complete supremacy in the advanced countries, and although the first steps towards the formation of the cartels were first taken by countries enjoying the protection of high tariffs (Germany, America), Great Britain, with her system of free trade, revealed the same basic phenomenon, only a little later, namely, the birth of monopoly out of the concentration of production.

Secondly, monopolies have stimulated the seizure of the most important sources of raw materials, especially for the basic and most highly cartelized industries in capitalist society: the coal and iron industries. The monopoly of the most important sources of raw materials has enormously increased the power of big capital, and has sharpened the antagonism between cartelized and non-cartelized industry.

Thirdly, monopoly has sprung from the banks. The banks have developed from humble middlemen enterprises into the monopolists of finance capital. Some three to five of the biggest banks in each of the foremost capitalist countries have achieved the 'personal union' of industrial and bank capital, and have concentrated in their hands the control of thousands upon thousands of millions which form the greater part of the capital and income of entire countries. A financial oligarchy, which throws a close network of dependence relationships over all the economic and political institutions of present-day bourgeois society without exception—such is the most striking manifestation of this monopoly.

Fourthly, monopoly has grown out of colonial policy. To the numerous 'old' motives of colonial policy, finance capital has added the struggle for the sources of raw materials, for the export of capital, for 'spheres of influence', i.e., for spheres for profitable deals, concessions, monopolist profits, and so on, and finally, for economic territory in general. When the colonies of the European powers in Africa, for instance, comprised only one-tenth of that territory (as was the case in 1876), colonial policy was able to develop by methods other than those of monopoly—by the 'free grabbing' of territories, so to speak. But when nine-tenths of Africa had been seized (by 1900), when the whole world had been divided up, there was inevitably ushered in the era of monopoly ownership of colonies and, consequently, of particularly intense struggle for the division and the redivision of the world.

The extent to which monopolist capital has intensified all the contradictions of capitalism is generally known. It is sufficient to mention the high cost of living and the tyranny of the cartels. This intensification of contradictions constitutes the most powerful driving force of the transitional period of history, which began from the time of the final victory of world finance capital.

Monopolies, oligarchy, the striving for domination instead of striving for liberty, the exploitation of an increasing number of small or weak nations by a handful of the richest or most powerful nations—all these have given birth to those distinctive characteristics of imperialism which compel us to define it as parasitic or decaying capitalism. More and more prominently there emerges as one of the tendencies of imperialism, the creation of the 'rentier state', the usurer state, in which the bourgeoisie to an ever-increasing degree lives on the proceeds of capital exports and by 'clipping coupons'. It would be a mistake to believe that this tendency to decay precludes the rapid growth of capitalism. It does not. In the epoch of imperialism, certain branches of industry, certain strata of the bougeoisie, and certain countries betray, to a greater or lesser degree, now one and now another of these tendencies. On the whole, capitalism is growing far more rapidly than before; but this growth is not only becoming more and more uneven in general, its unevenness also manifests itself, in particular, in the decay of the countries which are richest in capital (England).

In regard to the rapidity of Germany's economic development, Riesser, the author of the book on the big German banks, states: 'The progress of the preceding period (1848–70), which had not been exactly slow, stood in about the same ratio to the rapidity with which the whole of Germany's national economy, and with it German banking, progressed during this period (1870–1905) as the speed of the mail coach in the good old days stood to the speed of the present-day automobile ... Which is whizzing past so fast that it endangers not only innocent pedestrians in its path, but also the occupants of the car.' In its turn, this finance capital which has grown with such extraordinary rapidity is not unwilling, precisely because it has grown so quickly, to pass on to a more 'tranquil' possession of colonies which have to be seized—and not only by peaceful methods—from richer nations. In the United States, economic development in the last decades has been even more rapid than in Germany, *and for this very reason*, the parasitic features of modern American capitalism have stood out with particular prominence. On the other hand, a comparison of, say, the republican American bourgeoisie with the monarchist Japanese or German bourgeoisie shows that the most pronounced political distinction diminishes to an extreme degree in the epoch of imperialism—not because it is unimportant in general, but because in all these cases we are discussing a bourgeoisie which has definite features of parasitism.

The receipt of high monopoly profits by the capitalists in one of the numerous branches of industry, in one of the numerous countries, etc., makes it economically possible for them to bribe certain sections of the workers, and for a time a fairly considerable minority of them, and win them to the side of the bourgeoisie of a given industry or given nation against all the others. The intensification of antagonisms between imperialist nations for the division of the world increases this striving. And so there is created that bond between imperialism and opportunism, which revealed itself first and most clearly in England, owing to the fact that certain features of imperialist development were observable there much earlier than in other countries. Some writers, L. Martov, for example, are prone to wave aside the connection between imperialism and opportunism in the working-class movement—a particularly glaring fact at the present time—by resorting to 'official optimism' (*à la* Kautsky and Huysmans) like the following: the cause of the opponents of capitalism would be hopeless if it were precisely progressive capitalism that led to the increase of opportunism, or, if it were precisely the best-paid workers who were inclined towards opportunism, etc. We must have no illusions about 'optimism' of this kind. It is optimism in regard to opportunism; it is optimism which serves to conceal opportunism. As a matter of fact the extraordinary rapidity and the particularly revolting character of the development of opportunism are by no means a guarantee that its victory will be durable: the rapid growth of a malignant abscess on a healthy body can only cause it to burst more quickly and thus relieve the body of it. The most dangerous of all in this respect are those who do not wish to understand that the fight against imperialism is a sham and humbug unless it is inseparably bound up with the fight against opportunism.

From all that has been said in this book on the economic essence of imperialism, it follows that we must define it as capitalism in transition, or, more precisely, as moribund capitalism. It is very instructive in this respect to note that the bourgeois economists, in describing modern capitalism, frequently employ catchwords and phrases like 'interlocking', 'absence of isolation', etc.: 'in conformity with their functions and course of development', banks are 'not purely private business enterprises, they are more and more outgrowing the sphere of purely private business regulation'. And this very Riesser, who uttered the words just quoted, declares with all seriousness that the 'prophecy' of the Marxists concerning 'socialization' has 'not come true'!

What then does this catchword 'interlocking' express? It merely expresses the most striking feature of the process going on before our eyes. It shows that the observer counts the separate trees, but cannot see the wood. It slavishly copies the superficial, the fortuitous, the chaotic. It reveals the observer as one who is overwhelmed by the mass of raw material and is utterly incapable of appreciating its meaning and importance. Ownership of shares, the relations between owners of private property, 'interlock in a haphazard way'. But underlying this interlocking, its very base, are the changing social relations of production. When a big enterprise assumes gigantic proportions, and, on the basis of an exact computation of mass data, organizes according to plan the supply of primary raw materials to the extent of two-thirds, or three-fourths, of all that is necessary for tens of millions of people; when the raw materials are transported in a systematic and organized manner to the most suitable place of production, sometimes hundreds or thousands of miles; when a single centre directs all the consecutive stages of work right up to the manufacture of numerous varieties of finished articles; when these products are distributed according to a single plan among tens and hundreds of millions of consumers (the distribution of oil in America and Germany by the American 'oil trust')—then it becomes evident that we have socialization of production, and not mere 'interlocking'; that private economic and private property relations constitute a shell which no longer fits its contents, a shell which must inevitably decay if its removal by artificial means be delayed; a shell which may continue in a state of decay for a fairly long period (if, at the worst, the cure of the opportunist abscess is protracted), but which will inevitably be removed.

4. The State and Revolution

The proletariat needs the State—this is repeated by all the opportunists, social-chauvinists, and Kautskyites, who assure us that this is what Marx taught. But they *'forget'* to add that, in the first place, according to Marx, the proletariat needs only a State which is withering away, i.e., a State so constituted that it begins to wither away immediately, and cannot but wither away. And, secondly, the working people need a 'State, i.e., the proletariat organized as the ruling class'.

The State is a special organization of force: it is an organization of violence for the suppression of some class. What class must the proletariat suppress? Naturally, only the exploiting class, i.e., the bourgeoisie. The working people need a State only to suppress the resistance of the exploiters, and only the proletariat is in a position to direct this suppression, carry it out; for the proletariat is the only class that is consistently revolutionary, the only class that can unite all the working people and the exploited in the struggle against the bourgeoisie, in completely displacing it.

The exploiting classes need political rule in order to maintain exploitation, i.e., in the selfish interests of an insignificant minority against the vast majority of the people. The exploited classes need political rule in order to completely abolish all exploitation, i.e., in the interests of the vast majority of the people, and against the insignificant minority consisting of the modern slave owners—the landlords and the capitalists.

The petty-bourgeois democrats, those sham Socialists who have replaced class struggle by dreams of class harmony, even pictured the Socialist transformation in a dreamy fashion—not as the overthrow of the rule of the exploiting class, but as the peaceful submission of the minority to the majority which has become conscious of its aims. This petty-bourgeois Utopia, which is inseparably connected with the idea of the State being above classes, led in practice to the betrayal of the interests of the working classes, as was shown, for example, by the history of the French revolutions of 1848 and 1871, and by the experience of 'Socialist' participation in bourgeois cabinets in Britain, France, Italy, and other countries at the turn of the century.

Marx fought all his life against this petty-bourgeois Socialism—now resurrected in Russia by the Socialist Revolutionary and Menshevik parties. He applied his teaching on the class struggle consistently, down to the teaching on political power, the teaching on the State.

The overthrow of bourgeois rule can be accomplished only by the proletariat, as the particular class whose economic conditions of existence prepare it for this task and provide it with the possibility and the power to perform it. While the bourgeoisie breaks up and disintegrates the peasants and all the petty-bourgeois groups, it welds together, unites, and organizes the proletariat. Only the proletariat—by virtue of the economic role it plays in large-scale production—is capable of being the leader of *all* the working and exploited people, whom the bourgeoisie exploits, oppresses, and crushes often not less, but more, than it does the proletarians, but who are incapable of waging an *independent* struggle for their emancipation.

The teaching on the class struggle, when applied by Marx to the question of the State and of the socialist revolution, leads of necessity to the recognition of the *political rule* of the proletariat, of its dictatorship, i.e., of undivided power relying directly upon the armed force of the people. The overthrow of the bourgeoisie can be achieved only by the proletariat becoming transformed into the *ruling class*, capable of crushing the inevitable and desperate resistance of the bourgeoisie, and of organizing *all* the working and exploited people for the new economic system.

The proletariat needs State power, the centralized oganization of force, the organization of violence, both to crush the resistance of the exploiters and to *lead* the enormous mass of the population—the peasants, the petty bourgeoisie, the semi-proletarians—in the work of organizing Socialist economy.

By educating the workers' party, Marxism educates the vanguard of the proletariat which is capable of assuming power and *of leading the whole people* to Socialism, of directing and organizing the new order, of being the teacher, the guide, the leader of all the workers and exploited in the task of building up their social life without the bourgeoisie and against the bourgeoisie. As against this, the opportunism which now holds sway trains the membership of the workers' party to be the representatives of the better-paid workers, who lose touch with the rank and file, 'get along' fairly well under capitalism, and sell their birthright for a mess of pottage, i.e., renounce their role of revolutionary leaders of the people against the bourgeoisie.

Marx's theory on 'the State, i.e., the proletariat organized as the ruling class', is inseparably bound up with all he taught on the revolutionary role of the proletariat in history. The culmination of this role is the proletarian dictatorship, the political rule of the proletariat.

But if the proletariat needs a State as a *special* form of organization of violence *against* the bourgeoisie, the following conclusion suggests itself: is it conceivable that such an oganization can be created without first abolishing, destroying the State machine created by the bourgeoisie *for itself*? The *Communist Manifesto* leads straight to this conclusion, and it is of this conclusion that Marx speaks when summing up the experience of the Revolution of 1848–51. . . .

Marx continues:

Between capitalist and communist society lies the period of the revolutionary transformation of the one into the other. There corresponds to this also a political transition period in which the state can be nothing but *the revolutionary dictatorship of the proletariat.* . . .

Marx bases this conclusion on an analysis of the role played by the proletariat in modern capitalist society, on the facts concerning the development of this society, and on the irreconcilability of the antagonistic interests of the proletariat and the bourgeoisie.

Previously the question was put in this way: in order to achieve its emancipation, the proletariat must overthrow the bourgeoisie, win political power, and establish its revolutionary dictatorship.

Now the question is put somewhat differently: the transition from capitalist society—which is developing towards Communism—to a Communist society is impossible without a 'political transition period', and the State in this period can only be the revolutionary dictatorship of the proletariat.

What, then, is the relation of this dictatorship to democracy?

We have seen that the *Communist Manifesto* simply places side by side the two concepts: 'to raise the proletariat to the position of the ruling class' and 'to win the battle of democracy'. On the basis of all that has been said above, it is possible to determine more precisely how democracy changes in the transition from capitalism to Communism.

In capitalist society, providing it develops under the most favourable conditions, we have a more or less complete democracy in the democratic republic. But this democracy is always hemmed in by the narrow limits set by capitalist exploitation, and consequently always remains, in reality, a democracy for the minority, only for the propertied classes, only for the rich. Freedom in capitalist society always remains about the same as it was in the ancient Greek republics: freedom for the slave owners. Owing to the conditions of capitalist exploitation the modern wage slaves are so crushed by want and poverty that 'they cannot be bothered with democracy', 'they cannot be bothered with politics'; in the ordinary peaceful course of events the majority of the population is debarred from participation in public and political life.

The correctness of this statement is perhaps most clearly confirmed by Germany, precisely because in that country constitutional legality steadily endured for a remarkably long time—for nearly half a century (1871–1914)—and during this period Social Democracy there was able to achieve far more than in other countries in the way of 'utilizing legality', and organized a larger proportion of the workers into a political party than anywhere else in the world.

What is this largest proportion of politically conscious and active wage slaves that has so far been observed in capitalist society? One million members of the Social Democratic Party—out of fifteen million wage-workers! Three million organized in trade unions—out of fifteen million!

Democracy for an insignificant minority, democracy for the rich—that is the democracy of capitalist society. If we look more closely into the machinery of capitalist democracy, we shall see everywhere, in the 'petty'—supposedly petty—details of the suffrage (residential qualification, exclusion of women, etc.), in the technique of the representative institutions, in the actual obstacles to the right of assembly (public buildings are not for 'beggars'!), in the purely capitalist organization of the daily press, etc., etc.—we shall see restriction after restriction upon democracy. These restrictions, exceptions, exclusions, obstacles for the poor, seem slight, especially in the eyes of one who has never known want himself and has never been in close contact with the oppressed classes in their mass life (and nine-tenths, if not ninety-nine hundredths, of the bourgeois publicists and politicians are of this category); but in their sum total these restrictions exclude and squeeze out the poor from politics, from active participation in democracy.

Marx grasped this *essence* of capitalist democracy splendidly, when, in analysing the experience of the commune, he said that the oppressed are allowed once every few years to decide which particular representatives of the oppressing class shall represent and repress them in parliament!

But from this capitalist democracy—that is inevitably narrow, and stealthily pushes aside the poor, and is therefore hypocritical and false to the core—forward development does not proceed simply, directly, and smoothly towards 'greater and greater democracy', as the liberal professors and petty-bourgeois opportunists would have us believe. No, forward development, i.e., towards Communism, proceeds through the dictatorship of the proletariat, and cannot do otherwise, for the *resistance* of the capitalist exploiters cannot be *broken* by anyone else or in any other way.

And the dictatorship of the proletariat, i.e., the organization of the vanguard of the oppressed as the ruling class for the purpose of suppressing the oppressors, cannot result merely in an expansion of democracy. *Simultaneously* with an immense expansion of democracy for the people, and not democracy for the money-bags, the dictatorship of the proletariat imposes a series of restrictions on the freedom of the oppressors, the exploiters, the capitalists. We must suppress them in order to free humanity from wage slavery, their resistance must be crushed by force; it is clear that where there is suppression, where there is violence, there is no freedom and no democracy.

Engels expressed this splendidly in his letter to Bebel when he said, as the reader will remember, that 'the proletariat uses the state not in the interests of freedom but in order to hold down its adversaries, and as soon as it becomes possible to speak of freedom the state as such ceases to exist'.

Democracy for the vast majority of the people, and suppression by force, i.e., exclusion from democracy, of the exploiters and oppressors of the people—this is the change democracy undergoes during the *transition* from capitalism to Communism.

Only in Communist society, when the resistance of the capitalists has been completely crushed, when the capitalists have disappeared, when there are no classes (i.e., when there is no distinction between the members of society as regards their relation to the social means of production), only then 'the State ... ceases to exist', and it *'becomes possible to speak of freedom'*. Only then will a truly complete democracy become possible and be realized, a democracy without any restrictions whatever. And only then will democracy begin to *wither* away, owing to the simple fact that, freed from capitalist slavery, from the untold horrors, savagery, absurdities, and infamies of capitalist exploitation, people will gradually *become accustomed* to observing the elementary rules of social intercourse that have been known for centuries and repeated for thousands of years in all copybook maxims; they will become accustomed to observing them without force, without compulsion, without subordination, *without the special apparatus* for compulsion called the State.

The expression 'the State *withers away*' is very well chosen, for it indicates both the gradual and the spontaneous nature of the process. Only habit can, and undoubtedly will, have such an effect; for we see around us on millions of occasions how readily people become accustomed to observing the necessary rules of social intercourse when there is no exploitation, when there is nothing that rouses indignation, nothing that evokes protest and revolt and creates the need for *suppression*.

Thus, in capitalist society we have a democracy that is curtailed, wretched, false; a democracy only for the rich, for the minority. The dictatorship of the proletariat, the period of transition to Communism, will for the first time create democracy for the people, for the majority, along with the necessary suppression of the minority—the exploiters. Communism alone is capable of giving really complete democracy, and the more complete it is the more quickly will it become unnecessary and wither away of itself.

In other words: under capitalism we have the State in the proper sense of the word, that is, a special machine for the suppression of one class by another, and, what is more, of the majority by the minority. Naturally, to be successful, such an undertaking as the systematic suppression of the exploited majority by the exploiting minority calls for the utmost ferocity and savagery in the work of suppressing, it calls for seas of blood through which mankind has to wade in slavery, serfdom, and wage labour.

Furthermore, during the *transition* from capitalism to communism suppression is *still* necessary; but it is not the suppression of the exploiting minority by the exploited majority. A special apparatus, a special machine for suppression, the 'State', is *still* necessary, but this is now a transitional State; it is no longer a State in the proper sense of the word; for the suppresson of the minority of exploiters by the majority of the wage slaves of *yesterday* is comparatively so easy, simple, and natural a task that it will entail far less bloodshed than the suppression of the risings of slaves, serfs, or wage labourers and it will cost mankind far less. And it is compatible with the extension of democracy to such an overwhelming majority of the population that the need for a *special machine* of suppression will begin to disappear. The exploiters are naturally unable to suppress the people without a highly complex machine for performing this task, but *the people* can suppress the exploited even with a very simple 'machine', almost without a 'machine' without a special apparatus, by the simpler *organization of the armed people* (such as the Soviets of Workers' and Soldiers Deputies, we would remark, running ahead).

Lastly, only Communism makes the State absolutely unnecessary, for there is *nobody* to be suppressed—'nobody' in the sense of a *class*, in the sense of a systematic struggle against a definite section of the population. We are not Utopians, and do not in the least deny the possibility and inevitability of excesses on the part of *individual persons*, or the need to suppress *such* excesses. In the first place, however, no special machine, no special apparatus of suppression is needed for this; this will be done by the armed people themselves, as simply and as readily as any crowd of civilized people, even in modern society, interferes to put a stop to a scuffle or to prevent a woman from being assaulted. And, secondly, we know that the fundamental social cause of excesses, which consist in the violation of the rules of social intercourse, is the exploitation of the people, their want, and their poverty. With the removal of this chief cause, excesses will inevitably begin to '*wither away*'. We do not know how quickly and in what succession, but we know that they will wither away. With their withering away the state will also *wither away*.

Without indulging in Utopias, Marx defined more fully what can be defined *now* regarding this future, namely, the difference between the lower and higher phases (levels, states) of Communist society.

In the *Critique of the Gotha Programme*, Marx goes into detail to disprove Lassalle's idea that under Socialism the worker will receive the 'undiminished' or 'full product of his labour'. Marx shows that from the whole of the social labour of society there must be deducted a reserve fund, a fund for the expansion of production, for the replacement of the 'wear and tear' of machinery, and so on. Then, from the means of consumption must be deducted a fund for administration expenses, for schools, hospitals, old people's homes, and so on.

Instead of Lassalle's hazy, obscure, general phrase ('the full product of his labour to the worker') Marx makes a sober estimate of exactly how Socialist society will have to manage its affairs. Marx proceeds to make a *concrete* analysis of the conditions of life of a society in which there will be no capitalism, and says:

What we have to deal with here [in analysing the programme of the workers' party] is a Communist society, not as it has *developed* on its own foundations, but, on the contrary, just as it *emerges* from capitalist society; which is thus in every respect, economically, morally, and intellectually, still stamped with the birthmarks of the old society from whose womb it emerges.

And it is this Communist society—a society which has just emerged into the light of day out of the womb of capitalism and which in every respect, bears the birthmarks of the old society—that Marx terms the 'first', or lower phase of Communist society.

The means of production are no longer the private property of individuals. The means of production belong to the whole of society. Every member of society, performing a certain part of the socially necessary work, receives a certificate from society to the effect that he has done a certain amount of work. And with this certificate he receives from the public store of consumption articles a corresponding quantity of products. After a deduction is made of the amount of labour which goes to the public fund, every worker, therefore, receives from society as much as he has given to it.

'Equality' apparently reigns supreme.

But when Lassalle, having in view such a social order (usually called Socialism, but termed by Marx the first phase of Communism), says that this is 'equitable distribution', that this is 'the equal right of all members of society to an equal product of labour', Lassalle is mistaken and Marx exposes his error.

'Equal right', says Marx, we certainly do have here; but it is *still* a 'bourgeois right', which, like every right, *presupposes inequality*. Every right is an application of an *equal* measure to *different* people who in fact are not alike, are not equal to one another; that is why 'equal right' is really a violation of equality and an injustice. In fact, every man, having performed as much social labour as another, receives an equal share of the social product (after the above-mentioned deductions).

But people are not alike: one is strong, another is weak; one is married, another is not; one has more children, another has less, and so on. And the conclusion Marx draws is:

with an equal performance of labour, and hence an equal share in the social consumption fund, one will in fact receive more than another, one will be richer than another and so on. To avoid all these defects, right instead of being equal would have to be unequal. . . .

The first phase of Communism, therefore, cannot yet produce justice and equality: differences, and unjust differences, in wealth will still exist, but the *exploitation* of man by man will have become impossible, because it will be impossible to seize the *means of production*, the factories, machines, land, etc., as private property. While smashing Lassalle's petty-bourgeois, confused phrases about 'equality' and 'justice' *in general*, Marx shows the *course of development* of Communist Society, which is *compelled* to abolish at *first* only the 'injustice' of the means of production seized by individuals, and which is *unable* at once to eliminate the other injustice, which consists in the distribution of articles of consumption 'according to the amount of labour peformed' (and not according to needs).

The vulgar economists, including the bourgeois professors and 'our' Tugan among them, constantly reproach the Socialists with forgetting the inequality of people and with 'dreaming' of eliminating this inequality. Such a reproach, as we see, only proves the extreme ignorance of the bourgeois ideologists.

Marx not only most scrupulously takes account of the inevitable inequality of men, but he also takes into account the fact that the mere conversion of the means of production into the common property of the whole of society (commonly called 'Socialism') *does not remove* the defects of distribution and the inequality of 'bourgeois right' which *continues to prevail* as long as products are divided 'according to the amount of labour performed'. Continuing, Marx says:

But these defects are inevitable in the first phase of Communist society as it is when it has just emerged after prolonged birth pangs from capitalist society. Right can never be higher than the economic structure of society and its cultural development conditioned thereby....

And so, in the first phase of Communist society (usually called Socialism) 'bourgeois right' is *not* abolished in its entirety, but only in part, only in proportion to the economic revolution so far attained, i.e., only in respect of the means of production. 'Bourgeois right' recognizes them as the private property of individuals. Socialism converts them into *common* property. *To that extent*—and to that extent alone—'bourgeois right' disappears.

However, it continues to exist as far as its other part is concerned; it continues to exist in the capacity of regulator (determining factor) in the distribution of products and the allotment of labour among the members of society. The Socialist principle: 'He who does not work, neither shall he eat,' is *already* realized; the other Socialist principle: 'An equal amount of products for an equal amount of labour,' is also *already* realized. But this is not yet Communism, and it does not yet abolish 'bourgeois right', which gives to unequal individuals, in return for unequal (really unequal) amounts of labour, equal amounts of products.

This is a 'defect', says Marx, but it is unavoidable in the first phase of Communism; for if we are not to indulge in Utopianism, we must not think that having overthrown capitalism people will at once learn to work for society *without any standard of right*; and in fact the abolition of capitalism *does not immediately create* the economic premises for *such* a change.

And there is no other standard than that of 'bourgeois right'. To this extent, therefore, there still remains the need for a State, which, while safeguarding the common ownership of the means of production, would safeguard equality in labour and equality in the distribution of products.

The State withers away in so far as there are no longer any capitalists, any classes, and, consequently, no *class* can be *suppressed*.

But the State has not yet completely withered away, since there still remains the safeguarding of 'bourgeois right', which sanctifies actual inequality. For the state to wither away completely full Communism is necessary.

Marx continues:

In a higher phase of Communist society, after the enslaving subordination of the individual to the division of labour, and therewith also the antithesis between mental and physical labour, has vanished; after labour has become not only a means of life but life's prime want; after the productive forces have also increased with the all-round development of the individual, and all the springs of co-operative wealth flow more abundantly—only then can the narrow horizon of bourgeois right be crossed in its entirety and society inscribe on its banners: From each according to his ability, to each according to his needs!

Only now can we fully appreciate the correctness of Engels's remarks in which he mercilessly ridiculed the absurdity of combining the words 'freedom' and 'state'. So long as the State exists there is no freedom. When there will be freedom, there will be no state.

The economic basis for the complete withering away of the State is such a high stage of development of Communism when the antithesis between mental and physical labour disappears, when there, consequently, disappears one of the principal sources of modern *social* inequality—a source, moreover, which cannot on any account be removed immediately by the mere conversion of the means of production into public property, by the mere expropriation of the capitalists.

This expropriation will create *the possibility* of an enormous development of the productive forces. And when we see how incredibly capitalism is already *retarding* this development, when we see how much progress could be achieved on the basis of the level of technique now already attained, we are entitled to say with the fullest confidence that the expropriation of the capitalists will inevitably result in an enormous development of the productive forces of human society. But how rapidly this development will proceed, how soon it will reach the point of breaking away from the division of labour, of doing away with the antithesis between mental and physical labour, of transforming labour into 'the prime necessity of life'—we do not and *cannot* know.

That is why we are entitled to speak only of the inevitable withering away of the State, emphasizing the protracted nature of this process and its dependence upon the rapidity of development of the *higher phase* of communism, and leaving the question of the time required for, or the concrete forms of, the withering away quite open, because there is *no* material for answering these questions.

It will become possible for the State to wither away completely when society adopts the rule: From each according to his ability, to each according to his needs, i.e., when people have become so accustomed to observing the fundamental rules of social intercourse and when their labour becomes so productive that they will voluntarily work *according to their ability*. The narrow horizon of bourgeois right, which compels one to calculate with the cold-heartedness of a Shylock whether one has not worked half an hour more than somebody else, whether one is not getting less pay than somebody else—this narrow horizon will then be crossed. There will then be no need for society to regulate the quantity of products to be received by each; each will take freely 'according to his needs'.

From the bourgeois point of view, it is easy to declare that such a social order is 'sheer Utopia' and to sneer at the socialists for promising everyone the right to receive from society, without any control over the labour of the individual citizen, any quantity of truffles, cars, pianos, etc. Even to this day, most bourgeois 'savants' confine themselves to sneering in this way, thereby displaying both their ignorance and their mercenary defence of capitalism.

Ignorance—for it has never entered the head of any Socialist to 'promise' that the higher phase of the development of Communism will arrive; whereas the great Socialists, in *foreseeing* that it will arrive presuppose not the present productivity of labour *and not the present* ordinary run of people, who, like the seminary students in Pomyalovsky's stories, are capable of damaging the stocks of public wealth 'just for fun', and of demanding the impossible.

Until the 'higher' phase of Communism arrives, the Socialists demand the *strictest* control by society *and by the State* of the measure of labour and the measure of consumption; but this control must *start* with the expropriation of the capitalists, with the establishment of workers' control over the capitalists, and must be exercised not by a State of bureaucrats, but by a State of *armed workers*.

The mercenary defence of capitalism by the bourgeois ideologists (and their hangers-on, like the Tseretelis, Chernovs, and Co.) consists precisely in that they *substitute* controversies and discussions about the distant future for the vital and burning question of *present-day* politics, namely, the expropriation of the capitalists, the conversion of *all* citizens into workers and employees of *one* huge 'syndicate'—the whole State—and the complete subordination of the entire work of this syndicate to a genuinely democratic State, to *the state of the Soviets of Workers' and Soldiers' Deputies.*

In fact, when a learned professor, followed by the Philistine, followed in turn by the Tseretelis and Chernovs, talks of unreasonable Utopias, of the demagogic promises of the Bolsheviks, of the impossibility of 'introducing' Socialism, it is the higher stage or phase of Communism they have in mind, which no one has ever promised or even thought to 'introduce' because generally speaking it cannot be 'introduced'.

And this brings us to the question of the scientific distinction between Socialism and Communism, which Engels touched on in his above-quoted argument about the incorrectness of the name 'Social Democrat'. Politically the distinction between the first, or lower, and the higher phase of Communism will in time, probably, be tremendous; but it would be ridiculous to identify this distinction now, under capitalism, and only individual anarchists, perhaps, could invest it with primary importance (if there still remain people among the anarchists who have learned nothing from the 'Plekhanovite' conversion of the Kropotkins, the Graveses, the Cornelissens, and other 'stars' of anarchism into social-chauvinists or 'anarcho-trench-ists', as Ghe, one of the few anarchists who have still preserved a sense of honour and a conscience, has put it).

But the scientific distinction between Socialism and Communism is clear. What is usually called Socialism was termed by Marx the 'first' or lower phase of Communist society. In so far as the means of production become *common* property, the word 'Communism' is also applicable here, providing we do not forget that this is *not* full Communism. The great significance of Marx's explanations is that here, too, he consistently applies materialist dialectics, the theory of development, and regards Communism as something which develops *out of* capitalism. Instead of scholastically invented, 'concocted' definitions and fruitless disputes about words (What is Socialism? What is Communism?), Marx gives an analysis of what might be called the stages of the economic maturity of Communism.

In its first phase, or first stage, Communism *cannot* as yet be fully mature economically and entirely free from traditions or traces of capitalism. Hence the interesting phenomenon that Communism in its first phase retains 'the narrow horizon of *bourgeois* right'. Of course, bourgeois right in regard to the distribution of *consumption* goods inevitably presupposes the existence of the *bourgeois State*, for right is nothing without an apparatus capable of *enforcing* the observance of the standards of right.

It follows that under Communism there remains for a time not only bourgeois right, but even the bourgeois State without the bourgeoisie!

This may sound like a paradox or simply a dialectical conundrum, of which Marxism is often accused by people who do not take the slightest trouble to study its extraordinarily profound content.

But as a matter of fact, remnants of the old, surviving in the new, confront us in life at every step, both in nature and in society. And Marx did not arbitrarily insert a scrap of 'bourgeois' right into Communism, but indicated what is economically and politically inevitable in a society emerging *out of the womb* of capitalism.

Democracy is of enormous importance to the working class in its struggle against the capitalists for its emancipation. But democracy is by no means a boundary not to be overstepped; it is only one of the stages on the road from feudalism to capitalism, and from capitalism to Communism.

Democracy means equality. The great significance of the proletariat's struggle for equality and of equality as a slogan will be clear if we correctly interpret it as meaning the abolition of *classes*. But democracy means only *formal* equality. And as soon as equality is achieved for all members of society *in relation* to ownership of the means of production, that is, equality of labour and equality of wages, humanity will inevitably be confronted with the question of advancing farther, from formal equality to actual equality, i.e., to the operation of the rule, 'from each according to his ability, to each according to his needs'. By what stages, by means of what practical measures humanity will proceed to this supreme aim—we do not and cannot know. But it is important to realize how infinitely mendacious is the ordinary bourgeois conception of Socialism as something lifeless, petrified, fixed once and for all, whereas in reality *only* under Socialism will a rapid, genuine, really mass forward movement, embracing first the *majority* and then the whole of the population, commence in all spheres of public and personal life.

Democracy is a form of the State, one of its varieties. Consequently, it, like every State, represents on the one hand the organized, systematic use of violence against persons; but on the other hand it signifies the formal recognition of equality of citizens, the equal right of all to determine the structure of, and to administer, the State. This, in turn, results in the fact that, at a certain stage in the development of democracy, it first welds together the class that wages a revolutionary struggle against capitalism— the proletariat, and enables it to crush, smash to atoms, wipe off the face of the earth the bourgeois, even the republican bourgeois, State machine, the standing army, the police, and the bureaucracy. And it enables it to substitute for them a *more* democratic State machine, but a State machine nevertheless, in the shape of the armed masses of workers who form a militia in which the entire population takes part.

Here 'quantity turns into quality': *such* a degree of democracy implies overstepping the boundaries of bourgeois society, the beginning of its socialist reconstruction. If really *all* take part in the administration of the State, capitalism cannot retain its hold. And the development of capitalism, in turn, itself creates the *premisses* that *enable* really 'all' to take part in the administration of the State. Some of these premisses are: universal literacy, which has already been achieved in a number of the most advanced capitalist countries, then the 'training and disciplining' of millions of workers by the huge, complex, socialized apparatus of the postal service, railways, big factories, large-scale commerce, banking, etc., etc.

Given these *economic* premisses it is quite possible, after the overthrow of the capitalists and the bureaucrats, to proceed immediately, overnight, to replace them in the *control* of production and distribution, in the work of *keeping account* of labour and products by the armed workers, by the whole of the armed population. (The question of control and accounting should not be confused with the question of the scientifically trained staff of engineers, agronomists, and so on. These gentlemen are working today in obedience with the wishes of the capitalists; they will work even better tomorrow in obedience with the wishes of the armed workers.)

Accounting and control—that is the *main* thing required for 'arranging' the smooth working, the correct functioning of the *first phase* of communist society. *All* citizens are transformed here into hired employees of the State, which consists of the armed workers. *All* citizens become employees and workers of a *single* nation-wide State 'syndicate'. All that is required is that they should work equally, do their proper share of work, and get equally paid. The accounting and control necessary for this have been *simplified* by capitalism to the extreme and reduced to the extraordinarily simple operations—which any literate person can perform—of supervising and recording, knowledge of the four rules of arithmetic, and issuing appropriate receipts.

When the *majority* of the people begin independently and everywhere to keep such accounts and maintain such control over the capitalists (now converted into employees) and over the intellectual gentry who preserve their capitalist habits, this control will really become universal, general, popular; and there will be no way of getting away from it, there will be 'nowhere to go'.

The whole of society will have become a single office and a single factory, with equality of labour and equality of pay.

But this 'factory' discipline, which the proletariat, after defeating the capitalists, after overthrowing the exploiters, will extend to the whole of society, is by no means our ideal, or our ultimate goal. It is but a necessary *step* for the purpose of thoroughly cleaning society of all the infamies and abominations of capitalist exploitation, *and for further* progress.

From the moment all members of society, or even only the vast majority, have learned to administer the State *themselves*, have taken this work into their own hands, have 'set in motion' control over the insignificant minority of capitalists, over the gentry who wish to preserve their capitalist habits, and over the workers who have been profoundly corrupted by capitalism— from this moment the need for government of any kind begins to disappear altogether. The more complete the democracy, the nearer the moment approaches when it becomes unnecessary. The more democratic the 'State' which consists of the armed workers, and which is 'no longer a State in the proper sense of the word', the more rapidly *every form* of State begins to wither away.

For when *all* have learned to administer and actually do independently administer social production, independently keep accounts, and exercise control over the idlers, the gentlefolk, the swindlers, and suchlike 'guardians of capitalist traditions', the escape from this popular accounting and control will inevitably become so incredibly difficult, such a rare exception, and will probably be accompanied by such swift and severe punishment (for the armed workers are practical men and not sentimental intellectuals, and they will scarcely allow anyone to trifle with them), that the *necessity* of observing the simple, fundamental rules of human intercourse will very soon become a *habit*.

Then the door will be wide open for the transition from the first phase of Communist society to its higher phase, and with it to the complete withering away of the State.

5. Against 'Left-wing' Communism

Parliamentarism has become 'historically obsolete'. This is true in the propaganda sense. But everyone knows that is still a long way from propaganda to overcoming it *practically*. Capitalism could have been declared, and with full justice to be 'historically obsolete' many decades ago, but that does not at all remove the need for a very long and very persistent struggle *on the soil* of capitalism. Parliamentarism is 'historically obsolete' from the standpoint of *world history*, that is to say the *era* of bourgeois parliamentarism has come to an end and the *era* of the proletarian dictatorship has *begun*. That is incontestable. But world history reckons in decades. Ten or twenty years sooner or later makes no difference when measured by the scale of world history; from the standpoint of world history it is a trifle that cannot be calculated even approximately. But for that very reason it is a glaring theoretical blunder to apply the scale of world history to practical politics.

Is parliamentarism 'politically obsolete'? That is quite another matter. Were that true, the position of the 'lefts' would be a strong one. But it has to be proved by a most searching analysis and the 'lefts' do not even know how to approach it. In the 'Theses on Parliamentarism', published in the *Bulletin of the Provisional Bureau in Amsterdam of the Communist International*, No. 1, February 1920, and obviously expressing the Dutch-left or left-Dutch strivings, the analysis, as we shall see, is also hopelessly bad.

In the first place, contrary to the opinion of such outstanding political leaders as Rosa Luxemburg and Karl Liebknecht, the German 'lefts', as we know, considered parliamentarism to be 'politically obsolete' even in January 1919. We know that the 'lefts' were mistaken. This fact alone utterly destroys at a single stroke, the proposition that parliamentarism is 'politically obsolete'. The obligation falls upon the 'lefts' of proving why their error, indisputable at that time, has now ceased to be an error. They do not, and cannot, produce even a shadow of proof. The attitude of a political party towards its own mistakes is one of the most important and surest ways of judging how earnest the party is and how it *in practice* fulfils its obligations towards its *class* and the *working people*. Frankly admitting a mistake, ascertaining the reasons for it, analysing the conditions which led to it, and thoroughly discussing the means of correcting it—that is the hallmark of a serious party; that is the way it should perform its duties, that is the way it should educate and train the *class*, and then the *masses*. By failing to fulfil this duty, by failing to give the utmost attention, care and consideration to the study of their obvious mistake, the 'lefts' in Germany (and in Holland) have proved that they are not a *party of the class*, but a circle, not *a party of the masses*, but a group of intellectuals and of a few workers who imitate the worst features of intellectualism.

Secondly, in the same pamphlet of the Frankfurt group of 'lefts' that we have already cited in detail, we read:

The millions of workers who still follow the policy of the Centre [the Catholic 'Centre' Party] are counter-revolutionary. The rural proletarians provide the legions of counter-revolutionary troops. (page 3 of the pamphlet.)

Everything goes to show that this statement is much too sweeping and exaggerated. But the basic fact set forth here is incontrovertible, and its acknowledgement by the 'lefts' is particularly clear evidence of their mistake. How can one say that 'parliamentarism is politically obsolete', when 'millions' and 'legions' of *proletarians* are not only still in favour of parliamentarism in general, but are downright 'counter-revolutionary'! Clearly, parliamentarism in Germany is *not yet* politically obsolete. Clearly, the 'lefts' in Germany have mistaken *their desire*, their politico-ideological attitude, for objective reality. That is a most dangerous mistake for revolutionaries to make. In Russia—where, over a particularly long period and in particularly varied forms, the extremely fierce and savage yoke of tsarism produced revolutionaries of diverse shades, revolutionaries who displayed astonishing devotion, enthusiasm, heroism, and strength of will— in Russia we have observed this mistake of the revolutionaries very closely, we have studied it very attentively and have first-hand knowledge of it; and we can therefore see it especially clearly in others. Parliamentarism, of course, is 'politically obsolete' for the Communists in Germany; but—and that is the whole point—we must *not* regard what is obsolete *for us* as being obsolete *for the class*, as being obsolete *for the masses*. Here again we find that the 'lefts' do not know how to reason, do not know how to act as the party of the *class*, as the party of the *masses*. You must not sink to the level of the masses, to the level of the backward strata of the class. That is incontestable. You must tell them the bitter truth. You must call their bourgeois-democratic and parliamentary prejudices—prejudices. But at the same time you must *soberly* follow the *actual* state of class-consciousness and preparedness of the whole class (not only of its Communist vanguard), of all the *working people* (not only of their advanced elements).

Even if not 'millions' and 'legions', but only a fairly large *minority* of industrial workers follow the Catholic priests—and a similar minority of rural workers follow the landowners and kulaks (*Grossbauern*)—it *undoubtedly* follows that parliamentarism in Germany is *not yet* politically obsolete, that participation in parliamentary elections and in the struggle on the parliamentary rostrum is *obligatory* for the party of the revolutionary proletariat *specifically* for the purpose of educating the backward strata of *its own class*, for the purpose of awakening and enlightening the undeveloped, downtrodden, ignorant rural *masses*. As long as you are unable to disperse the bourgeois parliament and every other type of reactionary institution, you *must* work inside them because *it is there* you will still find workers who are doped by the priests and stultified by the conditions of rural life; otherwise you risk becoming mere babblers.

Thirdly, the 'left' Communists have a great deal to say in praise of us Bolsheviks. One sometimes feels like telling them to praise us less and try to understand the tactics of the Bolsheviks more, to become more familiar with them! We took part in the elections to the Russian bourgeois parliament, the Constituent Assembly, in September–November 1917. Were our tactics correct or not? If not, then this should be clearly stated and proved, for it is essential in working out correct tactics for international Communism. If they were correct, then you must draw certain conclusions. Of course, there can be no question of placing conditions in Russia on a par with conditions in Western Europe. But as regards the special question of the meaning of the concept that 'parliamentarism has become politically obsolete', it is essential to take careful account of our experience, for unless concrete experience is taken into account such concepts very easily turn into empty phrases. In September–November 1917, did not we, the Russian Bolsheviks, have *more* right than any Western Communists to consider that parliamentarism was politically obsolete in Russia? Of course we did, for the point is not whether bourgeois parliaments have existed for a long time or a short time, but how far the broad masses of the working people are *prepared* (ideologically, politically, and practically) to accept the Soviet system and to disperse the bourgeois-democratic parliament (or allow it to be dispersed). It is an absolutely incontestable and fully established historical fact that, owing to a number of special conditions the urban working class and the soldiers and peasants of Russia were in September–November 1917 exceptionally well prepared to accept the Soviet system and to disperse the most democratic of bourgeois parliaments. Nevertheless, the Bolsheviks did *not* boycott the Constituent Assembly, but took part in the elections both before the proletariat conquered political power *and after*. That these elections yielded exceedingly valuable (and for the proletariat, highly useful) political results I have proved, I make bold to hope, in the above-mentioned article, which analyses in detail the returns of the elections to the Constituent Assembly in Russia.

The conclusion which follows from this is absolutely incontrovertible; it has been proved that participation in a bourgeois-democratic parliament even a few weeks before the victory of a Soviet republic, and even *after* such a victory, not only does not harm the revolutionary proletariat, but actually helps it to *prove* to the backward masses why such parliaments deserve to be dispersed; it *helps* their successful dispersal, and *helps* to make bourgeois parliamentarism 'politically obsolete'. To refuse to heed this experience, and at the same time to claim affiliation to the Communist *International*, which must work out its tactics internationally (not as narrow or one-sided national tactics, but as international tactics), is to commit the gravest blunder and actually to retreat from internationalism while recognizing it in words.

Now let us examine the 'Dutch-left' arguments in favour of non-participation in parliaments. The following is the text of the most important of the above-mentioned 'Dutch' theses, Thesis No. 4:

When the capitalist system of production has broken down, and society is in a state of revolution, parliamentary activity gradually loses importance as compared with the action of the masses themselves. When, then, parliament becomes the centre and organ of the counter-revolution, whilst, on the other hand, the labouring class builds up the instruments of its power in the Soviets, it may even prove necessary to abstain from all and any participation in parliamentary action.

The first sentence is obviously wrong, since the action of the masses—a big strike, for instance—is more important than parliamentary activity at *all* times, and not only during a revolution or in a revolutionary situation. This obviously untenable and historically and politically incorrect argument only shows very clearly that the authors ignore completely both the general European experience (the French experience before the revolutions of 1848 and 1870; the German experience of 1878–90, etc.) and the Russian experience (see above) as to the importance of *combining* legal with illegal struggle. This question is of immense importance in general, and in particular, because in *all* civilized and advanced countries the time is rapidly approaching when such a combination will more and more become—in part it has already become—mandatory for the party of the revolutionary proletariat owing to the fact that civil war between the proletariat and the bourgeoisie is maturing and is imminent, and owing to the fierce persecution of the Communists by republican governments and bourgeois governments generally, which resort to any violation of legality (witness the example of America alone!), etc. This very important question the Dutch, and the lefts in general, have utterly failed to understand.

As for the second sentence—in the first place it is wrong historically. We Bolsheviks participated in the most counter-revolutionary parliaments, and experience has shown that this participation was not only useful but essential for the party of the revolutionary proletariat after the first bourgeois revolution in Russia (1905) in order to prepare the way for the second bourgeois revolution (February 1917), and then for the Socialist revolution (October 1917). In the second place, this sentence is amazingly illogical. If parliament becomes an organ and a 'centre' (in reality it never has been and never can be a 'centre' but that by the way) of counter-revolution, while the workers are building up the instruments of their power in the form of Soviets, it follows that the workers must prepare—ideologically, politically, and technically—for the struggle of the Soviets against parliament, for the dispersal of parliament by the Soviets. But it does not follow that this dispersal is hindered, or is not facilitated, by the presence of a Soviet opposition *within* the counter-revolutionary parliament. In the course of our victorious struggle against Denikin and Kilchak, we never found that the existence of a Soviet, proletarian opposition in their camp was immaterial to our victories. We know pefectly well that the dispersal of the Constituent Assembly on January 5, 1918, far from being hindered, was actually facilitated by the fact that within the counter-revolutionary Constituent Assembly about to be dispersed there was a consistent, Bolshevik, as well as an inconsistent, left Socialist Revolutionary, Soviet opposition. The authors of the theses are completely confused and have forgotten the experience of many, if not all, revolutions, which shows how very useful during a revolution is the *combination* of mass action outside the reactionary parliament with an opposition sympathetic to (or, better still, directly supporting) the revolution inside it. The Dutch, and the 'lefts' in general, *argue* like doctrinaire revolutionaries who have never taken part in a real revolution, or who have never deeply studied the history of revolutions, or who have naïvely mistaken the subjective 'rejection' of a certain reactionary institution for its actual destruction by the combined action of a number of objective factors. The surest way of discrediting and damaging a new political (and not only political) idea is to reduce it to absurdity on the plea of defending it. For every truth, if 'overdone' (as Dietzgen senior put it), if exaggerated, if carried beyond the limits of its actual applicability, can be reduced to absurdity, and is even bound to become an absurdity under these conditions. That is just the kind of backhanded service the Dutch and German lefts are rendering the new truth that the Soviet form of government is superior to bourgeois-democratic parliaments. It stands to reason that anyone who subscribed to the old view, or in general maintained that refusal to participate in bourgeois parliaments is impermissible under any circumstances, would be wrong. I cannot attempt to formulate here the conditions under which a boycott is useful, for the object of this pamphlet is far more modest, namely, to study Russian experience in connection with certain topical questions of international Communist tactics. Russian experience has given us one successful and correct (1905) and one incorrect (1906) example of the application of the boycott by the Bolsheviks. Analysing the first case, we see that we succeeded in *preventing* a reactionary government from *convening* a reactionary parliament in a situation in which extra-parliamentary, revolutionary mass action (strikes in particular) was developing with exceptional rapidity, when not a single section of the proletariat and the peasantry could support the reactionary government in any way, when the revolutionary proletariat was acquiring influence over the broad, backward masses through the strike struggle and through the agrarian movement. It is quite obvious that *this* experience is not applicable to present-day European conditions. It is likewise quite obvious—and the foregoing arguments bear this out—that the advocacy, even if with reservations, by the Dutch and other 'lefts' of refusal to participate in parliaments is fundamentally wrong and detrimental to the cause of the revolutionary proletariat.

In Western Europe and America parliament has become especially abhorrent to the advanced revolutionary members of the working class. That is incontestable. It is quite comprehensible, for it is difficult to imagine anything more vile, abominable, and treacherous than the behaviour of the vast majority of the socialist and Social Democratic parliamentary deputies during and after the war. But it would be not only unreasonable, but actually criminal, to yield to this mood when deciding *how* this generally recognized evil should be fought. In many countries of Western Europe the revolutionary mood, we might say, is at present a 'novelty', or a 'rarity', which had all too long been vainly and impatiently awaited; and perhaps that is why people so easily give way to it. Certainly, without a revolutionary mood among the masses, and without conditions facilitating the growth of this mood, revolutionary tactics would never be converted into action; but we in Russia have become convinced by very long, painful, and bloody experience of the truth that revolutionary tactics cannot be built on revolutionary moods alone. Tactics must be based on a sober and strictly objective appraisal of *all* class forces of the particular state (and of the states that surround it, and of all states the world over) as well as of the experience of revolutionary movements. To show how 'revolutionary' one is solely by hurling abuse at parliamentary opportunism, solely by repudiating participation in parliaments, is very easy; but just because it is too easy, it is not the solution for a difficult, a very difficult problem. It is much more difficult to create a really revolutionary parliamentary group in a European parliament than it was in Russia. That stands to reason. But it is only a particular expression of the general truth that it was easy for Russia, in the specific, historically very unique situation of 1917, to *start* the Socialist revolution, but it will be more difficult for Russia than for the European countries to *continue* the revolution and bring it to its consummation. I had occasion to point this out already at the beginning of 1918, and our experience of the past two years has entirely confirmed the correctness of this view. Certain specific conditions, (1) the possibility of linking up the Soviet revolution with the ending, as a consequence of this revolution, of the imperialist war, which had exhausted the workers and peasants to an incredible degree; (2) the possibility of taking advantage for a certain time of the mortal conflict between the world's two most powerful groups of imperialist robbers, who were unable to unite against their Soviet enemy; (3) the possibility of enduring a comparatively lengthy civil war, partly owing to the enormous size of the country and to the poor means of communication; (4) the existence of such a profound bourgeois-democratic revolutionary movement among the peasantry that the party of the proletariat was able to adopt the revolutionary demands of the peasant party (the Socialist Revolutionary Party, the majority of the members of which were definitely hostile to Bolshevism) and realize them at once, thanks to the conquest of political power by the proletariat—these specific conditions do not exist in Western Europe at present; and a repetition of such or similar conditions will not occur so easily. That, by the way, apart from a number of other causes, is why it will be more difficult for Western Europe to *start* a Socialist revolution than it was for us. To attempt to 'circumvent' this difficulty by 'skipping' the arduous job of utilizing reactionary parliaments for revolutionary purposes is absolutely childish. You want to create a new society, yet you fear the difficulties involved in forming a good parliamentary group, made up of convinced, devoted, heroic Communists, in a reactionary parliament! Is that not childish? If Karl Liebknecht in Germany and Z. Höglund in Sweden were able, even without mass support from below, to set examples in the truly revolutionary utilization of reactionary parliaments, how can one say that a rapidly growing revolutionary, mass party, in the midst of the post-war disillusionment and embitterment of the masses, cannot *forget* a Communist group in the worst of parliaments?! It is because the backward masses of the workers and—to an even greater degree—of the small peasants are in Western Europe much more imbued with bourgeois-democratic and parliamentary prejudices than they were in Russia—because of that, it is *only* from within such institutions as bourgeois parliaments that Communists can (and must) wage a long and persistent struggle, undaunted by any difficulties, to expose, dissipate and overcome these prejudices.

6.   On Soviet Bureaucracy

In the matter of improving our State apparatus, the Workers' and Peasants' Inspection should not, in my opinion, either strive after quantity or hurry. We have so far been able to devote so little thought and attention to the quality of our State apparatus that it would now be quite legitimate if we took special care to secure its thorough organization, and concentrated in the Workers' and Peasants' Inspection a staff of workers really abreast of the times, i.e., not inferior to the best West European standards. For a Socialist republic this condition is, of course, too modest. But our experience of the first five years has fairly crammed our heads with mistrust and scepticism. These qualities assert themselves involuntarily when, for example, we hear people dilating at too great length and too flippantly on 'proletarian culture'. For a start, we should be satisfied with real bourgeois culture, for a start, we should be glad to dispense with the cruder types of pre-bourgeois culture, i.e., bureaucratic culture or serf culture, etc. In matters of culture, haste and sweeping measures are most harmful. Many of our young writers and Communists should get this well into their heads.

Thus, in the matter of our State apparatus we should now draw the conclusion from our past experience that it would be better to proceed more slowly.

Our State apparatus is so deplorable, not to say disgusting, that we must first think very carefully how to combat its defects, bearing in mind that these defects are rooted in the past, which, although it has been overthrown, has not yet been overcome, has not yet reached the stage of a culture that has receded into the distant past. I say culture deliberately, because in these matters we can only regard as achieved what has become part and parcel of our culture, of our social life, our habits. We might say that the good in our social system has not been properly studied, understood, and taken to heart; it has been hastily grasped at; it has not been verified or tested or tried by experience, and not made durable, etc. Of course it could not be otherwise in a revolutionary epoch, when development proceeded at such breakneck speed that in a matter of five years we passed from tsarism to the Soviet system.

It is time we did something about it. We must show sound scepticism for too rapid progress, for boastfulness, etc. We must give thought to testing the steps forward we proclaim every hour, take every minute, and then prove every second that they are flimsy, superficial, and misunderstood. The most harmful thing here would be haste. The most harmful thing would be to rely on the assumption that we know at least something, or that we have any considerable number of elements necessary for the building of a really new State apparatus, one really worthy to be called Socialist, Soviet, etc.

No, we are ridiculously deficient of such an apparatus and even of the elements of it, and we must remember that we should not stint time on building it, and that it will take many, many years....

We must strive to build up a State in which the workers retain the leadership of the peasants, in which they retain the confidence of the peasants, and by exercising the greatest economy remove every trace of extravagance from our social relations.

We must reduce our State apparatus to the utmost degree of economy. We must banish from it all traces of extravagance, of which so much has been left over from tsarist Russia, from its bureaucratic capitalist State machine.

Will not this be a reign of peasant limitations?

No. If we see to it that the working class retains its leadership over the peasantry, we shall be able, by exercising the greatest possible economy in the economic life of our State, to use every saving we make to develop our large-scale machine industry, to develop electrification, the hydraulic extraction of peat, to complete the Volkhov power project, etc.

In this, and in this alone, lies our hope. Only when we have done this will we, speaking figuratively, be able to change horses, to change from the peasant, moujik horse of poverty, from the horse of an economy designed for a ruined peasant country, to the horse which the proletariat is seeking and must seek—the horse of large-scale machine industry, of electrification, of the Volkhov power station, etc.

That is how I link up in my mind the general plan of our work, of our policy, of our tactics, of our strategy, with the functions of the reorganized Workers' and Peasants' Inspection. This is what, in my opinion, justifies the exceptional care, the exceptional attention that we must devote to the Workers' and Peasants' Inspection in raising it to an exceptionally high place, in giving it a leadership with Central Committee rights, etc., etc.

And this justification is that only by thoroughly purging our government offices, by reducing to the utmost everything that is not absolutely essential in them, shall we be certain of being able to keep going. Moreover, we shall be able to keep going not on the level of a small-peasant country, not on the level of universal limitation, but on a level steadily advancing to large-scale machine industry.

These are the lofty tasks that I dream of for our Workers' and Peasants' Inspection. That is why I am planning for it the amalgamation of the most authoritative Party body with an 'ordinary' People's Commissariat.

Source: V. Lenin, *Selected Works* (Moscow, 1960), i. 156 f., 189 ff., 227 ff., 516 f., 521 f., 546 ff., 781 f., 797 f., 810 ff., ii. 320 ff., 371 f., iii. 406 ff., 829 f., 841 f.

FURTHER NOTES

WORKS

LENIN, V. I., *Collected Works*, 45 vols. (Moscow, 1960–70).
—— *Selected Works*, 1 vol. (Moscow, 1968); 3 vols. (Moscow, 1960–1).
—— *The Essential Works of Lenin*, ed. M. Christian (New York, 1966).

COMMENTARIES

FISCHER, LOUIS B., *The Life of Lenin* (London, 1965).

HARDING, N., *Lenin's Political Thought*, 2 vols. (New York and London, 1977–).

KRUPSKAYA, N., *Memories of Lenin*, with an Introduction by Andrew Rothstein (London, 1970).

LEWIN, MOSHE, *Lenin's Last Struggle* (London, 1969).

LIEBMAN, M., *Leninism Under Lenin* (London, 1975).

MEYER, A. G., *Leninism*, 2nd edn. (New York, 1962).

SCHAPIRO, L., and P. REDDAWAY (eds.), *Lenin: The Man, the Theorist, the Leader: A Reappraisal* (London, 1967).

SHUB, DAVID, *Lenin: A Biography* (New York, 1948).

SHUKMAN, HAROLD, *Lenin and the Russian Revolution* (New York, 1968).

ULAM, ADAM, *Lenin and the Bolsheviks* (New York, 1964).

WILSON, EDMUND, *To the Finland Station* (London, 1960).

WOLFE, BERTRAM D., *Three Who Made a Revolution* (New York, 1948).

8

LEON TROTSKY

As a matter of fact, all through history, mind limps after reality.
Literature and Revolution (Ann Arbor, 1960), p. 19.

LEON TROTSKY (1879–1940), *whose original name was Lev Davidovich Bronstein, was, with Lenin, the main architect of the Russian revolution of 1917. Trotsky joined the Russian Marxists in 1902 and was prominent in the 1905 revolution. It was around that time that he evolved two of his central ideas, expounded in the first two extracts below. The first idea— 'combined and uneven development'—states that a backward country does not develop by passing through all the stages already traversed by advanced countries: it can combine the most backward with the most advanced features. This leads to the second idea of 'permanent revolution': because of this uneven development and the lack of a long capitalist stage, an important proletariat could exist without a correspondingly powerful bourgeoisie and this revolutionary momentum could—like the stages of economic development—be telescoped and socialism be placed on the agenda of even comparatively backward countries such as Russia. Trotsky's prognosis of 1906 (contained in the first extract) was fulfilled in 1917, but the aftermath brought disillusion: ousted by the rise of Stalin during the 1920s, he spent the 1930s in exile and was eventually assassinated by one of Stalin's agents. Trotsky's explanation for the degeneration of the Russian revolution is contained in the third extract, from* The Revolution Betrayed *(1937): the growth of a parasitic bureaucracy consequent on the attempt to build socialism in a single country and the neglect of the vital international dimension of the revolution. The last extract, from the final pages of* Literature and Revolution *(1924), strikingly conveys the visionary side of the Bolshevik project.*

1. The Prospect of Revolution

THE proletariat can only achieve power by relying upon a national upsurge and national enthusiasm. The proletariat will enter the government as the revolutionary representative of the nation, as the recognized national leader in the struggle against absolutism and feudal barbarism. In taking power, however, it will open a new epoch, an epoch of revolutionary legislation, of positive policy, and in this connection it cannot at all be sure of retaining the role of the recognized expressor of the will of the nation. The first measures of the proletariat, cleansing the Augean stables of the old regime and driving out its inmates, will meet with the active support of the whole nation, in spite of what the liberal eunuchs may say about the tenacity of certain prejudices among the masses of the people.

This political cleansing will be supplemented by a democratic re-organization of all social and state relations. The workers' government will be obliged, under the influence of direct pressures and demands, to intervene decisively in all relationships and events ...

Its first task will have to be the dismissal from the army and admin-istration of all those who are stained with the blood of the people, and the cashiering or disbandment of the regiments which have most sullied themselves with crimes against the people. This will have to be done in the very first days of the revolution, that is, long before it is possible to introduce the system of elected and responsible officials and organize a national militia. But the matter will not end there. Workers' democracy will immedi-ately be confronted by questions of the length of the working day, the agrarian question, and the problem of unemployment.

One thing is clear. Every passing day will deepen the policy of the proletariat in power, and more and more define its *class character*. Side by side with that, the revolutionary ties between the proletariat and the nation will be broken, the class disintegration of the peasantry will assume political form, and the antagonism between the component sections will grow in proportion as the policy of the workers' government defines itself, ceasing to be a general-democratic and becoming a class policy.

Though the absence of accumulated bourgeois-individualistic traditions and anti-proletarian prejudices among the peasantry and intellectuals will assist the proletariat to come into power, it is necessary on the other hand to bear in mind that this absence of prejudices is due not to political consciousness but to political barbarism, social formlessness, primitiveness, and lack of character. None of these features can in any way create a reliable basis for a consistent, active proletarian policy.

The abolition of feudalism will meet with support from the *entire* peas-antry, as the burden-bearing estate. A progressive income tax will also be supported by the great majority of the peasantry. But any legislation carried through for the purpose of protecting the agricultural proletariat will not only not receive the active sympathy of the majority, but will even meet with the active opposition of a minority of the peasantry.

The proletariat will find itself compelled to carry the class struggle into the villages and in this manner destroy that community of interest which is undoubtedly to be found among all peasants, although within comparatively narrow limits. From the very first moment after its taking power, the proletariat will have to find support in the antagonisms between the village poor and village rich, between the agricultural proletariat and the agricultural bourgeoisie. While the heterogeneity of the peasantry creates difficulties and narrows the basis for a proletarian policy, the insufficient degree of class differentiation will create obstacles to the introduction among the peasantry of developed class struggle, upon which the urban proletariat could rely. The primitiveness of the peasantry turns its hostile face towards the proletariat.

The cooling-off of the peasantry, its political passivity, and all the more the active opposition of its upper sections, cannot but have an influence on a section of the intellectuals and the petty bourgeoisie of the towns.

Thus, the more definite and determined the policy of the proletariat in power becomes, the narrower and more shaky does the ground beneath its feet become. All this is extremely probable and even inevitable...

The two main features of proletarian policy which will meet opposition from the allies of the proletariat are *collectivism* and *internationalism*.

The primitiveness and petty-bourgeois character of the peasantry, its limited rural outlook, its isolation from world-political ties and allegiances, will create terrible difficulties for the consolidation of the revolutionary policy of the proletariat in power.

To imagine that it is the business of Social Democrats to enter a provisional government and lead it during the period of revolutionary-democratic reforms, fighting for them to have a most radical character, and relying for this purpose upon the organized proletariat—and then, after the democratic programme has been carried out, to leave the edifice they have constructed so as to make way for the bourgeois parties and themselves go into opposition, thus opening up a period of parliamentary politics, is to imagine the thing in a way that would compromise the very idea of a workers' government. This is not because it is inadmissible 'in principle'—putting the question in this abstract form is devoid of meaning—but because it is absolutely unreal, it is Utopianism of the worst sort—a sort of revolutionary-Philistine Utopianism.

For this reason:

The division of our programme into maximium and minimum programmes has a profound and tremendous principled significance during the period when power lies in the hands of the bourgeoisie. The very fact of the bourgeoisie being in power drives out of our minimum programme all demands which are incompatible with private property in the means of production. Such demands form the content of a Socialist revolution and presuppose a proletarian dictatorship.

Immediately, however, that power is transferred into the hands of a revolutionary government with a Socialist majority, the division of our programme into maximum and minimum loses all significance, both in principle and in immediate practice. A proletarian government under no circumstances can confine itself within such limits. Take the question of the eight-hour day. As is known, this by no means contradicts capitalist relations, and therefore it forms an item in the minimum programme of Social Democracy. But let us imagine the actual introduction of this measure during a period of revolution, in a period of intensified class passions; there is no question but that this measure would then meet the organized and determined resistance of the capitalists in the form, let us say, of lock-outs and the closing down of factories.

Hundreds of thousands of workers would find themselves thrown on the streets. What should the government do? A bourgeois government, however radical it might be, would never allow affairs to reach this stage because, confronted with the closing down of factories, it would be left powerless. It would be compelled to retreat, the eight-hour day would not be introduced, and the indignant workers would be suppressed.

Under the political domination of the proletariat, the introduction of an eight-hour day should lead to altogether different consequences. For a government that desires to rely upon the proletariat, and not on capital, as liberalism does, and which does not desire to play the role of an 'impartial' intermediary of bourgeois democracy, the closing down of factories would not of course be an excuse for increasing the working day. For a workers' government there would be only one way out: expropriation of the closed factories and the organization of production in them on a socialized basis.

Of course, one can argue in this way: we will suppose that the workers' government, true to its programme, issues a decree for an eight-hour day; if capital puts up a resistance which cannot be overcome by the resources of a democratic programme based on the preservation of private property, the Social Democrats will resign and appeal to the proletariat. Such a solution would be a solution only from the standpoint of the group constituting the membership of the government, but it would be no solution for the proletariat or for the development of the revolution. After the resignation of the Social Democrats, the situation would be exactly as it was at the time when they were compelled to take power. To flee before the organized opposition of capital would be a greater betrayal of the revolution than a refusal to take power in the first instance. It would really be far better for the working-class party not to enter the government than to go in so as to expose its own weakness and then to quit.

Let us take another example. The proletariat in power cannot but adopt the most energetic measures to solve the question of unemployment, because it is quite obvious that the representatives of the workers in the government cannot reply to the demands of unemployed workers with arguments about the bourgeois character of the revolution.

But if the government undertakes to maintain the unemployed—it is not important for us at the moment in what form—this would mean an immediate and quite substantial shift of economic power to the side of the proletariat. The capitalists, who in their oppression of the workers always relied upon the existence of a reserve army of labour, would feel themselves *economically* powerless while the revolutionary government, at the same time, doomed them to *political* impotence.

In undertaking the maintenance of the unemployed, the government thereby undertakes the maintenance of strikers. If it does not do *that*, it immediately and irrevocably undermines the basis of its own existence.

There is nothing left for the capitalists to do then but to resort to the lock-out, that is, to close the factories. It is quite clear that the employers can stand the closing down of production much longer than the workers, and therefore there is only one reply that a workers' government can give to a general lock-out: the expropriation of the factories and the introduction in at least the largest of them of State or communal production.

Similar problems arise in agriculture by the mere fact of the expropriation of the land. In no way must it be supposed that a proletarian government, on expropriating the privately owned estates carrying on production on a large scale, would break these up and sell them for exploitation to small producers. The only path open to it in this sphere is the organization of co-operative production under communal control or organized directly by the State. But this is the path to Socialism.

All this quite clearly shows that Social Democrats cannot enter a revolutionary government, giving the workers in advance an undertaking not to *give way* on the minimum programme, and at the same time promising the bourgeoisie not to *go beyond* it. Such a bilateral undertaking is absolutely impossible to realize. The very fact of the proletariat's representatives entering the government, not as powerless hostages, but as the leading force, destroys the border-line between maximum and minimum programme; that is to say, it *places collectivism on the order of the day*. The point at which the proletariat will be held up in its advance in this direction depends upon the relation of forces, but in no way upon the original intentions of the proletarian party.

For this reason there can be no talk of any sort of special form of

proletarian dictatorship in the bourgeois revolution, of *democratic* proletarian dictatorship (or dictatorship of the proletariat and the peasantry).
The working class cannot preserve the democratic character of its dictatorship without refraining from overstepping the limits of its democratic
programme. Any illusions on this point would be fatal. They would compromise Social Democracy from the very start.

The proletariat, once having taken power, will fight for it to the very
end. While one of the weapons in this struggle for the maintenance and the
consolidation of power will be agitation and organization, especially in the
countryside, another will be a policy of collectivism. Collectivism will
become not only the inevitable way forward from the position in which the
party in power will find itself, but will also be a means of preserving this
position with the support of the proletariat.

When the idea of uninterrupted revolution was formulated in the Socialist
press—an idea which connected the liquidation of absolutism and feudalism
with a Socialist revolution, along with growing social conflicts, uprisings
of new sections of the masses, unceasing attacks by the proletariat upon
the economic and political privileges of the ruling classes—our 'progressive'
press raised a unanimous howl of indignation. 'Oh!' it cried, 'we have put
up with a lot, but we cannot allow this. Revolution', it cried, 'is not a road
that can be "legalized". The application of exceptional measures is only
permissible under exceptional circumstances. The aim of the movement for
emancipation is not to make revolution permanent but to lead it as soon
as possible into the channel of *law*,' etc., etc.

The more radical representatives of this same democracy do not risk
taking up a stand against revolution even from the point of view of already-
secured constitutional 'gains'. For them this parliamentary cretinism, preceding the rise of parliamentarism itself, does not constitute a strong weapon
in the struggle against the proletarian revolution. They choose another
path. They take their stand not on the basis of law but on what seems to
them the basis of facts—on the basis of historical 'possibility', on the basis
of political 'realism' and, finally … finally, even on the basis of 'Marxism'.
And why not? That pious Venetian bourgeois, Antonio, very aptly said:
'The devil can quote Scripture to his purpose.'

These radical democrats not only regard the idea of a workers' government in Russia as fantastic, but they even deny the possibility of a Socialist
revolution in Europe in the historical epoch immediately ahead. 'The
prerequisites of revolution', they say, 'are not yet visible.' Is that true?
Certainly there is no question of appointing a dateline for the Socialist
revolution, but it is necesseary to point out its real historical prospects. …

Without the direct State support of the European proletariat the working class of Russia cannot remain in power and convert its temporary domination into a lasting Socialist dictatorship. Of this there cannot for one moment be any doubt. But on the other hand there cannot be any doubt that a Socialist revolution in the West will enable us directly to convert the temporary domination of the working class into a Socialist dictatorship....

The influence of the Russian revolution upon the European proletariat is tremendous. Besides destroying Russian absolutism, the main force of European reaction, it will create the necessary prerequisites for revolution in the consciousness and temper of the European working class.

The function of the Socialist parties was and is to revolutionize the consciousness of the working class, just as the development of capitalism revolutionized social relations. But the work of agitation and organization amongst the ranks of the proletariat has an internal inertia. The European Socialist Parties, particularly the largest of them, the German Social Democratic Party, have developed their conservatism in proportion as the great masses have embraced Socialism and the more these masses have become organized and disciplined. As a consequence of this, Social Democracy as an organization embodying the political experience of the proletariat may at a certain moment become a direct obstacle to open conflict between the workers and bourgeois reaction. In other words, the propagandist-Socialist conservatism of the proletarian parties may at a certain moment hold back the direct struggle of the proletariat for power. The tremendous influence of the Russian revolution indicates that it will destroy party routine and conservatism, and place the question of an open trial of strength between the proletariat and capitalist reaction on the order of the day. The struggle for universal suffrage in Austria, Saxony, and Prussia has become acute under the direct influence of the October strikes in Russia. The revolution in the East will infect the Western proletariat with a revolutionary idealism and rouse a desire to speak to their enemies 'in Russian'. Should the Russian proletariat find itself in power, if only as the result of a temporary conjuncture of circumstances in our bourgeois revolution, it will encounter the organized hostility of world reaction, and on the other hand will find a readiness on the part of the world proletariat to give organized support.

Left to its own resources, the working-class of Russia will inevitably be crushed by the counter-revolution the moment the peasantry turns its back on it. It will have no alternative but to link the fate of its political rule, and, hence, the fate of the whole Russian revolution, with the fate of the Socialist revolution in Europe. That colossal State-political power given it by a temporary conjuncture of circumstances in the Russian bourgeois revolution it will cast into the scales of the class struggle of the entire capitalist world. With state power in its hands, with counter-revolution behind it, and European reaction in front of it, it will send forth to its comrades the world over the old rallying cry, which this time will be a call for the last attack: *Workers of all countries, unite!*

2. Permanent Revolution

The permanent revolution, in the sense which Marx attached to this concept, means a revolution which makes no compromise with any single form of class rule, which does not stop at the democratic stage, which goes over to Socialist measures and to war against reaction from without; that is, a revolution whose every successive stage is rooted in the preceding one and which can end only in the complete liquidation of class society.

To dispel the chaos that has been created around the theory of the permanent revolution, it is necessary to distinguish three lines of thought that are united in this theory.

First, it embraces the problem of the transition from the democratic revolution to the Socialist. This is in essence the historical origin of the theory.

The concept of the permanent revolution was advanced by the great Communists of the middle of the nineteenth century, Marx and his co-thinkers, in opposition to the democratic ideology which, as we know, claims that with the establishment of a 'rational' or democratic state all questions can be solved peacefully by reformist or evolutionary measures. Marx regarded the bourgeois revolution of 1848 as the direct prelude to the proletarian revolution. Marx 'erred'. Yet his error has a factual and not a methodological character. The Revolution of 1848 did not turn into the Socialist revolution. But that is just why it also did not achieve democracy. As to the German revolution of 1918, it was no democratic completion of the bourgeois revolution, it was a proletarian revolution decapitated by the Social Democrats; more correctly, it was a bourgeois counter-revolution, which was compelled to preserve pseudo-democratic forms after its victory over the proletariat.

Vulgar 'Marxism' has worked out a pattern of historical development according to which every bourgeois society sooner or later secures a democratic regime, after which the proletariat, under conditions of democracy, is gradually organized and educated for Socialism. The actual transition to Socialism has been variously conceived: the avowed reformists pictured this transition as the reformist filling of democracy with a socialist content (Jaurès): the formal revolutionists acknowledged the inevitability of applying revolutionary violence in the transition to Socialism (Guesde). But both

the former and the latter considered democracy and Socialism, for all people
and countries, as two stages in the development of society which are not
only entirely distinct but also separated by great distances of time from
each other. This view was predominant also among those Russian Marxists
who, in the period of 1905, belonged to the left wing of the Second
International. Plekhanov, the brilliant progenitor of Russian Marxism,
considered the idea of the dictatorship of the proletariat a delusion in
contemporary Russia. The same standpoint was defended not only by the
Mensheviks but also by the overwhelming majority of the leading Bolsheviks, in particular by those present party leaders, without exception, who
in their day were resolute revolutionary democrats but for whom the
problems of the Socialist revolution, not only in 1905 but also on the eve
of 1917, still signified the vague music of a distant future.

The theory of the permanent revolution, which originated in 1905,
declared war upon these ideas and moods. It pointed out that the democratic
tasks of the backward bourgeois nations lead directly, in our epoch, to the
dictatorship of the proletariat and that the dictatorship of the proletariat
puts Socialist tasks on the order of the day. Therein lay the central idea of
the theory. While the traditional view was that the road to the dictatorship
of the proletariat led through a long period of democracy, the theory of the
permanent revolution established the fact that for backward countries the
road to democracy passed through the dictatorship of the proletariat. Thus
democracy is not a regime that remains self-sufficient for decades, but is
only a direct prelude to the Socialist revolution. Each is bound to the other
by an unbroken chain. Thus there is established between the democratic
revolution and the Socialist reconstruction of society a permanent state of
revolutionary development.

The second aspect of the 'permanent theory' has to do with the Socialist
revolution as such. For an indefinitely long time and in constant internal
struggle, all social relations undergo transformation. Society keeps on
changing its skin. Each stage of transformation stems directly from the
preceding. This process necessarily retains a political character, that is, it
develops through collisions between various groups in the society which is
in transformation. Outbreaks of civil war and foreign wars alternate with
periods of 'peaceful' reform. Revolutions in economy, technique, science,
the family, morals, and everyday life develop in complex reciprocal action
and do not allow society to achieve equilibrium. Therein lies the permanent
character of the Socialist revolution as such.

The international character of the Socialist revolution, which constitutes
the third aspect of the theory of the permanent revolution, flows from the
present state of economy and the social structure of humanity. Inter-

nationalism is no abstract principle but a theoretical and political reflection of the character of world economy, of the world development of productive forces, and the world scale of the class struggle. The socialist revolution begins on national foundations—but it cannot be completed within these foundations. The maintenance of the proletarian revolution within a national framework can only be a provisional state of affairs, even though, as the experience of the Soviet Union shows, one of long duration. In an isolated proletarian dictatorship, the internal and external contradictions grow inevitably along with the successes achieved. If it remains isolated, the proletarian state must finally fall victim to these contradictions. The way out for it lies only in the victory of the proletariat of the advanced countries. Viewed from this standpoint, a national revolution is not a self-contained whole: it is only a link in the international chain. The international revolution constitutes a permanent process, despite temporary declines and ebbs.

The struggle of the epigones is directed, even if not always with the same clarity, against all three aspects of the theory of the permanent revolution. And how could it be otherwise, when it is a question of three inseparably connected parts of a whole? The epigones mechanically separate the *democratic* and the *Socialist* dictatorships. They separate the *national* Socialist revolution from the *international*. They consider that, in essence, the conquest of power within national limits is not the initial act but the final act of the revolution; after that follows the period of reforms that lead to the national Socialist society. In 1905, they did not even grant the idea that the proletariat could conquer power in Russia earlier than in Western Europe. In 1917, they preached the self-sufficing democratic revolution in Russia and spurned the dictatorship of the proletariat. In 1925–7, they steered a course toward national revolution in China under the leadership of the national bourgeoisie. Subsequently, they raised the slogan for China of the democratic dictatorship of the workers and peasants in opposition to the slogan of the dictatorship of the proletariat. They proclaimed the possibility of the construction of an isolated and self-sufficient Socialist society in the Soviet Union. The world revolution became for them, instead of an indispensable condition for victory, only a favourable circumstance. This profound breach with Marxism was reached by the epigones in the process of permanent struggle against the theory of the permanent revolution. . . .

I hope that the reader will not object if, to end this book, I attempt, without fear of repetition, to formulate succinctly my principal conclusions.

1. The theory of the permanent revolution now demands the greatest

attention from every Marxist, for the course of the class and ideological struggle has fully and finally raised this question from the realm of reminiscences over old differences of opinion among Russian Marxists, and converted it into a question of the character, the inner connections, and methods of the international revolution in general.

2. With regard to countries with a belated bourgeois development, especially the colonial and semi-colonial countries, the theory of the permanent revolution signifies that the complete and genuine solution of their tasks of achieving *democracy and national emancipation* is conceivable only through the dictatorship of the proletariat as the leader of the subjugate nation, above all of its peasant masses.

3. Not only the agrarian, but also the national question assigns to the peasantry—the overwhelming majority of the population in backward countries—an exceptional place in the democratic revolution. Without an alliance of the proletariat with the peasantry the tasks of the democratic revolution cannot be solved, nor even seriously posed. But the alliance of these two classes can be realized in no other way than through an irreconcilable struggle against the influence of the national-liberal bourgeoisie.

4. No matter what the first episodic stages of the revolution may be in the individual countries, the realization of the revolutionary alliance between the proletariat and the peasantry is conceivable only under the political leadership of the proletarian vanguard, organized in the Communist Party. This in turn means that the victory of the democratic revolution is conceivable only through the dictatorship of the proletariat which bases itself upon the alliance with the peasantry and solves first of all the tasks of the democratic revolution.

5. Assessed historically, the old slogan of Bolshevism—'the democratic dictatorship of the proletariat and peasantry'—expressed precisely the above-characterized relationship of the proletariat, the peasantry, and the liberal bourgeoisie. This has been confirmed by the experience of October. But Lenin's old formula did not settle in advance the problem of what the reciprocal relations would be between the proletariat and the peasantry within the revolutionary bloc. In other words, the formula deliberately retained a certain algebraic quality, which had to make way for more precise arithmetical quantities in the process of historical experience. However, the latter showed, and under circumstances that exclude any kind of misinterpretation, that no matter how great the revolutionary role of the peasantry may be, it nevertheless cannot be an independent role and even less a leading one. The peasant follows either the worker or the bourgeois. This means that the 'democratic dictatorship of the proletariat and peas-

antry' is only conceivable as a *dictatorship of the proletariat that leads the peasant masses behind it.*

6. A democratic dictatorship of the proletariat and peasantry, as a regime that is distinguished from the dictatorship of the proletariat by its class content, might be realized only in a case where an *independent* revolutionary party could be constituted, expressing the interest of the peasants and in general of petty-bourgeois democracy—a party capable of conquering power with this or that degree of aid from the proletariat, and of determining its revolutionary programme. As all modern history attests—especially the Russian experience of the last twenty-five years—an insurmountable obstacle on the road to the creation of a peasants' party is the petty bourgeoisie's lack of economic and political independence and its deep internal differentiation. By reason of this the upper sections of the petty bourgeoisie (of the peasantry) go along with the big bourgeoisie in all decisive cases, especially in war and in revolution; the lower sections go along with the proletariat; the intermediate section being thus compelled to choose between the two extreme poles. Between Kerenskyism and the Bolshevik power, between the Kuomintang and the dictatorship of the proletariat, there is not and cannot be any intermediate state, that is, no democratic dictatorship of the workers and peasants.

7. The Comintern's endeavour to foist upon the Eastern countries the slogan of the democratic dictatorship of the proletariat and peasantry, finally and long ago exhausted by history, can have only a reactionary effect. In so far as this slogan is counterposed to the slogan of the dictatorship of the proletariat, it contributes politically to the dissolution of the proletariat in the petty-bourgeois masses and thus creates the most favourable conditions for the hegemony of the national bourgeoisie and consequently for the collapse of the democratic revolution. The introduction of this slogan into the programme of the Comintern is a direct betrayal of Marxism and of the October tradition of Bolshevism.

8. The dictatorship of the proletariat which has risen to power as the leader of the democratic revolution is inevitably and very quickly confronted with tasks, the fulfilment of which is bound up with deep inroads into the rights of bourgeois property. The democratic revolution grows over directly into the Socialist revolution and thereby becomes a *permanent* revolution.

9. The conquest of power by the proletariat does not complete the revolution, but only opens it. Socialist construction is conceivable only on the foundation of the class struggle, on a national and international scale. This struggle, under the conditions of an overwhelming predominance of capitalist relationships on the world arena, must inevitably lead to explosions, that is, internally to civil wars and externally to revolutionary

wars. Therein lies the permanent character of the Socialist revolution as such, regardless of whether it is a backward country that is involved, which only yesterday accomplished its democratic revolution, or an old capitalist country which already has behind it a long epoch of democracy and parliamentarism.

10. The completion of the Socialist revolution within national limits is unthinkable. One of the basic reasons for the crisis in bourgeois society is the fact that the productive forces created by it can no longer be reconciled with the framework of the national state. From this follow, on the one hand, imperialist wars, on the other, the Utopia of a bourgeois United States of Europe. The Socialist revolution begins on the national arena, it unfolds on the international arena, and is completed on the world arena. Thus, the Socialist revolution becomes a permanent revolution in a newer and broader sense of the word; it attains completion only in the final victory of the new society on our entire planet.

11. The above-outlined sketch of the development of the world revolution eliminates the question of countries that are 'mature' or 'immature' for Socialism in the spirit of that pedantic, lifeless classification given by the present programme of the Comintern. In so far as capitalism has created a world market, a world division of labour, and world productive forces, it has also prepared world economy as a whole for Socialist transformation.

Different countries will go through this process at different tempos. Backward countries may, under certain conditions arrive at the dictatorship of the proletariat sooner than advanced countries, but they will come later than the latter to Socialism.

A backward colonial or semi-colonial country, the proletariat of which is insufficiently prepared to unite the peasantry and take power, is thereby incapable of bringing the democratic revolution to its conclusion. Contrariwise, in a country where the proletariat has power in its hands as the result of the democratic revolution, the subsequent fate of the dictatorship and Socialism depends in the last analysis not only and not so much upon the national productive forces as upon the development of the international Socialist revolution.

12. The theory of Socialism in one country, which rose on the yeast of the reaction against October, is the only theory that consistently and to the very end opposes the theory of the permanent revolution.

The attempt of the epigones, under the lash of our criticism, to confine the application of the theory of Socialism in one country exclusively to Russia, because of its specific characteristics (its vastness and its natural resources), does not improve matters but only makes them worse. The break with the internationalist position always and invariably leads to national *messianism*, that is, to attributing special superiorities and qualities to one's own country, which allegedly permit it to play a role to which other countries cannot attain.

The world division of labour, the dependence of Soviet industry upon foreign technology, the dependence of the productive forces of the advanced countries of Europe upon Asiatic raw materials, etc., etc., make the construction of an independent Socialist society in any single country in the world impossible.

13. The theory of Stalin and Bukharin, running counter to the entire experience of the Russian revolution, not only sets up the democratic revolution mechanically in contrast to the Socialist revolution, but also makes a breach between the national revolution and the international revolution.

This theory imposes upon revolutions in backward countries the task of establishing an unrealizable regime of democratic dictatorship, which it counterposes to the dictatorship of the proletariat. Thereby this theory introduces illusions and fictions into politics, paralyses the struggle for power of the proletariat in the East, and hampers the victory of the colonial revolution.

The very seizure of power by the proletariat signifies, from the standpoint of the epigones' theory, the completion of the revolution ('to the extent of nine-tenths', according to Stalin's formula) and the opening of the epoch of national reforms. The theory of the kulak growing into Socialism and the theory of the 'neutralization' of the world bourgeoisie are consequently inseparable from the theory of Socialism in one country. They stand or fall together.

By the theory of national Socialism, the Communist International is downgraded to an auxiliary weapon useful only for the struggle against military intervention. The present policy of the Comintern, its regime, and the selection of its leading personnel correspond entirely to the demotion of the Communist International to the role of an auxiliary unit which is not destined to solve independent tasks.

14. The programme of the Comintern created by Bukharin is eclectic through and through. It makes the hopeless attempt to reconcile the theory of Socialism in one country with Marxist internationalism, which is, however, inseparable from the permanent character of the world revolution. The struggle of the Communist Left Opposition for a correct policy and a healthy regime in the Communist International is inseparably bound up with the struggle for the Marxist programme. The question of the programme is in turn inseparable from the question of the two mutually exclusive theories: the theory of permanent revolution and the theory of Socialism in one

country. The problem of the permanent revolution has long ago outgrown
the episodic differences of opinion between Lenin and Trotsky, which were
completely exhausted by history. The struggle is between the basic ideas of
Marx and Lenin on the one side and the eclectism of the centrists on the
other.

3.   The Revolution Betrayed

The historian of the Soviet Union cannot fail to conclude that the policy of
the ruling bureaucracy upon great questions has been a series of contradictory zigzags. The attempt to explain or justify them by 'changing
circumstances' obviously won't hold water. To guide means at least in
some degree to exercise foresight. The Stalin faction have not in the slightest
degree foreseen the inevitable results of the development; they have been
caught napping every time. They have reacted with mere administrative
reflexes. The theory of each successive turn has been created after the fact,
and with small regard for what they were teaching yesterday. On the basis
of the same irrefutable facts and documents, the historian will be compelled
to conclude that the so-called 'left opposition' offered an immeasurably
more correct analysis of the processes taking place in the country, and far
more truly foresaw their further development.

This assertion is contradicted at first glance by the simple fact that the
faction which could not see ahead was steadily victorious, while the more
penetrating group suffered defeat after defeat. That kind of objection, which
comes automatically to mind, is convincing, however, only for those who
think rationalistically, and see in politics a logical argument or a chess
match. A political struggle is in its essence a struggle of interests and forces,
not of arguments. The quality of the leadership is, of course, far from a
matter of indifference for the outcome of the conflict, but it is not the only
factor, and in the last analysis is not decisive. Each of the struggling camps,
moreover, demands leaders in its own image.

The February revolution raised Kerensky and Tseretelli to power, not
because they were 'cleverer' or 'more astute' than the ruling tsarist clique,
but because they represented, at least temporarily, the revolutionary masses
of the people in their revolt against the old regime. Kerensky was able to
drive Lenin underground and imprison other Bolshevik leaders, not because
he excelled them in personal qualifications, but because the majority of the
workers and soldiers in those days were still following the patriotic petty
bourgeoisie. The personal 'superiority' of Kerensky, if it is suitable to
employ such a word in this connection, consisted in the fact that he did not
see farther than the overwhelming majority. The Bolsheviks in their turn

conquered the petty bourgeois democrats, not through the personal superiority of their leaders, but through a new correlation of social forces. The proletariat had succeeded at least in leading the discontented peasantry against the bourgeoisie.

The consecutive stages of the great French Revolution, during its rise and fall alike, demonstrate no less convincingly that the strength of the 'leaders' and 'heroes' who replaced each other consisted primarily in their correspondence to the character of those classes and strata which supported them. Only this correspondence, and not any irrelevant superiorities whatever, permitted each of them to place the impress of his personality upon a certain historic period. In the successive supremacy of Mirabeau, Brissot, Robespierre, Barras, and Bonaparte, there is an obedience to objective law incomparably more effective than the special traits of the historic protagonists themselves.

It is sufficiently well known that every revolution up to this time has been followed by a reaction, or even a counter-revolution. This, to be sure, has never thrown the nation all the way back to its starting-point, but it has always taken from the people the lion's share of their conquests. The victims of the first reactionary wave have been, as a general rule, those pioneers, initiators, and instigators who stood at the head of the masses in the period of the revolutionary offensive. In their stead people of the second line, in league with the former enemies of the revolution, have been advanced to the front. Beneath this dramatic duel of 'coryphées' on the open political scene shifts have taken place in the relations between classes, and, no less important, profound changes in the psychology of the recently revolutionary masses.

Answering the bewildered questions of many comrades as to what has become of the activity of the Bolshevik Party and the working class—where is its revolutionary initiative, its spirit of self-sacrifice and plebeian pride— why, in place of all this, has appeared so much vileness, cowardice, pusillanimity, and careerism—Rakovsky referred to the life story of the French Revolution of the eighteenth century, and offered the example of Babeuf, who on emerging from the Abbaye Prison likewise wondered what had become of the heroic people of the Parisian suburbs. A revolution is a mighty devourer of human energy, both individual and collective. The nerves give way. Consciousness is shaken and characters are worn out. Events unfold too swiftly for the flow of fresh forces to replace the loss. Hunger, unemployment, the death of the revolutionary cadres, the removal of the masses from administration, all this led to such a physical and moral impoverishment of the Parisian suburbs that they required three decades before they were ready for a new insurrection.

The axiom-like assertions of the Soviet lierature, to the effect that the laws of bourgeois revolutions are 'inapplicable' to a proletarian revolution, have no scientific content whatever. The proletarian character of the October revolution was determined by the world situation and by a special correlation of internal forces. But the classes themselves were formed in the barbarous circumstances of tsarism and backward capitalism, and were anything but made to order for the demands of a Socialist revolution. The exact opposite is true. It is for the very reason that a proletariat still backward in many respects achieved in the space of a few months the unprecedented leap from a semi-feudal monarchy to a Socialist dictatorship that the reaction in its ranks was inevitable. This reaction has developed in a series of consecutive waves. External conditions and events have vied with each other in nourishing it. Intervention followed intervention. The revolution got no direct help from the West. Instead of the expected prosperity of the country an ominous destitution reigned for long. Moreover, the outstanding representatives of the working class either died in the civil war or rose a few steps higher and broke away from the masses. And thus, after an unexampled tension of forces, hopes and illusions, there came a long period of weariness, decline, and sheer disappointment in the results of the revolution. The ebb of the 'plebeian pride' made room for a flood of pusillanimity and careerism. The new commanding caste rose to its place upon this wave.

The demobilization of the Red Army of five million played no small role in the formation of the bureaucracy. The victorious commanders assumed leading posts in the local Soviets, in economy, in education, and they persistently introduced everywhere that regime which had ensured success in the civil war. Thus on all sides the masses were pushed away gradually from actual participation in the leadership of the country.

The reaction within the proletariat caused an extraordinary flush of hope and confidence in the petty-bourgeois strata of town and country, aroused as they were to new life by the NEP, and growing bolder and bolder. The young bureaucracy, which had arisen at first as an agent of the proletariat, began now to feel itself a court of arbitration between the classes. Its independence increased from month to month.

The international situation was pushing with mighty forces in the same direction. The Soviet bureaucracy became more self-confident, the heavier the blows dealt to the working class. Between these two facts there was not only a chronological, but a causal connection, and one which worked in two directions. The leaders of the bureaucracy promoted the proletarian defeats; the defeats promoted the rise of the bureaucracy. The crushing of the Bulgarian insurrection and the inglorious retreat of the German workers' party in 1923, the collapse of the Estonian attempt at insurrection in 1924, the treacherous liquidation of the General Strike in England, and the unworthy conduct of the Polish workers' party at the installation of Pilsudski in 1926, the terrible massacre of the Chinese revolution in 1927, and, finally, the still more ominous recent defeats in Germany and Austria— these are the historic catastrophes which killed the faith of the Soviet masses in world revolution, and permitted the bureaucracy to rise higher and higher as the sole light of salvation.

As to the causes of the defeat of the world proletariat during the last thirteen years, the author must refer to his other works where he has tried to expose the ruinous part played by the leadership in the Kremlin, isolated from the masses and profoundly conservative as it is, in the revolutionary movement of all countries. Here we are concerned primarily with the irrefutable and instructive fact that the continual defeats of the revolution in Europe and Asia, while weakening the international position of the Soviet Union, have vastly strengthened the Soviet bureaucracy. Two dates are especially significant in this historic series. In the second half of 1923 the attention of the Soviet workers was passionately fixed upon Germany, where the proletariat, it seemed, had stretched out its hand to power. The panicky retreat of the German Communist Party was the heaviest possible disappointment to the working masses of the Soviet Union. The Soviet bureaucracy straightway opened a campaign against the theory of 'permanent revolution', and dealt the left opposition its first cruel blow. During the years 1926 and 1927 the population of the Soviet Union experienced a new tide of hope. All eyes were now directed to the East, where the drama of the Chinese revolution was unfolding. The left opposition had recovered from the previous blows and was recruiting a phalanx of new adherents. At the end of 1927 the Chinese revolution was massacred by the hangman Chiang Kai-shek, into whose hands the Communist International had literally betrayed the Chinese workers and peasants. A cold wave of disappointment swept over the masses of the Soviet Union. After an unbridled baiting in the Press and at meetings, the bureaucracy finally, in 1928, ventured upon mass arrests among the left opposition.

To be sure, tens of thousands of revolutionary fighters gathered around the banner of the Bolshevik-Leninists. The advanced workers were indubitably sympathetic to the opposition, but that sympathy remained passive. The masses lacked faith that the situation could be seriously changed by a new struggle. Meantime the bureaucracy asserted: 'For the sake of an international revolution the opposition proposes to drag us into a revolutionary war. Enough of shake-ups! We have earned the right to rest. We will build the Socialist society at home. Rely upon us, your leaders!' This

gospel of repose firmly consolidated the *apparatchiki* and the military and
State officials and indubitably found an echo among the weary workers,
and still more the peasant masses. Can it be, they asked themselves, that
the opposition is actually ready to sacrifice the interests of the Soviet Union
for the idea of 'permanent revolution'? In reality, the struggle had been
about the life interests of the Soviet State. The false policy of the International in Germany resulted ten years later in the victory of Hitler—that
is, in a threatening war danger from the West. And the no less false policy
in China reinforced Japanese imperialism and brought very much nearer
the danger in the East. But periods of reaction are characterized above all
by a lack of courageous thinking.

The opposition was isolated. The bureaucracy struck while the iron was
hot, exploiting the bewilderment and passivity of the workers, setting their
more backward strata against the advanced, and relying more and more
boldly upon the kulak and the petty bourgeois ally in general. In the course
of a few years the bureaucracy thus shattered the revolutionary vanguard
of the proletariat.

It would be naïve to imagine that Stalin, previously unknown to the
masses, suddenly issued from the wings fully armed with a complete
strategical plan. No, indeed. Before he felt out his own course the bureaucracy felt out Stalin himself. He brought it all the necessary guarantees: the
prestige of an old Bolshevik, a strong character, narrow vision, and close
bonds with the political machine as the sole source of his influence. The
success which fell upon him was a surprise at first to Stalin himself. It was
the friendly welcome of the new ruling group, trying to free itself from the
old principles and from the control of the masses, and having need of a
reliable arbiter in its inner affairs. A secondary figure before the masses and
in the events of the revolution, Stalin revealed himself as the indubitable
leader of the Thermidorian bureaucracy, as first in its midst.

The new ruling caste soon revealed its own ideas, feelings, and, more
important, its interests. The overwhelming majority of the older generation
of the present bureaucracy had stood on the other side of the barricades
during the October revolution. (Take, for example, the Soviet ambassadors
only: Troyanovsky, Maisky, Potemkin, Suritz, Khinchuk, etc.) Or at best
they had stood aside from the struggle. Those of the present bureaucrats
who were in the Bolshevik camp in the October days played in the majority
of cases no considerable role. As for the younger bureaucrats, they have
been chosen and educated by the elders, frequently from among their own
offspring. These people could not have achieved the October revolution,
but they were perfectly suited to exploit it.

Personal incidents in the interval between these two historic chapters were not, of course, without influence. Thus the sickness and death of Lenin undoubtedly hastened the denouement. Had Lenin lived longer, the pressure of the bureaucratic power would have developed, at least during the first years, more slowly. But as early as 1926 Krupskaya said, in a circle of left oppositionists: 'If Ilich were alive he would probably already be in prison.' The fears and alarming prophecies of Lenin himself were then still fresh in her memory, and she cherished no illusions as to his personal omnipotence against opposing historic winds and currents.

The bureaucracy conquered something more than the left opposition. It conquered the Bolshevik Party. It defeated the programme of Lenin, who had seen the chief danger in the conversion of the organs of the State 'from the servants of society to lords over society'. It defeated all these enemies, the Opposition, the party, and Lenin, not with ideas and arguments, but with its own social weight. The leaden rump of the bureaucracy outweighed the head of the revolution. That is the secret of the Soviet's Thermidor.

The Bolshevik Party prepared and insured the October victory. It also created the Soviet State, supplying it with a sturdy skeleton. The degeneration of the party became both cause and consequence of the bureaucratization of the State. It is necessary to show at least briefly how this happened.

The inner regime of the Bolshevik Party was characterized by the method of *democratic centralism*. The combination of these two concepts, democracy and centralism, is not in the least contradictory. The party took watchful care not only that its boundaries should always be strictly defined, but also that all those who entered these boundaries should enjoy the actual right to define the direction of the party policy. Freedom of criticism and intellectual struggle was an irrevocable content of the party democracy. The present doctrine that Bolshevism does not tolerate factions is a myth of the epoch of decline. In reality the history of Bolshevism is a history of the struggle of factions. And, indeed, how could a genuinely revolutionary organization, setting itself the task of overthrowing the world and uniting under its banner the most audacious iconoclasts, fighters, and insurgents, live and develop without intellectual conflicts, without groupings and temporary factional formations? The far-sightedness of the Bolshevik leadership often made it possible to soften conflicts and shorten the duration of factional struggle, but no more than that. The Central Committee relied upon this seething democratic support. From this it derived the audacity to make decisions and give orders. The obvious correctness of the leadership at all critical stages gave it that high authority which is the priceless moral capital of centralism.

The regime of the Bolshevik Party, especially before it came to power, stood thus in complete contradiction to the regime of the present sections of the Communist International, with their 'leaders' appointed from above, making complete changes of policy at a word of command, with their uncontrolled apparatus, haughty in its attitude to the rank and file, servile in its attitude to the Kremlin. But in the first years after the conquest of power also, even when the administrative rust was already visible on the party, every Bolshevik, not excluding Stalin, would have denounced as a malicious slanderer any one who should have shown him on a screen the image of the party ten or fifteen years later.

The very centre of Lenin's attention and that of his colleagues was occupied by a continual concern to protect the Bolshevik ranks from the vices of those in power. However, the extraordinary closeness and at times actual merging of the party with the State apparatus had already in those first years done indubitable harm to the freedom and elasticity of the party regime. Democracy had been narrowed in proportion as difficulties increased. In the beginning the party had wished and hoped to preserve freedom of political struggle within the framework of the Soviets. The civil war introduced stern amendments into this calculation. The opposition parties were forbidden one after the other. This measure, obviously in conflict with the spirit of Soviet democracy, the leaders of Bolshevism regarded not as a principle but as an episodic act of self-defence.

The swift growth of the ruling party, with the novelty and immensity of its tasks, inevitably gave rise to inner disagreements. The underground oppositional currents in the country exerted a pressure through various channels upon the sole legal political organization, increasing the acuteness of the factional struggle. At the moment of completion of the civil war this struggle took such sharp forms as to threaten to unsettle the State power. In March 1921, in the days of the Kronstadt revolt, which attracted into its ranks no small number of Bolsheviks, the Tenth Congress of the party thought it necessary to resort to a prohibition of factions—that is, to transfer the political regime prevailing in the State to the inner life of the ruling party. This forbidding of factions was again regarded as an exceptional measure to be abandoned at the first serious improvement in the situation. At the same time the Central Committee was extremely cautious in applying the new law, concerning itself most of all lest it lead to a strangling of the inner life of the party.

However, what was in its original design merely a necessary concession to a difficult situation proved perfectly suited to the taste of the bureaucracy, which had then begun to approach the inner life of the party exclusively from the viewpoint of convenience in administration. Already in 1922, during a brief improvement in his health, Lenin, horrified at the threatening growth of bureaucratism, was preparing a struggle against the faction of Stalin, which had made itself the axis of the party machine as a first step toward capturing the machinery of State. A second stroke and then death prevented him from measuring forces with this internal reaction.

The entire effort of Stalin, with whom at this time Zinoviev and Kamenev were working hand in hand, was thenceforth directed to freeing the party machine from the control of the rank-and-file members of the party. In this struggle for 'stability' of the Central Committee Stalin proved the most consistent and reliable among his colleagues. He had no need to tear himself away from international problems; he had never been concerned with them. The petty-bourgeois outlook of the new ruling stratum was his own outlook. He profoundly believed that the task of creating Socialism was national and administrative in its nature. He looked upon the Communist International as a necessary evil which should be used so far as possible for the purposes of foreign policy. His own party kept a value in his eyes merely as a submissive support for the machine.

Together with the theory of Socialism in one country there was put into circulation by the bureaucracy a theory that in Bolshevism the Central Committee is everything and the party nothing. This second theory was in any case realized with more success than the first. Availing itself of the death of Lenin, the ruling group announced a 'Leninist levy'. The gates of the party, always carefully guarded, were now thrown wide open. Workers, clerks, petty officials, flocked through in crowds. The political aim of this manœuvre was to dissolve the revolutionary vanguard in raw human material without experience, without independence, and yet with the old habit of submitting to the authorities. The scheme was successful. By freeing the bureaucracy from the control of the proletarian vanguard, the 'Leninist levy' dealt a death-blow to the party of Lenin. The machine had won the necessary independence. Democratic centralism gave place to bureaucratic centralism. In the party apparatus itself there now took place a radical reshuffling of personnel from top to bottom. The chief merit of a Bolshevik was declared to be obedience. Under the guise of a struggle with the opposition there occurred a sweeping replacement of revolutionists with *chinovniks*. The history of the Bolshevik Party became a history of its rapid degeneration.

The political meaning of the developing struggle was darkened for many by the circumstance that the leaders of all three groupings, Left, Centre, and Right, belonged to one and the same staff in the Kremlin, the Politburo. To superficial minds it seemed to be a matter of personal rivalry, a struggle for the 'heritage' of Lenin. But in the conditions of iron dictatorship social antagonisms could not show themselves at first except through the institutions of the ruling party. Many Thermidorians emerged in their day from the circle of the Jacobins. Bonaparte himself belonged to that circle in his early years, and subsequently it was from among former Jacobins that the First Consul and emperor of France selected his most faithful servants. Times change and the Jacobins with them, not excluding the Jacobins of the twentieth century.

Of the Politburo of Lenin's epoch there now remains only Stalin. Two of its members, Zinoviev and Kamenev, collaborators of Lenin throughout many years as émigrés, are enduring ten-year prison terms for a crime which they did not commit. Three other members, Rykov, Bukharin, and Tomsky, are completely removed from the leadership, but as a reward for submission occupy secondary posts. And, finally, the author of these lines is in exile. The widow of Lenin, Krupskaya, is also under the ban, having proved unable with all her efforts to adjust herself completely to the Thermidor.

The members of the present Politburo occupied secondary posts throughout the history of the Bolshevik Party. If anybody in the first years of the revolution had predicted their future elevation they would have been the first in surprise, and there would have been no false modesty in their surprise. For this very reason the rule is more stern at present that the Politburo is always right, and in any case that no man can be right against the Politburo. But, moreover, the Politburo cannot be right against Stalin, who is unable to make mistakes and consequently cannot be right against himself.

Demands for party democracy were through all this time the slogans of all the oppositional groups, as insistent as they were hopeless. The above-mentioned platform of the left opposition demanded in 1927 that a special law be written into the Criminal Code 'punishing as a serious State crime every direct or indirect persecution of a worker for criticism'. Instead of this there was introduced into the Criminal Code an article against the left opposition itself.

Of party democracy there remained only recollections in the memory of the older generation. And together with it had disappeared the democracy of the Soviets, the trade unions, the co-operatives, the cultural and athletic organizations. Above each and every one of them there reigns an unlimited hierarchy of party secretaries. The regime had become 'totalitarian' in character several years before this word arrived from Germany. 'By means of demoralizing methods, which convert thinking communists into machines, destroying will, character, and human dignity,' wrote Rakovsky in 1928, 'the ruling circles have succeeded in converting themselves into an unremovable and inviolate oligarchy, which replaces the class and the party.' Since those indignant lines were written, the degeneration of the regime has gone immeasurably further. The GPU has become the decisive factor in the inner life of the party. If Molotov in March 1936 was able to boast to a French journalist that the ruling party no longer contains any factional struggle, it is only because disagreements are now settled by the automatic intervention of the political police. The old Bolshevik Party is dead, and no force will resurrect it. . . .

We have defined the Soviet Thermidor as a triumph of the bureaucracy over the masses. We have tried to disclose the historic conditions of this triumph. The revolutionary vanguard of the proletariat was in part devoured by the administrative apparatus and gradually demoralized, in part annihilated in the civil war, and in part thrown out and crushed. The tired and disappointed masses were indifferent to what was happening on the summits. These conditions, however, important as they may have been in themselves, are inadequate to explain why the bureaucracy succeeded in raising itself above society and getting its fate firmly into its own hands. Its own will to this would in any case be inadequate: the arising of a new ruling stratum must have deep social causes.

The victory of the Thermidorians over the Jacobins in the eighteenth century was also aided by the weariness of the masses and the demoralization of the leading cadres, but beneath these essentially incidental phenomena a deep organic process was taking place. The Jacobins rested upon the lower petty bourgeoisie lifted by the great wave. The revolution of the eighteenth century, however, corresponding to the course of development of the productive forces, could not but bring the great bourgeoisie to political ascendancy in the long run. The Thermidor was only one of the stages in this inevitable process. What similar social necessity found expression in the Soviet Thermidor? We have tried already in one of the preceding chapters to make a preliminary answer to the question why the gendarme triumphed. We must now prolong our analysis of the conditions of the transition from capitalism to Socialism, and the role of the State in this process. Let us again compare theoretic prophecy with reality. 'It is still necessary to suppress the bourgeoisie and its resistance,' wrote Lenin in 1917, speaking of the period which should begin immediately after the conquest of power, 'but the organ of suppression here is now the majority of the population, and not the minority, as has heretofore always been the case. . . . In that sense the State *is beginning to die away*.' In what does this dying away express itself? Primarily in the fact that 'in place of special institutions of a privileged minority (privileged officials, commanders of a standing army) the majority itself can directly carry out' the functions of suppression. Lenin follows this with a statement axiomatic and unanswerable. 'The more universal becomes the very fulfilment of the functions of the State power, the less need is there of this power. The annulment of private property in the means of production removes the principal task of the historic State—defence of the proprietary privileges of the minority against the overwhelming majority.

The dying away of the State begins, then, according to Lenin, on the very day after the expropriation of the expropriators—that is, before the new regime has had time to take up its economic and cultural problems. Every success in the solution of these problems means a further step in the liquidation of the State, its dissolution in the Socialist society. The degree of this dissolution is the best index of the depth and efficacy of the Socialist structure. We may lay down approximately this sociological theorem: The strength of the compulsion exercised by the masses in a workers' State is directly proportional to the strength of the exploitive tendencies, or the danger of a restoration of capitalism, and inversely proportional to the strength of the social solidarity and the general loyalty to the new regime. Thus the bureaucracy—that is, the 'privileged officials and commanders of a standing army'—represents a special kind of compulsion which the masses cannot or do not wish to exercise, and which, one way or another, is directed against the masses themselves.

If the democratic Soviets had preserved to this day their original strength and independence, and yet were compelled to resort to repressions and compulsions on the scale of the first years, this circumstance might of itself give rise to serious anxiety. How much greater must be the alarm in view of the fact that the mass Soviets have entirely disappeared from the scene, having turned over the function of compulsion to Stalin, Yagoda, and company. And what forms of compulsion! First of all we must ask ourselves: What social cause stands behind this stubborn virility of the State and especially behind its policification? The importance of this question is obvious. In dependence upon the answer we must either radically revise our traditional views of the Socialist society in general, or as radically reject the official estimates of the Soviet Union.

Let us now take from the latest number of a Moscow newspaper a stereotyped characterization of the present Soviet regime, one of those which are repeated throughout the country from day to day and which schoolchildren learn by heart: 'In the Soviet Union the parasitical classes of capitalists, landlords, and kulaks are completely liquidated, and thus is forever ended the exploitation of man by man. The whole national economy has become socialistic, and the growing Stakhanov movement is preparing the conditions for a transition from Socialism to Communism.' (*Pravda*, 4 April 1936.) The world Press of the Communist International, it goes without saying, has no other thing to say on this subject. But if exploitation is 'ended forever', if the country is really now on the road from Socialism, that is, the lowest stage of Communism, to its higher stage, then there remains nothing for society to do but to throw off at last the strait-jacket of the State. In place of this—it is hard even to grasp this contrast with the mind—the Soviet State has acquired a totalitarian bureaucratic character.

The same fatal contradiction finds illustration in the fate of the party. Here the problem may be formulated approximately thus: Why from 1917 to 1921, when the old ruling classes were still fighting with weapons in their hands, when they were actively supported by the imperialists of the whole world, when the kulaks in arms were sabotaging the army and food supplies of the country, why was it possible to dispute openly and fearlessly in the party about the most critical questions of policy? Why now, after the cessation of intervention, after the shattering of the exploiting classes, after the indubitable successes of industrialization, after the collectivization of the overwhelming majority of the peasants, is it impossible to permit the slightest word of criticism of the unremovable leaders? Why is it that any Bolshevik who should demand a calling of the congress of the party in accordance with its constitution would be immediately expelled, any citizen who expressed out loud a doubt of the infallibility of Stalin would be tried and convicted almost as though a participant in a terrorist plot? Whence this terrible, monstrous, and unbearable intensity of repression and of the police apparatus?

Theory is not a note which you can present at any moment to reality for payment. If a theory proves mistaken we must revise it or fill out its gaps. We must find out those real social forces which have given rise to the contrast between Soviet reality and the traditional Marxian conception. In any case we must not wander in the dark, repeating ritual phrases, useful for the prestige of the leaders, but which nevertheless slap the living reality in the face. We shall now see a convincing example of this.

In a speech at a session of the Central Executive Committee in January 1936 Molotov, the president of the Council of People's Commissars, declared: 'The national economy of the country has become Socialistic. (*Applause.*) In that sense [?] we have solved the problem of the liquidation of classes. (*Applause.*)' However, there still remain from the past 'elements in their nature hostile to us', fragments of the former ruling classes. More-over, among the collectivized farmers, State employees, and sometimes also the workers, 'petty speculators' are discovered, 'grafters in relation to the collective and State wealth, anti-Soviet gossips, etc.' And hence results the necessity of a further reinforcement of the dictatorship. In opposition to Engels, the workers' State must not 'fall asleep', but on the contrary become more and more vigilant.

The picture drawn by the head of the Soviet government would be reassuring in the highest degree were it not murderously self-contradictory. Socialism completely reigns in the country: 'In that sense' classes are abolished. (If they are abolished in that sense, then they are in every other.) To be sure, the social harmony is broken here and there by fragments and remnants of the past, but it is impossible to think that scattered dreamers of a restoration of capitalism, deprived of power and property, together with 'petty speculators' (not even *speculators!*) and 'gossips' are capable of overthrowing the classless society. Everything is getting along, it seems, the very best you can imagine. But what is the use then of the iron dictatorship of the bureaucracy?

Those reactionary dreamers, we must believe, will gradually die out. The 'petty speculators' and 'gossips' might be disposed of with a laugh by the super-democratic Soviets. 'We are not Utopians', responded Lenin in 1917 to the bourgeois and reformist theoreticians of the bureaucratic State, and 'by no means deny the possibility and inevitability of excesses on the part of *individual persons*, and likewise the necessity for suppressing *such* excesses. But ... for this there is no need of a special machine, a special apparatus of repression. This will be done by the armed people themselves with the same simplicity and ease with which any crowd of civilized people even in contemporary society separate a couple of fighters or stop an act of violence against a woman.' Those words sound as though the author had especially foreseen the remarks of one of his successors at the head of the government. Lenin is taught in the public schools of the Soviet Union, but apparently not in the Council of People's Commissars. Otherwise it would be impossible to explain Molotov's daring to resort without reflection to the very construction against which Lenin directed his well-sharpened weapons. The flagrant contradiction between the founder and his epigone is before us! Whereas Lenin judged that even the liquidation of the exploiting classes might be accomplished without a bureaucratic apparatus, Molotov, in explaining why *after* the liquidation of classes the bureaucratic machine has strangled the independence of the people, finds no better pretext than a reference to the 'remnants' of the liquidated classes.

To live on these 'remnants' becomes, however, rather difficult since, according to the confession of authoritative representatives of the bureaucracy itself, yesterday's class enemies are being successfully assimilated by the Soviet society. Thus Postyshev, one of the secretaries of the Central Committee of the party, said in April 1936, at a congress of the League of

Communist Youth: 'Many of the sabotagers ... have sincerely repented and joined the ranks of the Soviet people.' In view of the successful carrying out of collectivization, 'the children of kulaks are not to be held responsible for their parents'. And yet more: 'The kulak himself now hardly believes in the possibility of a return to his former position of exploiter in the village.' Not without reason did the government annul the limitations connected with social origin! But if Postyshev's assertion, wholly agreed to by Molotov, makes any sense, it is only this: Not only has the bureaucracy become a monstrous anachronism, but State compulsion in general has nothing whatever to do in the land of the Soviets. However, neither Molotov nor Postyshev agrees with that immutable inference. They prefer to hold the power even at the price of self-contradiction.

In reality, too, they cannot reject the power. Or, to translate this into objective language: The present Soviet society cannot get along without a State, not even—within limits—without a bureaucracy. But the cause of this is by no means the pitiful remnants of the past, but the mighty forces and tendencies of the present. The justification for the existence of a Soviet State as an apparatus of compulsion lies in the fact that the present transitional structure is still full of social contradictions, which in the sphere of *consumption*—most close and sensitively felt by all—are extremely tense, and forever threaten to break over into the sphere of production. The triumph of Socialism cannot be called either final or irrevocable.

The basis of bureaucratic rule is the poverty of society in objects of consumption, with the resulting struggle of each against all. When there is enough goods in a store the purchasers can come whenever they want to. When there is little goods the purchasers are compelled to stand in line. When the lines are very long, it is necessary to appoint a policeman to keep order. Such is the starting-point of the power of the Soviet bureaucracy. It 'knows' who is to get something and who has to wait.

A raising of the material and cultural level ought, at first glance, to lessen the necessity of privileges, narrow the sphere of application of 'bourgeois law', and thereby undermine the standing ground of its defenders, the bureaucracy. In reality the opposite thing has happened: the growth of the productive forces has been so far accompanied by an extreme development of all forms of inequality, privilege, and advantage, and therewith of bureaucratism. That too is not accidental.

In its first period, the Soviet regime was undoubtedly far more equalitarian and less bureaucratic than now. But that was an equality of general poverty. The resources of the country were so scant that there was no opportunity to separate out from the masses of the population any broad privileged strata. At the same time the 'equalizing' character of wages, destroying personal interestedness, became a brake upon the development of the productive forces. Soviet economy had to lift itself from its poverty to a somewhat higher level before fat deposits of privilege became possible. The present state of production is still far from guaranteeing all necessities to everybody. But it is already adequate to give significant privileges to a minority, and convert inequality into a whip for the spurring on of the majority. That is the first reason why the growth of production has so far strengthened not the Socialist, but the bourgeois, features of the State.

But that is not the sole reason. Alongside the economic factor dictating capitalistic methods of payment at the present stage there operates a parallel political factor in the person of the bureaucracy itself. In its very essence it is the planter and protector of inequality. It arose in the beginning as the bourgeois organ of a workers' State. In establishing and defending the advantages of a minority, it of course draws off the cream for its own use. Nobody who has wealth to distribute ever omits himself. Thus out of a social necessity there has developed an organ which has far outgrown its socially necessary function, and become an independent factor and therewith the source of great danger for the whole social organism.

The social meaning of the Soviet Thermidor now begins to take form before us. The poverty and cultural backwardness of the masses has again become incarnate in the malignant figure of the ruler with a great club in his hand. The deposed and abused bureaucracy, from being a servant of society, has again become its lord. On this road it has attained such a degree of social and moral alienation from the popular masses that it cannot now permit any control over either its activities or its income.

The bureaucracy's seemingly mystic fear of 'petty speculators, grafters, and gossips' thus finds a wholly natural explanation. Not yet able to satisfy the elementary needs of the population, the Soviet economy creates and resurrects at every step tendencies to graft and speculation. On the other side, the privileges of the new aristocracy awaken in the masses of the population a tendency to listen to anti-Soviets 'gossip'—that is, to any one who, albeit in a whisper, criticizes the greedy and capricious bosses. It is a question, therefore, not of spectres of the past, not of the remnants of what no longer exists, not in short, of the snows of yesteryear, but of new, mighty, and continually reborn tendencies to personal accumulation. The first still very meagre wave of prosperity in the country, just because of its meagreness, has not weakened, but strengthened, these centrifugal tendencies. On the other hand, there has developed simultaneously a desire of the unprivileged to slap the grasping hands of the new gentry. The social struggle again grows sharp. Such are the sources of the power of the bureaucracy. But from those same sources comes also a threat to its power....

Classes are characterized by their position in the social system of economy, and primarily by their relation to the means of production. In civilized societies property relations are validated by laws. The nationalization of the land, the means of industrial production, transport, and exchange, together with the monopoly of foreign trade, constitute the basis of the Soviet social structure. Through these relations, established by the proletarian revolution, the nature of the Soviet Union as a proletarian State is for us basically defined.

In its intermediary and regulating function, its concern to maintain social ranks, and its exploitation of the State apparatus for person goals, the Soviet bureaucracy is similar to every other bureaucracy, especially the fascist. But it is also in a vast way different. In no other regime has a bureaucracy ever achieved such a degree of independence from the dominating class. In bourgeois society the bureaucracy represents the interests of a possessing and educated class, which has at its disposal innumerable means of everyday control over its administration of affairs. The Soviet bureaucracy has risen above a class which is hardly emerging from destitution and darkness, and has no tradition of dominion or command. Whereas the fascists, when they find themselves in power, are united with the big bourgeoisie by bonds of common interest, friendship, marriage, etc., the Soviet bureaucracy takes on bourgeoisie customs without having beside it a national bourgeoisie. In this sense we cannot deny that it is something more than a bureaucracy. It is in the full sense of the word the sole privileged and commanding stratum in the Soviet society.

Another difference is no less important. The Soviet bureaucracy has expropriated the proletariat politically in order by methods of *its own* to defend the social conquests. But the very fact of its appropriation of political power in a country where the principal means of production are in the hands of the State creates a new and hitherto unknown relation between the bureaucracy and the riches of the nation. The means of production belong to the State. But the State, so to speak, 'belongs' to the bureaucracy. If these as yet wholly new relations should solidify, become the norm, and be legalized, whether with or without resistance from the workers, they would, in the long run, lead to a complete liquidation of the social conquests of the proletarian revolution. But to speak of that now is at least premature. The proletariat has not yet said its last word. The bureaucracy has not yet created social supports for its dominion in the form of special types of property. It is compelled to defend State property as the source of its power and its income. In this aspect of its activity it still remains a weapon of proletarian dictatorship.

The attempt to represent the Soviet bureaucracy as a class of 'State capitalists' will obviously not withstand criticism. The bureaucracy has neither stocks nor bonds. It is recruited, supplemented, and renewed in the manner of an administrative hierarchy, independently of any special property relations of its own. The individual bureaucrat cannot transmit to his heirs his rights in the exploitation of the State apparatus. The bureaucracy enjoys its privileges under the form of an abuse of power. It conceals its income; it pretends that as a special social group it does not even exist. Its appropriation of a vast share of the national income has the character of social parasitism. All this makes the position of the commanding Soviet stratum in the highest degree contradictory, equivocal, and undignified, notwithstanding the completeness of its power and the smokescreen of flattery that conceals it.

Bourgeois society has in the course of its history displaced many political regimes and bureaucratic castes, without changing its social foundations. It has preserved itself against the restoration of feudal and guild relations by the superiority of its productive methods. The State power has been able either to co-operate with capitalist development or put brakes on it. But in general the productive forces, upon a basis of private property and competition, have been working out their own destiny. In contrast to this the property relations which issued from the Socialist revolution are indivisibly bound up with the new State as their repository. The predominance of Socialist over petty-bourgeois tendencies is guaranteed not by the automatism of the economy—we are still far from that—but by the political measures taken by the dictatorship. The character of the economy as a whole thus depends upon the character of the State power.

A collapse of the Soviet regime would lead inevitably to the collapse of the planned economy, and thus to the abolition of State property. The bond of compulsion between the trusts and the factories within them would fall away. The more successful enterprises would succeed in coming out on the road of independence. They might convert themselves into stock companies, or they might find some other transitional form of property—one, for example, in which the workers should participate in the profits. The collective farms would disintegrate at the same time, and far more easily. The fall of the present bureaucratic dictatorship, if it were not replaced by a new socialist power, would thus mean a return to capitalist relations with a catastrophic decline of industry and culture.

But if a Socialist government is still absolutely necessary for the preservation and development of the planned economy, the question is all the more important, upon whom the present Soviet government relies, and in what measure the Socialist character of its policy is guaranteed. At the Eleventh Party Congress in March 1922 Lenin, in practically bidding farewell to the party, addressed these words to the commanding group: 'History knows transformations of all sorts. To rely upon conviction, devotion, and other excellent spiritual qualities—that is not to be taken seriously in politics.' Being determines consciousness. During the last fifteen years the government has changed its social composition even more deeply than its ideas. Since of all the strata of Soviet society the bureaucracy has best solved its own social problem, and is fully content with the existing situation, it has ceased to offer any subjective guarantee whatever of the Socialist direction of its policy. It continues to preserve State property only to the extent that it fears the proletariat. This saving fear is nourished and supported by the illegal party of Bolshevik-Leninists, which is the most conscious expression of the Socialist tendencies opposing that bourgeois reaction with which the Thermidorian bureaucracy is completely saturated. As a conscious political force the bureaucracy has betrayed the revolution. But a victorious revolution is fortunately not only a programme and a banner, not only political institutions, but also a system of social relations. To betray it is not enough. You have to overthrow it. The October revolution has been betrayed by the ruling stratum, but not yet overthrown. It has a great power of resistance, coinciding with the established property relations, with the living force of the proletariat, the consciousness of its best elements, the impasse of world capitalism, and the inevitability of world revolution.

In order better to understand the character of the present Soviet Union let us make two different hypotheses about its future. Let us assume first that the Soviet bureaucracy is overthrown by a revolutionary party having all the attributes of the old Bolshevism, enriched moreover by the world experience of the recent period. Such a party would begin with the restoration of democracy in the trade unions and the Soviets. It would be able to, and would have to, restore freedom of Soviet parties. Together with the masses, and at their head, it would carry out a ruthless purgation of the State apparatus. It would abolish ranks and decorations, all kinds of privileges, and would limit inequality in the payment of labour to the life necessities of the economy and the State apparatus. It would give the youth free opportunity to think independently, learn, criticize, and grow. It would introduce profound changes in the distribution of the national income in correspondence with the interests and will of the worker and peasant masses. But so far as concerns property relations, the new power would not have to resort to revolutionary measures. It would retain and further develop the experiment of planned economy. After the political revolution—that is, the deposing of the bureaucracy—the proletariat would have to introduce in the economy a series of very important reforms, but not another social revolution.

If—to adopt a second hypothesis—a bourgeois party were to overthrow the ruling Soviet caste, it would find no small number of ready servants among the present bureaucrats, administrators, technicians, directors, party secretaries, and privileged upper circles in general. A purgation of the State apparatus would, of course, be necessary in this case too. But a bourgeois restoration would probably have to clean out fewer people than a revolutionary party. The chief task of the new power would be to restore private property in the means of production. First of all it would be necessary to create conditions for the development of strong farmers from the weak collective farms, and for converting the strong collectives into producers' co-operatives of the bourgeois type—into agricultural stock companies. In the sphere of industry denationalization would begin with the light industries and those producing food. The planning principle would be converted for the transitional period into a series of compromises between State power and individual 'corporations'—potential proprietors, that is, among the Soviet captains of industry, the *émigré* former proprietors, and foreign capitalists. Notwithstanding that the Soviet bureaucracy has gone far toward preparing a bourgeois restoration, the new regime would have to introduce in the matter of forms of property and methods of industry not a reform, but a social revolution.

Let us assume—to take a third variant—that neither a revolutionary nor a counter-revolutionary party seizes power. The bureaucracy continues at the head of the State. Even under these conditions social relations will not jellify. We cannot count upon the bureaucracy's peacefully and voluntarily renouncing itself on behalf of Socialist equality. If at the present time, notwithstanding the too obvious inconveniences of such an operation, it has considered it possible to introduce ranks and decorations, it must inevitably, in future stages seek supports for itself in property relations. One may argue that the big bureaucrat cares little what are the prevailing forms of property, provided only they guarantee him the necessary income. This argument ignores not only the instability of the bureaucrat's own rights, but also the question of his descendants. The new cult of the family has not fallen out of the clouds. Privileges have only half their worth, if they cannot be transmitted to one's children. But the right of testament is inseparable from the right of property. It is not enough to be the director of a trust; it is necessary to be a stockholder. The victory of the bureaucracy in this decisive sphere would mean its conversion into a new possessing class. On the other hand, the victory of the proletariat over the bureaucracy would ensure a revival of the Socialist revolution. The third variant consequently brings us back to the two first, with which, in the interests of clarity and simplicity we set out.

To define the Soviet regime as transitional, or intermediate, means to abandon such finished social categories as *capitalism* (and therewith 'State capitalism') and also *Socialism*. But besides being completely inadequate in itself such a definition is capable of producing the mistaken idea that from the present Soviet regime *only* a transition to Socialism is possible. In reality a backslide to capitalism is wholly possible. A more complete definition will of necessity be complicated and ponderous.

The Soviet Union is a contradictory society half-way between capitalism and socialism, in which: (*a*) the productive forces are still far from adequate to give the State property a Socialist character; (*b*) the tendency toward primitive accumulation created by want breaks out through innumerable pores of the planned economy; (*c*) norms of distribution preserving a bourgeois character lie at the basis of a new differentiation of society; (*d*) the economic growth, while slowly bettering the situation of the toilers, promotes a swift formation of privileged strata; (*e*) exploiting the social antagonisms, a bureaucracy has converted itself into an uncontrolled caste alien to Socialism; (*f*) the social revolution, betrayed by the ruling party, still exists in property relations and in the consciousness of the toiling masses; (*g*) a further development of the accumulating contradictions can as well lead to Socialism as back to capitalism; (*h*) on the road to capitalism the counter-revolution would have to break the resistance of the workers; (*i*) on the road to Socialism the workers would have to overthrow the bureaucracy. In the last analysis, the question will be decided by a struggle of living social forces, both on the national and the world arena.

Doctrinaires will doubtless not be satisfied with this hypothetical definition. They would like categorical formulae: yes—yes, and no—no. Sociological problems would certainly be simpler, if social phenomena had always a finished character. There is nothing more dangerous, however, than to throw out of reality, for the sake of logical completeness, elements which today violate your scheme and tomorrow may wholly overturn it. In our analysis we have above all avoided doing violence to dynamic social formations which have had no precedent and have no analogies. The scientific task, as well as the political, is not to give a finished definition to an unfinished process, but to follow all its stages, separate its progressive from its reactionary tendencies, expose their mutual relations, foresee

possible variants of development, and find in this foresight a basis for
action. . . .

The increasingly insistent deification of Stalin is, with all its elements of
caricature, a necessary element of the regime. The bureaucracy has need of
an inviolable super-arbiter, a first consul if not an emperor, and it raises
upon its shoulders him who best responds to its claim for lordship. That
'strength of character' of the leader which so enraptures the literary dilett-
antes of the West is in reality the sum total of the collective pressure of a
caste which will stop at nothing in defence of its position. Each one of them
at his post is thinking: *L'Etat—c'est moi.* In Stalin each one easily finds
himself. But Stalin also finds in each one a small part of his own spirit.
Stalin is the personification of the bureaucracy. That is the substance of his
political personality.

Caesarism, or its bourgeois form, Bonapartism, enters the scene in those
moments of history when the sharp struggle of two camps raises the State
power, so to speak, above the nation, and guarantees it, in appearance, a
complete independence of classes—in reality, only the freedom necessary
for a defence of the privileged. The Stalin regime, rising above a politically
atomized society, resting upon a police and officers' corps, and allowing of
no control whatever, is obviously a variation of Bonapartism—a Bonapartism of a new type not before seen in history.

Caesarism arose upon the basis of a slave society shaken by inward strife.
Bonapartism is one of the political weapons of the capitalist regime in its
critical period. Stalinism is a variety of the same system, but upon the basis
of a workers' State torn by the antagonism between an organized and armed
Soviet aristocracy and the unarmed toiling masses.

As history testifies, Bonapartism gets along admirably with a universal,
and even a secret, ballot. The democratic ritual of Bonapartism is the
plebiscite. From time to time the question is presented to the citizens: *for*
or *against* the leader? And the voter feels the barrel of a revolver between
his shoulders. Since the time of Napoleon III, who now seems a provincial
dilettante, this technique has received an extraordinary development. The
new Soviet constitution which established *Bonapartism on a plebiscite basis*
is the veritable crown of the system.

In the last analysis, Soviet Bonapartism owes its birth to the belatedness
of the world revolution. But in the capitalist countries the same cause gave
rise to Fascism. We thus arrive at the conclusion, unexpected at first glance,
but in reality inevitable, that the crushing of Soviet democracy by an all-
powerful bureaucracy and the extermination of bourgeois democracy by
Fascism were produced by one and the same cause: the dilatoriness of the

world proletariat in solving the problems set for it by history. Stalinism and Fascism, in spite of a deep difference in social foundations, are symmetrical phenomena. In many of their features they show a deadly similarity. A victorious revolutionary movement in Europe would immediately shake not only Fascism, but Soviet Bonapartism. In turning its back to the international revolution, the Stalinist bureaucracy was, from its own point of view, right. It was merely obeying the voice of self-preservation.

4. The Socialist Society of the Future

The personal dreams of a few enthusiasts today for making life more dramatic and for educating man himself rhythmically find a proper and real place in this outlook. Having rationalized his economic system, that is, having saturated it with consciousness and planfulness, man will not leave a trace of the present stagnant and worm-eaten domestic life. The care for food and education, which lies like a millstone on the present-day family, will be removed, and will become the subject of social initiative and of an endless collective creativeness. Woman will at last free herself from her semi-servile condition. Side by side with technique, education, in the broad sense of the psycho-physical moulding of new generations, will take its place as the crown of social thinking. Powerful 'parties' will form themselves around pedagogic systems. Experiments in social education and an emulation of different methods will take place to a degree which has not been dreamed of before. Communist life will not be formed blindly, like coral islands, but will be built consciously, will be tested by thought, will be directed and corrected. Life will cease to be elemental, and for this reason stagnant. Man, who will learn how to move rivers and mountains, how to build peoples' palaces on the peaks of Mont Blanc and at the bottom of the Atlantic, will not only be able to add to his own life richness, brilliancy, and intensity, but also a dynamic quality of the highest degree. The shell of life will hardly have time to form before it will be burst open again under the pressure of new technical and cultural inventions and achievements. Life in the future will not be monotonous.

More than that. Man at last will begin to harmonize himself in earnest. He will make it his business to achieve beauty by giving the movement of his own limbs the utmost precision, purposefulness, and economy in his work, his walk, and his play. He will try to master first the semi-conscious and then the subconscious processes in his own organism, such as breathing, the circulation of the blood, digestion, reproduction, and, within necessary limits, he will try to subordinate them to the control of reason and will. Even purely physiologic life will become subject to collective experiments.

The human species, the coagulated *Homo sapiens*, will once more enter into a state of radical transformation, and, in his own hands, will become an object of the most complicated methods of artificial selection and psycho-physical training. This is entirely in accord with evolution. Man first drove the dark elements out of industry and ideology, by displacing barbarian routine by scientific technique, and religion by science. Afterwards he drove the unconscious out of politics, by overthrowing monarchy and class with democracy and rationalist parliamentarianism and then with the clear and open Soviet dictatorship. The blind elements have settled most heavily in economic relations, but man is driving them out from there also, by means of the Socialist organization of economic life. This makes it possible to reconstruct fundamentally the traditional family life. Finally, the nature of man himself is hidden in the deepest and darkest corner of the unconscious, of the elemental, of the subsoil. Is it not self-evident that the greatest efforts of investigative thought and of creative initiative will be in that direction? The human race will not have ceased to crawl on all fours before God, kings, and capital, in order later to submit humbly before the dark laws of heredity and a blind sexual selection! Emancipated man will want to attain a greater equilibrium in the work of his organs and a more proportional developing and wearing out of his tissues, in order to reduce the fear of death to a rational reaction of the organism towards danger. There can be no doubt that man's extreme anatomical and physiological disharmony, that is, the extreme disproportion in the growth and wearing out of organs and tissues, give the life instinct the form of a pinched, morbid, and hysterical fear of death, which darkens reason and which feeds the stupid and humiliating fantasies about life after death.

Man will make it his purpose to master his own feelings, to raise his instincts to the heights of consciousness, to make them transparent, to extend the wires of his will into hidden recesses, and therby to raise himself to a new plane, to create a higher social biologic type, or, if you please, a superman.

It is difficult to predict the extent of self-government which the man of the future may reach or the heights to which he may carry his technique. Social construction and psycho-physical self-education will become two aspects of one and the same process. All the arts—literature, drama, paint-ing, music, and architecture—will lend this process beautiful form. More correctly, the shell in which the cultural construction and self education of Communist man will be enclosed will develop all the vital elements of contemporary art to the highest point. Man will become immeasurably stronger, wiser, and subtler; his body will become more harmonized, his movements more rhythmic, his voice more musical. The forms of life will become dynamically dramatic. The average human type will rise to the heights of an Aristotle, a Goethe, or a Marx. And above this ridge new peaks will rise.

Sources: L. Trotsky, *Results and Prospects and The Permanent Revolution* (New York, 1969), pp. 75 ff., 105, 114 ff., 130 ff., 276 ff. L. Trotsky, *The Revolution Betrayed* (London, 1937), pp. 88 ff., 103 ff., 235 ff., 262 ff. L. Trotsky, *Literature and Revolution* (Ann Arbor, 1960), pp. 253 ff.

FURTHER NOTES

WORKS

TROTSKY, LEON, *The Age of Permanent Revolution: A Trotsky Anthology*, ed. Isaac Deutscher (New York, 1964).
—— *Basic Writings*, ed. Irving Howe (London, 1964).
—— *The Essential Trotsky* (London, 1963).

COMMENTARIES

BEILHARZ, P., *Trotsky, Trotskyism and the Transition to Socialism* (London, 1987).
DAY, RICHARD B., *Leon Trotsky and the Politics of Economic Isolation* (Cambridge, 1973).
DEUTSCHER, ISAAC, *The Prophet Armed: Trotsky 1879–1921* (Oxford, 1954).
HANSON, JOSEPH, et al., *Leon Trotsky: The Man and His Work* (New York, 1969).
HOWE, IRVING, *Trotsky* (London and New York, 1978).
KNEI-PAZ, BARUCH, *The Social and Political Thought of Leon Trotsky* (Oxford, 1978).
KRASSO, NICHOLAS (ed.), *Trotsky: The Great Debate Renewed* (St Louis, 1972).
SMITH, IRVING H. (ed.), *Trotsky* (Englewood Cliffs, NJ, 1973).

9

NIKOLAI BUKHARIN

Theoretically there are two possible alternatives: either the workers' movement, in the same way as all the organizations of the bourgeoisie, becomes an organization of the universal State and concerts itself into a mere appendage of the State apparatus, or it outgrows the limitations of the State and blows it up from the inside, organizing its own State power (dictatorship).

Quoted in N. Harding, *Lenin's Political Thought*, vol. ii (London, 1981), p. 109.

NIKOLAI BUKHARIN (1888–1938) *was one of the most intellectually gifted of the Bolshevik leaders. Elected to the party's Central Committee just before the October 1917 revolution, he remained a member until 1934. Before 1917, Bukharin wrote two works which strongly influenced Lenin: the first, on imperialism, maintained that competition between capitalist states was more important than internal competition; the second argued that Marxists should be hostile in principle to any conception of the State. (For Lenin's assimilation of these ideas, see the extracts above.) The first of the extracts below is taken from Bukharin's* Economics of the Transformation Period (1920). *With its admission that proletarian revolution implied an initial disintegration of the productive forces and its insistence at this stage on the all-encompassing coercive role of the State, this reflects the heroic and desperate period of War Communism. With the introduction of the New Economic Policy in 1921, Bukharin moderated his views: in the extracts below from his 1924 primer on* Historical Materialism, *he explains Marxist dialectics in terms of the establishment and breakdown of equilibrium and discusses the relation between class, State, and future communist society. As an advocate of a gradualist economic strategy in Russia, Bukharin became one of the main opponents of Stalin and was executed in 1938. His ideas have, however, recently been revived, particularly among Western Communist parties.*

1. The Communist Revolutionary State

THE Communist revolution of the proletariat is accompanied, as every revolution, by a *reduction* in productive powers. The civil war, still in the powerful dimensions of modern class wars, since not only the bourgeoisie, but also the proletariat, is organized as state power, signifies a net minus economically speaking and from the point of view of the subsequent *reproduction of cycles*. Nevertheless, we have already seen in the example of crises and of capitalist wars that a consideration from this point of view is a limited consideration; one must examine the roles of the relevant phenomenon, starting with *successive cycles of reproduction in its broad historical scale*. Then the costs of the revolution and the civil war appear as a temporary reduction of productive powers, through which, however, the basis for their powerful development is given by the *restructuring of production relations according to a new basic design*.

The restructuring of production relations has as presupposition: the 'power of the proletariat', its 'commando' in the State apparatus as well as in the army as part of this apparatus, and in production.

In the process of the struggle for power and of the civil war, in the period of proletarian dictatorship, the curve of productive powers sinks further and further, *at the same time that the forms of organization grow*. This growth of organizational forms occurs under the resistance (above all, so-called sabotage) of the 'officers of industry', i.e. of the technicians who do not want to be in a *different* hierarchical system in which they previously were. *But the resistance of this class is much less dangerous for the growing new system than the resistance of the working class for the system of capitalist relationships.* From the point of view of the preservation and development of human society, therefore, only Socialist production relations can signify the way out, for only these *are capable of creating the conditions for a relatively flexible equilibrium of the social system of production*....

We want to dwell on the general question of the structure of the administrative-economic and administrative-technological apparatus of State power of the proletariat. Under State capitalist production relations, all organizations of the bourgeoisie (syndicates, trusts, cartels, etc.) are subjected to the State power and fused with it. With the destruction of bourgeois dictatorship and the formation of proletarian dictatorship, these administrative apparatus are also destroyed. The organizations of trusts, the State organs of regulation of the old society, etc. decay. As a rule (theoretically we have proven it in the last chapter) they cannot be taken as 'entire apparatus'. But this does not mean that they have not fulfilled their historical role, for the entire, most highly complicated total series of these sometimes very subtle organizations, which encompass all of social economic life with their feelers, played, taken objectively, the role of the screw, which increased and accelerated the process of centralization of the means of production and of the proletariat. On the other side, upon the decay of these apparatus, their materially technological, objective skeleton remains. And, looking at the question on a general scale, just as the proletariat takes first of all the centralized means of production, i.e. the materially technological bone-and-muscle system of capitalist production, which mainly finds its expression in a system of machines and as Marx said in a 'vascular system' of apparatus, likewise does the proletariat seize not the personal element but the *essential element* of the old administrative system (buildings, chancelleries, offices, typewriters, the entire inventory in general, books by which one can orientate himself more easily, and finally all possible materially symbolical devices, such as diagrams, models, etc.). After it has seized these along with the other 'centralized means of production', it builds its apparatus, the base of which is made up of the *workers' organizations*.

The working class has at its disposal the following organizations: the *councils of workers delegates*, which transform themselves from a tool for the takeover of power into tools of the government; the *party of Communist overthrow*, the *spiritus rector* of proletarian action; the *trade-union associations*, which transform themselves from tools of the struggle against the employers into organs of the administration of production; the *partnerships*, which are transformed from means of the struggle against commercial transaction into an organization of the State apparatus for the purpose of general distribution; the *factory committees* and similar organizations ('factory councils' in Germany, 'workers' committees' and 'shop stewards' committees' in England), which transform themselves from local organs of the workers' struggle against the employers into elements of the general administration of production.

The network of these, as well as of *totally new* organizations *specially created* on their base, constitutes the organizational ground floor of the new apparatus.

Under the given conditions we have before us above all a dialectical *change of functions* of the workers' organizations. It is completely clear that it cannot be otherwise with the *transposition* of relations of domination, because the working class, which has seized State power, must inevitably become the power which appears as *organizer of production*.

Now we must raise the question as to the general principle of the system of organization of the proletarian apparatus, i.e. as to the interchanging relationships between different forms of the proletarian organizations. It is clear that the same method is formally necessary for the working class as for the bourgeoisie at the time of State capitalism. This organizational method exists in the co-ordination of all proletarian organizations with one all-encompassing organization, i.e. with the State organization of the working class, with the *soviet State of the proletariat*. The 'nationalization' of the trade unions and the effectual nationalization of all mass organizations of the proletariat result from the internal logic of the process of transformation itself. The minutest cells of the labour apparatus must transform themselves into agents of the general process of organization, which is systematically directed and led by the collective reason of the working class, which finds its material embodiment in the highest and most all-encompassing organization, in its State apparatus. Thus the system of State capitalism dialectically transforms itself into its own inversion, into the State form of workers' socialism.

No new structure can be born before it has become an objective necessity. Capitalist development and the collapse of capitalism have led society to a dead end, have halted the process of production, the base of existence of society. The renewal of the production process will only be possible under the domination of the proletariat, and therefore its dictatorship is an objective necessity.

A stability of the emerging new society can only be achieved with the greatest possible union, contact, and working out in common of all organizational powers. And therefore that general form of labour apparatus, of which we spoke above, is so necessary. Out of the bloody passion of war, out of the chaos and the ruins, out of misery and destruction, arises the structure of the new, harmonic society....

It is obvious that this element of compulsion, which is here the self-compulsion of the working class, grows from the crystallized centre towards the significantly more amorphous and dispersed periphery. *This is the conscious power of cohesion of the little parts of the working class, which power represents for some categories, subjectively, an external pressure, which constitutes for the entire working class, objectively, its accelerated self-organization.*

In *Communist* society there will be an absolute freedom of 'personality', and any kind of external regulation of relationships between people will be absent; that is, self-activity without force will exist. In *capitalist* society there was no self-activity for the working class and only force *from the side of the enemy class*. In the *transition period* the self-activity of the working class is present along with the force which the working class, as a class for itself, introduces for all its parts. The contradiction between force and self-activity expresses the self-contradictory character of the transition period itself, since the proletariat has already abandoned the framework of capitalist compulsion, but has not *yet* become a member of Communist society.

One of the main forms of compulsion of the new kind which operates in the sphere of the working class itself is the abolition of the so-called freedom to work. 'Freedom to work' in capitalist society means one of the numerous fictions of this society, for in reality the monopolizing of the means of production by the capitalists *forces* the workers to sell their labour power. This 'freedom' was based on the following: first, on a relative possibility of *choosing* one's master (moving from one factory to another), the possibility 'of leaving' and of being 'dismissed'; second, this 'freedom' meant *competition among the workers themselves*. In this last sense, 'freedom to work' was already partly conquered in the period of capitalism by the *workers' organizations*, when unions partly eliminated competition between workers by uniting them, organizing the splintered parts of the class, joining them, and making them stronger in their struggle against the capitalist class. The unions made the demand that only members of the unions be admitted to the factory; they declared a boycott (i.e. used force) against strike-breakers, this living incarnation of the bourgeois 'freedom to work', etc. Under the dictatorship of the proletariat, the question of the 'master' falls aside, since the 'expropriators' are 'expropriated'. On the other hand, the remnants of disorganization, unsolidarity, individualism, guild restrictions, the depravity of capitalist society in not recognizing the *universal proletarian tasks*, which receive their concentrated expression in the tasks and demands of the soviet dictatorship, of the workers' state. Since these tasks must be mastered at any price, it is understandable that from the point of view of the proletariat, in the very name of actual and non-fictitious freedom of the working class, an abolition of the so-called 'freedom to work' is required. For the latter no longer agrees with the regularly organized 'planned economy' and a corresponding distribution of labour powers. Consequently, the regime of compulsory labour and State distribution of labour powers in the dictatorship of the proletariat already expresses a relatively high degree of organization of the entire apparatus and the stability of proletarian power on the whole.

In the capitalist regime, compulsion was defended in the name of the 'interests of the totality', while it was in reality in the interests of capitalist groups. Under proletarian dictatorship, compulsion is for the first time really the tool of the majority in the interest of this majority.

The proletariat as class is the only class to which prejudices of possession are on the whole foreign. But the proletariat must operate side by side with the often very large *peasantry*. If the large farmers actively struggle against the measures of the proletarian dictatorship, then the 'concentrated force' of the proletariat must more or less offer resistance to the Vendée of the kulaks. But the masses of middle and sometimes even poor peasantry vacillate constantly, now led by the hate of capitalist exploitation by the large-estate owners, a hate which drives them into the arms of Communism, and now motivated by the feeling of being owner (and consequently in the period of famine also that of being a *black-market dealer*), which drives them into the army of the reaction. The latter is expressed in the resistance to the State monopoly of grain and in the struggle for free trade which is speculation, as well as for speculation which is free trade—in resistance to the system of compulsory labour and in general to any forms of State restraint of the economic anarchy. These stimuli appear especially when the exhausted cities can offer no equivalent in the first period for the grain and for the sacrifices to the 'common pot'. Therefore, compulsion constitutes an absolute and peremptory necessity.

Therefore: in relation to the previous *bourgeois groups*, compulsion from the side of the proletarian dictatorship is a compulsion from the side of another class which wages a class struggle against the object of its compulsion; in relation to the non-exploitative mass of farmers, compulsion from the side of the proletariat is a class struggle in so far as the farmer is owner and speculator; compulsion signifies union of the peasantry with the labour organization, education and attracting to the building of Communism in so far as the farmer is a working man and non-exploiter, an enemy of capitalism; finally, in relation to the *proletariat* itself, compulsion is a method of organization which is introduced by the working class itself, i.e. a method of forced, accelerated *self-organization*.

From a broader point of view, i.e. from the point of view of a historical scale of greater scope, proletarian compulsion in all its forms, from executions to compulsory labour, constitutes, as paradoxical as this may sound, a method of the formation of a new Communist humanity from the human material of the capitalist epoch. In actuality, the epoch of proletarian dictatorship is at the same time an epoch of *class deformation*. Capitalism was followed by a more or less progressive social splintering of the society: it disintegrated the peasantry, annihilated the 'middle class', and drove class contradictions to their extreme. The dictatorship of the proletariat, which in the first period expresses the crassest division of the capitalist world, *begins*, after the commencement of a certain equilibrium, *to re-collect humanity*. The previous bourgeoisie, which has now been conquered, struck, held down, and impoverished, learns physical labour, undergoes mental changes, and is re-educated. A part of the bourgeoisie is destroyed in the civil war, but that part which survives already represents another social category. Likewise, the intellectuals. The peasantry, which in the general stream maintains more position than the others, is nevertheless drawn into the universal channel, and slowly but surely experiences a transformation. The proletariat itself likewise changes its 'own nature'. In this way specific class distinctions are blurred; the classes begin to decay as classes and to become like the proletariat. A period of class *deformation* begins. The lever of this deformation is the proletarian dictatorship. As concentrated application of force, the dictatorship finally abolishes any kind of force. As the highest expression of the class, it abolishes all classes. As regime of the class which has organized itself as State power, it prepares the extinction of every state. By carrying on the struggle for its existence, it annihilates its own existence. In classless, State-less, Communist society, where in place of discipline from without, the simple joy of working on the part of the normal social human being will have appeared, the external norms of human behaviour will lose all meaning. Compulsion, no matter what its form, will disappear once and for all.

2. Dialectics and Revolution

The basis of all things is therefore the law of change, the law of constant motion. Two philosophers particularly (the ancient Heraclitus and the modern Hegel, as we have already seen) formulated this law of change, but they did not stop there. They also set up the question of the manner in which the process operates. The answer they discovered was that changes are produced by constant internal contradictions, internal struggle. Thus, Heraclitus declared: 'Conflict is the mother of all happenings,' while Hegel said: 'Contradiction is the power that moves things.'

There is no doubt of the correctness of this law. A moment's thought will convince the reader. For, if there were no conflict, no clash of forces, the world would be in a condition of unchanging, stable equilibrium, i.e., complete and absolute permanence, a state of rest precluding all motion. Such a state of rest would be conceivable only in a system whose component parts and forces would be so related as not to permit of the introduction of any conflicts, as to preclude all mutual interaction, all disturbances. As we already know that all things change, all things are 'in flux', it is certain that such an absolute state of rest cannot possibly exist. We must therefore reject a condition in which there is no 'contradiction between opposing and colliding forces', no disturbance of equilibrium, but only an absolute immutability. Let us take up this matter somewhat more in detail.

In biology, when we speak of adaptation, we mean that process by which one thing assumes a relation toward another thing that enables the two to exist simultaneously. An animal that is 'adapted' to its environment is an animal that has achieved the means of living in that environment. It is suited to its surroundings, its qualities are such as to enable it to continue to live. The mole is 'adapted' to conditions prevailing under the earth's surface; the fish, to conditions in the water; either animal transferred to the other's environment will perish at once.

A similar phenomenon may be observed also in so called 'inanimate' nature: the earth does not fall into the sun, but revolves around it 'without mishap'. The relation between the solar system and the universe which surrounds it, enabling both to exist side by side, is a similar relation. In the latter case we commonly speak, not of the adaptation, but of the equilibrium between bodies, or systems of such bodies, etc. We may observe the same state of things in society. Whether we like it or not, society lives within nature: is therefore in one way or another in equilibrium with nature. And the various parts of society, if the latter is capable of surviving, are so adapted to each other as to enable them to exist side by side: capitalism, which included both capitalists and workers, had a very long existence!

In all these examples it is clear that we are dealing with one phenomenon, that of *equilibrium*. This being the case, where do the contradictions come in? For there is no doubt that conflict is a *disturbance* of equilibrium. It must be recalled that such equilibrium as we observe in nature and in society is *not* an absolute, unchanging equilibrium, but an equilibrium *in flux*, which means that the equilibrium may be established and destroyed, may be re-established on a new basis, and again disturbed.…

In other words, the world consists of forces, acting in many ways, opposing each other. These forces are balanced for a moment in exceptional cases only. We then have a state of 'rest', i.e., their actual 'conflict' is concealed. But if we change only one of these forces, immediately the 'internal contradictions' will be revealed, equilibrium will be disturbed, and if a new equilibrium is again established, it will be on a new basis, i.e., with a new combination of forces, etc. It follows that the 'conflict', the 'contradiction', i.e., antagonism of forces acting in various directions, determines the motion of the system.

On the other hand, we have here also the *form* of this process: in the first place, the condition of equilibrium; in the second place, a disturbance of this equilibrium; in the third place, the re-establishment of equilibrium on a *new* basis. And then the story begins all over again: the new equilibrium is the point of departure for a new disturbance, which in turn is followed by another state of equilibrium, etc., *ad infinitum*. Taken all together, we are dealing with a process of motion based on the development of internal contradictions.

Hegel observed this characteristic of motion and expressed it in the following manner: he called the original condition of equilibrium the *thesis*, the disturbance of equilibrium the *antithesis*, the re-establishment of equilibrium on a new basis the *synthesis* (the unifying proposition reconciling the contradictions). The characteristic of motion present in all things, expressing itself in this tripartite formula (or triad), he called *dialectic....*

Any object, a stone, a living thing, a human society, etc., may be considered as a whole consisting of parts (elements) related with each other; in other words, this whole may be regarded as a *system*. And no such system exists in empty space; it is surrounded by other natural objects, which, with reference to it, may be called the *environment*. For the tree in the forest, the environment means all the other trees, the brook, the earth, the ferns, the grass, the bushes, together with all their properties. Man's environment is society, in the midst of which he lives; the environment of human society is external nature. There is a constant relation between environment and system, and the latter, in turn, acts upon the environment. We must first of all investigate the fundamental question as to the nature of the relations between the environment and the system; how are they to be defined; what are their forms; what is their significance for their system. Three chief types of such relations may be distinguished.

1. *Stable equilibrium.* This is present when the mutual action of the environment and the system results in an unaltered condition, or in a disturbance of the first condition which is again re-established in the original state. For example, let us consider a certain type of animals living in the steppes. The environment remains unchanged. The quantity of food available for this type of beast neither increases nor decreases; the number of animals preying upon them also remains the same; all the diseases, all the microbes (for all must be included in the 'environment'), continue to exist in the original proportions. What will be the result? viewed as a whole, the number of our animals will remain the same; some of them will die or be destroyed by beasts of prey, others will be born, but the given type and the given conditions of the environment will remain the same as they were before. This means a condition of rest due to an unchanged *relation* between the system (the given type of animals) and the environment, which is equivalent to stable equilibrium. Stable equilibrium is not always a complete absence of motion; there may be motion, but the resulting disturbance is followed by a re-establishment of equilibrium on the former basis. The contradiction between the environment and the system is constantly being reproduced *in the same quantitative relation*.

We shall find the case the same in a society of the stagnant type (we shall go into this question more in detail later). If the relation between society and nature remains the same; i.e., if society extracts from nature, by the process of production, precisely as much energy as it consumes, the contradiction between society and nature will again be reproduced in the former shape; the society will mark time, and there results a state of stable equilibrium.

2. *Unstable equilibrium with positive (favourable) indication (and expanding system)*. In actual fact, however, stable equilibrium does not exist. It constitutes merely an imaginary, sometimes termed the 'ideal', case. As a matter of fact, the relation between environment and the system is never reproduced in precisely the same proportions; the disturbance of equilibrium never actually leads to its re-establishment on exactly the same basis as before, but a new equilibrium is created on a new basis. For example, in the case of the animals mentioned above, let us assume that the number of beasts of prey opposing them decreases for some reason, while the available food increases. There is no doubt that the number of our animals would then also increase; our 'system' will then grow; a new equilibrium is established on a better basis; this means *growth*. In other words, the contradiction between the environment and the system has become quantitatively different.

If we consider human society, instead of these animals, and assume that the relation between it and nature is altered in such manner that society—by means of production—extracts more energy from nature than is consumed by society (either the soil becomes more fruitful, or new tools are devised, or both), this society will *grow* and not merely mark time. The new equilibrium will in each case be actually new. The contradiction between society and nature will in each case be reproduced on a new and 'higher' basis, a basis on which society will increase and develop. This is a case of unstable equilibrium with positive indication.

3. *Unstable equilibrium with negative indication (a declining system)*. Now let us consider the quite different case of a new equilibrium being established on a 'lower' basis. Let us suppose, for example, that the quantity of food available to our beasts has decreased, or that the number of beasts of prey has for some reason increased. Our animals will die out. The equilibrium between the system and the environment will in each case be established on the basis of the extinction of a portion of this system. The contradiction will be re-established on a new basis, with a negative indication. Or, in the case of society, let us assume that the relation between it and nature has been altered in such manner that society is obliged to consume more and more and obtain less and less (the soil is exhausted, technical methods become poorer, etc.). New equilibrium will here be

established in each case on a lowered basis, by reason of the destruction of a portion of society. We are now dealing with a declining society, a disappearing system, in other words, with motion having a negative indication.

Every conceivable case will fall under one of these three heads. At the basis of the motion, as we have seen, there is in fact the contradiction between the environment and the system, which is constantly being re-established.

But the matter has another phase also. Thus far we have spoken only of the contradictions between the environment and the system, i.e., the *external* contradictions. But there are also internal contradictions, those that are within the system. Each system consists of its component parts (elements), united with each other in one way or another. Human society consists of people; the forests, of trees and bushes; the pile of stones, of the various stones; the herd of animals, of the individual animals, etc. Between them there are a number of contradictions, differences, imperfect adaptations, etc. In other words, here also there is no absolute equilibrium. If there can be, strictly speaking, no absolute equilibrium between the environment and the system, there can also be no such equilibrium between the elements of the system itself.

This may be seen best by the example of the most complicated system, namely, human society. Here we encounter an endless number of contradictions; we find the struggle between classes, which is the sharpest expression of 'social contradictions', and we know that 'the struggle between classes is the motive force of history'. The contradictions between the classes, between groups, between ideals, between the quantity of labour performed by individuals and the quantity of goods distributed to them, the planlessness in production (the capitalist 'anarchy' in production), all these constitute an endless chain of contradictions, all of which are within the system and grow out of its contradictory *structure* ('structural contradictions'). But these contradictions do not of themselves destroy society. They *may* destroy it (if, for example, both opposing classes in a civil war destroy each other), but it is also possible they may at times not destroy it.

In the latter case, there will be an unstable equilibrium between the various elements of society. We shall later discuss the nature of this equilibrium; for the present we need not go into it. But we must not regard society stupidly, as do so many bourgeois scholars, who overlook its internal contradictions. On the contrary, a scientific consideration of society requires that we consider it from the point of view of the contradictions present within it. *Historical 'growth' is the development of contradictions.*

We must again point out a fact with which we shall have to deal more than once in this book. We have said that these contradictions are of two kinds: between the environment and this system, and between the elements of the system and the system itself. Is there any relation between these two phenomena? A moment's thought will show us that such a relation exists.

It is quite clear that the internal structure of the system (its internal equilibrium) must change together with the relation existing between the system and its environment. The latter relation is the decisive factor; for the entire situation of the system, the fundamental forms of its motion (decline, prosperity, or stagnation) are determined by this relation only.

Let us consider the question in the following form: we have seen above that the character of the equilibrium between society and nature determines the fundamental course of the motion of society. Under these circumstances, could the internal structure continue for long to develop in the opposite direction? Of course not. In the case of a growing society, it would not be possible for the internal structure of society to continue *constantly* to grow worse. If, *in a condition of growth*, the structure of society should become poorer, i.e., its internal disorders grow worse, this would be equivalent to the appearance of a new contradiction: a contradiction between the external and the internal equilibrium, which would require the society, if it is to continue growing, to undertake a reconstruction, i.e., its internal structure must adapt itself to the character of the external equilibrium. Consequently, *the internal (structural) equilibrium is a quantity which depends on the external equilibrium (is a 'function' of this external equilibrium)*.

We have now to consider the final phase of the dialectic method, namely, the theory of sudden changes. No doubt it is a widespread notion that 'nature makes no sudden jumps' (*natura non facit saltus*). This wise saying is often applied in order to demonstrate 'irrefutably' the impossibility of revolution, although revolutions have a habit of occurring in spite of the moderation of our friends the professors. Now, is nature really so moderate and considerate as they pretend?

In his *Science of Logic (Wissenschaft der Logik)*, Hegel says:

It is said that there are no sudden changes in nature, and the common view has it (*meint*) that when we speak of a growth or a destruction (*Entstehen oder Vergehen*), we always imagine a *gradual* growth (*Hervorgehen*) or disappearance (*Verschwinden*). Yet we have seen cases in which the alteration of existence (*des Seins*) involves not only a transition from one proportion to another, but also a transition, by a sudden leap, into a *quantitatively*, and, on the other hand, also *qualitatively* different thing (*Anderswerden*); an interruption of the gradual process (*ein*

Abbrechen des Allmählichen), differing qualitatively from the preceding, the former, state. [the italics are mine.—N.B.]

Hegel speaks of a transition of quantity into quality; there is a very simple illustration of such a transition. If we should heat water, we should find that throughout the process of heating, before a temperature of 100°C. (212°F.) is reached, the water will not boil and turn into steam. Portions of the water will move faster and faster, but they will not bubble on the surface in the form of steam. The change thus far is merely *quantitative*; the water moves faster, the temperature rises, but the water remains water, having all the properties of water. Its quantity is changing gradually; its quality remains the same. But when we have heated it to 100°C., we have brought it to the 'boiling-point'. At once it begins to boil, at once the particles that have been madly in motion burst apart and leap from the surface in the form of little explosions of *steam*. The water has ceased to be water; it becomes *steam*, a gas. The former quality is lost; we now have a new quality, with new properties. We have thus learned two important peculiarities in the process of change.

In the first place, having reached a certain stage in motion, the quantitative changes call forth qualitative changes (or, in more abbreviated form, 'quantity becomes quality'); in the second place, this transition from quantity to quality is accomplished in a sudden leap, which constitutes an interruption in the gradual continuous process. The water was not constantly changing, with gradual deliberateness, into a little steam at a time, with the quantity of steam *constantly* increasing. For a long time it did not boil at all. But having reached the 'boiling-point', it began to boil. We must consider this a *sudden* change.

The transformation of quantity into quality is one of the fundamental laws in the motion of matter; it may be traced literally at every step both in nature and society. Hang a weight at the end of a string, and gradually add slight additional weights, each weight being as small as you like; up to a certain limit, the string 'will hold'. But once this limit has been exceeded, it will suddenly break. Force steam into a boiler; all will go well for a while; only the pressure indicator will show increases in the pressure of the steam against the walls of the boiler. But when the dial has exceeded a certain limit, the boiler will explode. The pressure of the steam exceeded—perhaps by a very little—the power of resistance offered by the walls of the boiler. Before this moment, the quantitative changes had not led to a 'cataclysm', to a *qualitative* change, but at that 'point' the boiler exploded.

Several men are unable to lift a stone. Another joins them; they are still unable to do it. A weak old woman joins them—and their united strength raises the stone. Here, but a slight additional force was needed, and as soon as this force was added the job was done. Let us take another example. Leo Tolstoi wrote a story called 'Three Rolls and a Cookie'. The point of the story is the following: a man, to appease his hunger, ate one roll after another, for each still left him hungry; in fact, after his third roll, he was still hungry; then he ate a little cookie, and his hunger was appeased. He then cursed his folly for not having eaten the cookie first: for then he would not have had to eat the rolls. Of course, we are aware of his mistake; we are dealing here with a *qualitative* change, the transition from the feeling of hunger to that of satiation, which transition was accomplished in one bound (after eating the cookie). But this qualitative difference *ensued after the quantitative differences*: the cookie would have been of no use without the rolls.

We thus find that it is foolish to deny the existence of sudden changes; and to admit only a deliberate gradual process. Sudden leaps are often found in nature, and the notion that nature permits of no such violent alterations is merely a reflection of the fear of such shifts in society, i.e., of the fear of revolution.

The denial of the contradictory character of evolution by bourgeois scholars is based on their fear of the class struggle and on their concealment of social contradictions. Their fear of sudden changes is based on their fear of revolution; all their wisdom is contained in the following reasoning: there are no violent changes in nature, there cannot be any such violent changes anywhere; therefore, you proletarians, do not dare make a revolution! Yet here it becomes exceptionally evident that bourgeois science is in contradiction with the most fundamental requirements of all science. Everybody knows that there have been many revolutions in human society. Will anyone deny that there was an English Revolution, or a French Revolution, or a Revolution of 1848, or the Revolution of 1917? If these violent changes have taken place in society, and are still taking place, science should not 'deny' them, refusing to recognize facts, but should *understand* these sudden shifts, and *explain* them.

Revolutions in society are of the same character as the violent changes in nature. They do not suddenly 'fall from the sky'. They are prepared by the entire preceding course of development, as the boiling of water is prepared by the preceding process of heating or as the explosion of a steam-boiler is prepared by the increasing pressure of the steam against its walls. A revolution in society means its reconstruction, 'a structual alteration of the system'. Such a revolution is an inevitable consequence of the contradictions between the structure of society and the demands for its development....

Considered as a whole, we find that the process of reproduction is a process of constant disturbance and re-establishment of equilibrium between society and nature.

Marx distinguishes between *simple* reproduction and reproduction *on an extending scale*.

Let us first consider the case of simple reproduction. We have seen that in the process of production, the means of production are used up (the raw material is worked over, various auxiliary substances are required, such as lubricating oil, rags, etc.; the machines themselves, and the buildings in which the work is done, as well as all kinds of instruments and their parts, wear out); on the other hand, labour power is also exhausted (when people work, they also deteriorate, their labour power is used up, and a certain expenditure must be incurred in order to re-establish this labour power). In order that the process of production may continue, it is necessary to reproduce in it and by means of it the substances that it consumes. For example, in textile production, cotton is consumed as a raw material, while the weaving machinery deteriorates. In order that production may continue, cotton must continue to be raised somewhere, and looms to be manufactured. At one point the cotton disappears by reason of its transformation into fabrics, at another point, fabrics disappear (workers, etc., use them) and cotton reappears. At one point, looms are being slowly wiped out, while at another they are being produced. In other words, the necessary elements of production required in one place must be produced somewhere else; there must be a constant replacement of everything needed in production; if this replacement proceeds smoothly and at the same rate as the disappearance, we have a case of simple reproduction, which corresponds to a situation in which the productive social labour remains uniform, with the productive forces unchanging, and society moving neither forward nor backward. It is clear that this is a case of stable equilibrium between society and nature. It involves constant disturbances of equilibrium (disappearance of products in consumption and deterioration) and a constant re-establishment of equilibrium (the products reappear); but this re-establishment is always on the old basis: just as much as produced as has been consumed; and again just as much is consumed as has been produced, etc., etc. The process of reproduction is here a dance to the same old tune.

But where the productive forces are increasing, the case is different. Here, as we have seen, a portion of the social labour is liberated and devoted to an extension of social production (new production branches; extension of old branches). This involves not only a replacement of the formerly existing elements of production, but also the insertion of new elements into the new cycle of production. Production here does not continue on the same path, moving in the same cycle all the time, but increases in scope. This is *production on an extending scale*, in which case equilibrium is always established on a new basis; simultaneously with a certain consumption proceeds a larger production; consumption consequently also increases, while production increases still further. Equilibrium results in each case in a wider basis; we are now dealing with *unstable equilibrium with positive indication*.

The third case, finally, is that of a decline in the productive forces. In this case, the process of reproduction falls asleep: smaller and smaller quantities are reproduced. A certain quantity is consumed, but reproduction involves a smaller quantity still; less is consumed, and still less is reproduced, etc. Here again, reproduction does not repeat the same old cycle in each case; its sphere grows narrower and narrower; society's condition of life becomes poorer and poorer. The equilibrium between society and nature is re-established on a level that goes lower and lower each time....

... The various situations of the classes must result in a difference in their interests, aspirations, struggles, even in their death struggles. It is interesting to observe the nature of the equilibrium existing in the *structure* of a society of classes. The fact that such a society, in which, in the words of an English statesman, there are in reality two 'nations' (classes), can exist at all, without danger of disintegrating at any moment, is of itself very striking.

Yet there is no doubt of the existence of class societies. In some way or other, a unifying bond has been attained in such societies, a sort of hoop holding together the staves of the barrel; this hoop is the State, an organization of all society, with its threads, retaining them all in the system of its tentacles. If we should ask how the State originates, we should not be satisfied with any answer attributing a supernatural origin to the State, nor with any declaration that the State stands beyond all classes; for the simple reason that classless persons do not exist in a class society. There would therefore be no material with which to construct an organization standing outside of all classes or above all classes, no matter how often this may be asserted by bourgeois scholars. The organization of the State is altogether an organization of the 'ruling class'.

It now becomes of interest to determine which is the ruling class, for we shall then understand which class is represented by the State power, which subjugates all the other classes by means of its strength, its force, its mental system, its widely ramified apparatus. The question is not difficult to answer. In capitalist society, we find the capitalist class dominant in production; it would be absurd to expect to find the proletariat permanently dominant in the State, for one of the fundamental conditions of equilibrium would now be lacking; either the proletariat would also seize control of production, or the bourgeoisie would seize the State power. The existence of a society with a specific economic structure also involves the adaptation of its State organization; in other words, the economic structure of society also determines its State and political structure. The State, furthermore, is a huge organization embracing an entire nation and ruling many millions of men. This organization needs a whole army of employees, officials, soldiers, officers, legislators, jurists, ministers, judges, generals, etc., etc., and embraces great layers of human beings, one superimposed on the other. This structure is a precise reflection of the conditions in production. In capitalist society, for example, the bourgeoisie is in control of production, and therefore also of the State. Following upon the manufacturer comes the factory superintendent himself, often a capitalist; the same is true of the ministers of a capitalist state, its politicians in high places. From these circles are recruited the generals for the army; the intermediate positions in production are filled by the technical specialist, the engineer, the technical mental worker; these mental workers occupy the posts of intermediate officials in the State apparatus; they often furnish the army officers. The lower employees, as well as the soldiers, are furnished by the working class. Of course, there are many fluctuations, but the structure of the State authority corresponds closely, on the whole, to the structure of society.

If we should assume, for a moment, that by a miracle the lower employees had raised themselves above the higher employees, our assumption would involve a loss of equilibrium in the whole of society, i.e., a revolution. But such a revolution also cannot take place unless corresponding alterations have already been accomplished in production. Here also it is apparent that the structure of the State apparatus itself reflects the economic structure, i.e., the same classes occupy relatively the same positions. . . .

. . . The peculiarity of the State is its centralized administration; therefore, the anarchists tell us, any centralized administration is a State authority. Therefore, even the most advanced Communist society, if it has a systematic economy, will also be a State. This reasoning is based entirely on the naïve bourgeois error: bourgeois science, instead of perceiving *social* relations, perceives relations between *things*, or *technical* relations. But it is obvious that the 'essence' of the State is not in the *thing* but in the *social* relation; not in the centralized administration as such, but in the *class envelope* of the centralized administration. As capital is not a thing (as is, for instance, a machine), but a social relation between workers and employers, a relation expressed by means of a thing, so centralization *per se* by no means necessarily signifies a State organization; it does not become a State organization until it expresses a class relation.

Here we encounter a question that has been but little discussed in Marxian literature. We have seen that the class rules through the party, the party through its leaders; each class and each party therefore having its staff of officers. This staff is technically necessary, for we have seen that it is the result of the lack of uniformity within the class and the inequality of the party members. Each class therefore has its organizers. Viewing the evolution of society from this point of view, we may reasonably ask the following question: Is—in general—the communist classless society, of which Marxists speak, a possibility?

It is. We know that the classes themselves have risen organically as Engels described, from the division of labour, from the organizational functions that had become technically necessary for the further evolution of society. Obviously, in the society of the future, such organizational work will also be necessary. One might object that the society of the future will not involve private property, or the formation of such private property, and it is precisely this private property that constitutes this basis of the class.

But this argument need not remain unanswered. Professor Robert Michels, in his very interesting book, *Zur Soziologie des Parteiwesens in der modernen Demokràtie* (Leipzig, 1910, p. 370), says: 'Doubts again arise on this point, however, whose consistent application would lead to an outright denial of the possibility of a classless State [the author should not have said 'State' but 'society'.—N.B.]. Their administration of boundless capital [i.e., means of production.—N.B.] assigns at least as much power to the administrators as would possession of their own private property.' Viewed from this point of view, the entire evolution of society seems to be nothing more than a substitution of one group of leaders for another. Accordingly, Vilfredo Pareto speaks of a 'theory of the circulation of élites' (*théorie de la circulation des élites*). If this view is a correct one, Michels must also be correct in his conclusion, i.e., *Socialists* may be victorious, but not *Socialism*. An example will show Michels's error. When the bourgeoisie is in power, it is by reason of the power—as we know—not of all the members of the class, but of its leaders. Yet it is evident that this condition does not result in a class stratification *within* the bourgeoisie. The landlords in Russia ruled their high officials, constituting an entire staff, an entire stratum, but this stratum did not set itself up as a class against the other landlords. The reason was that these other landlords did not have a lower standard of living than that of the former; furthermore, their cultural level was about the same, on the whole, and the rulers were constantly recruited from this class.

Engels was therefore right when he said that the classes up to a certain moment are an outgrowth of the insufficient evolution of the productive forces; administration is necessary, but there is not sufficient bread for all, so to speak. Parallel with the growth of the socially necessary organizational functions, we therefore have also a growth of private property. But Communist society is a society with highly developed, increased productive forces. Consequently, it can have no economic basis for the creation of its peculiar ruling class. For—even assuming the power of the administrators to be stable, as does Michels—this power will be the power of specialists over machines, not over men. How could they, in fact, realize this power with regard to men? Michels neglects the fundamental decisive fact that each administratively dominant position has hitherto been an envelope for economic exploitation. This economic exploitation may not be subdivided. But there will not even exist a stable, close corporation, dominating the machines, for the fundamental basis for the formation of monopoly groups will disappear; what constitutes an eternal category in Michels's presentation, namely, the 'incompetence of the masses', will disappear, for this incompetence is by no means a necessary attribute of every system; it likewise is a product of the economic and technical conditions, expressing themselves in the general cultural being and in the educational conditions. We may state that in the society of the future there will be a colossal overproduction of organizers, which will nullify the *stability* of the ruling groups.

But the question of the *transition period* from capitalism to Socialism, i.e., the period of the proletarian dictatorship, is far more difficult. The working class achieves victory, although it is not and cannot be a unified mass. It attains victory while the productive forces are going down and the great masses are materially insecure. There will inevitably result a *tendency* to 'degeneration', i.e., the excretion of a leading stratum in the form of a class-germ. This tendency will be retarded by two opposing tendencies; first, by the *growth of the productive forces*; second, by the abolition of the *educational monopoly*. The increasing reproduction of technologists and of organizers in general, out of the working class itself, will undermine this possible new class alignment. The outcome of the struggle will depend on which tendencies turn out to be the stronger.

The working class, having in its possession so fine an instrument as the Marxian theory, must be mindful of this fact: by its hands an order of society will be put through and ultimately established, differing in principle from all the preceding formations; namely, from the primitive communist horde by the fact that it will be a society of highly cultivated persons, conscious of themselves and others; and from the class forms of society by the fact that for the first time the conditions for a human existence will be realized, not only for individual groups, but for the entire aggregate of humanity, a mass which will have ceased to be a mass, and will become a single, harmoniously constructed human society.

Sources: N. Bukharin, *Economics of the Transformation Period* (New York, 1971), pp. 58 f., 77 ff., 156 ff. N. Bukharin, *Historical Materialism* (New York, 1924), pp. 72 ff., 118 f., 151 f., 303 ff.

FURTHER NOTES

WORKS

BUKHARIN, N. *Imperialism and the World Economy* (New York, 1973).
—— *Economics of the Transformation Period* (New York, 1971).
—— *Historical Materialism: A System of Sociology* (New York, 1924).
—— and E. PREOBRAZHENSKY, *ABC of Communism* (Baltimore, 1969).

COMMENTARIES

COATES, K., *The Case of Nicolai Bukharin* (Nottingham, 1978).
COHEN, S., *Bukharin and the Bolshevik Revolution* (New York, 1974).
HARDING, N., *Lenin's Political Thought* (New York, 1981), esp. chs. 3 and 5.
HEITMAN, S., 'Between Lenin and Stalin: Nicolai Bukharin', in L. Labedz (ed.), *Revisionism* (New York, 1972).
LEWIN, M., *Political Undercurrents in Soviet Economic Debates: From Bukharin to Modern Reformers* (Princeton, 1975).

10

GEORG LUKACS

It is not the primacy of economic motives in historical explanation that constitutes the decisive difference between Marxism and bourgeois thought, but the point of view of totality.... The primacy of the category of totality is the bearer of the principle of revolution in science.

History and Class Consciousness (London, 1971), p. 27.

GEORG LUKACS (1885–1971) *was a leading Hungarian Communist who had an immense influence on the development of Marxist theory in the West. After joining the Hungarian Communist Party at the end of the First World War, Lukacs played a prominent role in the revolutionary government of 1919. On its collapse, Lukacs went into exile and lived mainly in the Soviet Union until his return to Hungary in 1945. Most of his voluminous writings are in the area of literary criticism and, towards the end of his life, in social ontology. By far his most influential work is* History and Class Consciousness, *published in 1921, from which the extracts below are taken. This book is a collection of articles written during the heady years of the October revolution in Russia and its immediate aftermath. His emphasis on class-consciousness and its embodiment in a revolutionary party whose theory and practice would break through the reified forms of bourgeois society obviously reflected the politics of Lenin. Lukacs's book was condemned by the Soviet authorities in 1924 and he himself wrote a Preface in 1967 (see below) in which he retracted his 'revolutionary Messianism'. But this aspect of his work continued to exercise considerable influence on such writers as Marcuse and student radicals of the late 1960s.*

History and Class-consciousness

(a) FROM THE 1967 PREFACE

I MUST confine myself here to a critique of *History and Class Consciousness*, but this is not to imply that this deviation from Marxism was less pronounced in the case of other writers with a similar outlook. In my book this deviation has immediate consequences for the view of economics I give there and fundamental confusions result, as in the nature of the case economics must be crucial. It is true that the attempt is made to explain all ideological phenomena by reference to their basis in economics but, despite this, the purview of economics is narrowed down because its basic Marxist category, labour as the mediator of the metabolic interaction between society and nature, is missing. Given my basic approach, such a consequence is quite natural. It means that the most important real pillars of the Marxist view of the world disappear and the attempt to deduce the ultimate revolutionary implications of Marxism in as radical a fashion as possible is deprived of a genuinely economic foundation. It is self-evident that this means the disappearance of the ontological objectivity of nature upon which this process of change is based. But it also means the disappearance of the interaction between labour as seen from a genuinely materialist standpoint and the evolution of the men who labour. Marx's great insight that 'even production for the sake of production means nothing more than the *development of the productive energies of man, and hence the development of the wealth of human nature as an end in itself*' lies outside the terrain which *History and Class Consciousness* is able to explore. Capitalist exploitation thus loses its objective revolutionary aspect and there is a failure to grasp the fact that 'although this evolution of the species Man is accomplished at first at the expense of the majority of individual human beings and of certain human classes, it finally overcomes this antagonism and coincides with the evolution of the particular individual. Thus the higher development of individuality is only purchased by a historical process in which individuals are sacrificed.' In consequence, my account of the contradictions of capitalism as well as of the revolutionization of the proletariat is unintentionally coloured by an overriding subjectivism.

This has a narrowing and distorting effect on the book's central concept of praxis. With regard to this problem, too, my intention was to base myself on Marx and to free his concepts from every subsequent bourgeois distortion and to adapt them to the requirements of the great revolutionary upsurge of the present. Above all I was absolutely convinced of one thing: that the purely contemplative nature of bourgeois thought had to be radically overcome. As a result the conception of revolutionary praxis in this book takes on extravagant overtones that are more in keeping with the current Messianic Utopianism of the Communist left than with authentic Marxist doctrine. Comprehensibly enough in the context of the period, I attacked the bourgeois and opportunistic currents in the workers' movement that glorified a conception of knowledge which was ostensible objective but was in fact isolated from any sort of praxis: with considerable justice I directed my polemics against the over-extension and over-valuation of contemplation. Marx's critique of Feuerbach only reinforced my convictions. What I failed to realize, however, was that in the absence of a basis in real praxis, in labour as its original form and model, the over-extension of the concept of praxis would lead to its opposite: a relapse into idealistic contemplation. My intention, then, was to chart the correct and authentic class-consciousness of the proletariat, distinguishing it from 'public opinion surveys' (a term not yet in currency), and to confer upon it an indisputably practical objectivity. I was unable, however, to progress beyond the notion of an 'imputed' [*zugerechnet*] class-consciousness. By this I meant the same thing as Lenin in *What is to be done?* when he maintained that socialist class-consciousness would differ from the spontaneously emerging trade-union consciousness in that it would be implanted in the workers 'from outside', i.e. 'from outside the economic struggle and the sphere of the relations between workers and employers'. Hence, what I had intended subjectively, and what Lenin had arrived at as the result of an authentic Marxist analysis of a practical movement, was transformed in my account into a purely intellectual result and thus into something contemplative. In my presentation it would indeed be a miracle if this 'imputed' consciousness could turn into revolutionary praxis. . . .

Already Bernstein had wished to eliminate everything reminiscent of Hegel's dialectics in the name of 'science'. And nothing was further from the mind of his philosophical opponents, and above all Kautsky, than the wish to undertake the defence of this tradition. For anyone wishing to return to the revolutionary traditions of Marxism the revival of the Hegelian traditions was obligatory. *History and Class Consciousness* represents what was perhaps the most radical attempt to restore the revolutionary nature of Marx's theories by renovating and extending Hegel's dialectics and method. The task was made even more important by the fact the bourgeois philosophy at the time showed signs of a growing interest in Hegel. Of course they never succeeded in making Hegel's breach with Kant the foundation of their analysis and, on the other hand, they were influenced by Dilthey's attempts to construct theoretical bridges between Hegelian dialectics and modern irrationalism. A little while after the appearance of *History and Class Consciousness* Kroner described Hegel as the greatest irrationalist of all time and in Löwith's later studies Marx and Kierkegaard were to emerge as parallel phenomena out of the dissolution of Hegelianism. It is by contrast with all these developments that we can best see the relevance of *History and Class Consciousness*. Another fact contributing to its importance to the ideology of the radical workers' movement was that whereas Plekhanov and others had vastly overestimated Feuerbach's role as an intermediary between Hegel and Marx, this was relegated to the background here. Anticipating the publication of Lenin's later philosophical studies by some years, it was nevertheless only somewhat later, in the essay on Moses Hess, that I explicitly argued that Marx followed directly from Hegel. However, this position is contained implicitly in many of the discussions in *History and Class Consciousness*.

In a necessarily brief summary it is not possible to undertake a concrete criticism of all the issues raised by the book, and to show how far the interpretation of Hegel it contained was a source of confusion and how far it pointed towards the future. The contemporary reader who is qualified to criticize will certainly find evidence of both tendencies. To assess the impact of the book at that time, and also its relevance today, we must consider one problem that surpasses in its importance all questions of detail. This is the question of alienation, which for the first time since Marx, is treated as central to the revolutionary critique of capitalism and which has its theoretical and methodological roots in the Hegelian dialectic. Of course the problem was in the air at the time. Some years later, following the publication of Heidegger's *Being and Time* (1927), it moved into the centre of philosophical debate. Even today it has not lost this position, largely because of the influence of Sartre, his followers, and his opponents. The philosophical problem raised above all by Lucien Goldmann when he interpreted Heidegger's work in part as a polemical reply to mine—which however was not mentioned explicitly—can be left on one side here. The statement that the problem was in the air is perfectly adequate, particularly as it is not possible to discuss the reasons for this here and to lay bare the mixture of Marxist and Existentialist ideas that were so influential after World War II, especially in France. The question of who was first and who influenced whom is not particularly interesting here. What is important is that the alienation of man is a crucial problem of the age in which we live and is recognized as such by both bourgeois and proletarian thinkers, by commentators on both right and left. Hence *History and Class Consciousness* had a profound impact in youthful intellectual circles; I know of a whole host of good Communists who were won over to the movement by this very fact. Without a doubt the fact that this Marxist and Hegelian question was taken up by a Communist was one reason why the impact of the book went far beyond the limits of the party.

As to the way in which the problem was actually dealt with, it is not hard to see today that it was treated in purely Hegelian terms. In particular its ultimate philosophical foundation is the identical subject-object that realizes itself in the historical process. Of course, in Hegel it arises in a purely logical and philosophical form when the highest stage of absolute spirit is attained in philosophy by abolishing alienation and by the return of self-consciousness to itself, thus realizing the identical subject-object. In *History and Class Consciousness*, however, this process is socio-historical and it culminates when the proletariat reaches this stage in its class-consciousness, thus becoming the identical subject-object of history. This does indeed appear to 'stand Hegel on his feet'; it appears as if the logico-metaphysical construction of the *Phenomenology of Mind* had found its authentic realization in the existence and the consciousness of the proletariat. And this appears in turn to provide a philosophical foundation for the proletariat's efforts to form a classless society through revolution and to conclude the 'prehistory' of mankind. But is the identical subject-object here anything more in truth than a purely metaphysical construct? Can a genuinely identical subject-object be created by self-knowlege, however adequate, and however truly based on an adequate knowledge of society, i.e. however perfect that self-knowledge is? We need only formulate the question precisely to see that it must be answered in the negative. For even when the content of knowledge is referred back to the knowing subject, this does not mean that the act of cognition is thereby freed of its alienated nature. In the *Phenomenology of Mind* Hegel rightly dismisses the notion of a mystical and irrationalistic realization of the identical subject-object, of Schelling's 'intellectual intuition', calling instead for a philosophical and rational solution to the problem. His healthy sense of reality induced him to leave the matter at this juncture; his very general system does indeed culminate in the vision of such a realization but he never shows in concrete terms how it might be achieved. Thus the proletariat seen as the identical subject-object of the real history of mankind is no materialist consummation that overcomes the constructions of idealism. It is rather an attempt to out-Hegel Hegel, it is an edifice boldly erected above every possible reality and thus attempts objectively to surpass the master himself.

Hegel's reluctance to commit himself on this point is the product of the wrong-headedness of his basic concept. For it is in Hegel that we first encounter alienation as the fundamental problem of the place of man in the world and *vis-à-vis* the world. However, in the term alienation he includes every type of objectification. Thus 'alienation' when taken to its logical conclusion is identical with objectification. Therefore, when the identical subject-object transcends alienation it must also transcend objectification at the same time. But as, according to Hegel, the object, the thing exists only as an alienation from the self-consciousness, to take it back into the subject would mean the end of objective reality and thus of any reality at all. *History and Class Consciousness* follows Hegel in that it too equates alienation with objectification [*Vergegenständlichung*] (to use the term employed by Marx in the *Economic-Philosophical Manuscripts*). This fundamental and crude error has certainly contributed greatly to the success enjoyed by *History and Class Consciousness*. The unmasking of alienation by philosophy was in the air, as we have remarked, and it soon became a central problem in the type of cultural criticism that undertook to scrutinize the condition of man in contemporary capitalism. In the philosophical, cultural criticism of the bourgeoisie (and we need look no further than Heidegger), it was natural to sublimate a critique of society into a purely philosophical problem, i.e. to convert an essentially social alienation into an eternal 'condition humaine', to use a term not coined until somewhat later. It is evident that *History and Class Consciousness* met such attitudes half-way, even though its intentions had been different and indeed opposed to them. For when I identified alienation with objectification I meant this as a societal category—socialism would after all abolish alienation—but its irreducible presence in class society and above all its basis in philosophy brought it into the vicinity of the 'condition humaine'.

(b) FROM THE 1921 TEXT

Great disunity has prevailed even in the 'Socialist' camp as to what constitutes the essence of Marxism, and which theses it is 'permissible' to criticize and even reject without forfeiting the right to the title of 'Marxist'. In consequence it came to be thought increasingly 'unscientific' to make scholastic exegeses of old texts with a quasi-biblical status, instead of fostering an 'impartial' study of the 'facts'. These texts, it was argued, had long been 'superseded' by modern criticism and they should no longer be regarded as the sole fount of truth.

If the question were really to be formulated in terms of such a crude antithesis it would deserve at best a pitying smile. But in fact it is not (and never has been) quite so straightforward. Let us assume for the sake of argument that recent research had disproved once and for all every one of Marx's individual theses. Even if this were to be proved, every serious 'orthodox' Marxist would still be able to accept all such modern findings without reservation and hence dismiss all of Marx's theses *in toto*—without having to renounce his orthodoxy for a single moment. Orthodox Marxism, therefore, does not imply the uncritical acceptance of the results of Marx's investigations. It is not the 'belief' in this or that thesis, nor the exegesis of a 'sacred' book. On the contrary, orthodoxy refers exclusively to *method*. It is the scientific conviction that dialectical materialism is the road to truth and that its methods can be developed, expanded, and deepened only along the lines laid down by its founders. It is the conviction, moreover, that all attempts to surpass or 'improve' it have led and must lead to over-simplification, triviality, and eclecticism....

Economic fatalism and the reformation of Socialism through ethics are intimately connected. It is no accident that they reappear in similar form in Bernstein, Tugan-Baranovsky, and Otto Bauer. This is not merely the result of the need to seek and find a subjective substitute for the objective path to revolution that they themselves have blocked. It is the logical consequence of the vulgar-economic point of view and of methodological individualism. The 'ethical' reformation of Socialism is the subjective side of the missing category of totality which alone can provide an overall view. For the individual, whether capitalist or proletarian, his environment, his social milieu (including Nature which is the theoretical reflection and projection of that milieu), must appear the servant of a brutal and senseless fate which is eternally alien to him. This world can only be understood by means of a theory which postulates 'eternal laws of nature'. Such a theory endows the world with a rationality alien to man and human action can neither penetrate nor influence the world if man takes up a purely contemplative and fatalistic stance....

For the destruction of a totalizing point of view disrupts *the unity of theory and practice*. Action, praxis—which Marx demanded before all else in his *Theses on Feuerbach*—is in essence the penetration and trans-formation of reality. But reality can only be understood and penetrated as a totality, and only a subject which is itself a totality is capable of this penetration. It was not for nothing that the young Hegel erected his philosophy upon the principle that 'truth must be understood and expressed not merely as substance, but also as subject'. With this he exposed the deepest and the ultimate limitation of classical German philosophy. However, his own philosophy failed to live up to this precept and for much of the time it remained enmeshed in the same snares as those of his predecessors.

It was left to Marx to make the concrete discovery of 'truth as the subject' and hence to establish the unity of theory and practice. This he achieved by focusing the known totality upon the reality of the historical process and by confining it to this. By this means he determined both the knowable totality and the totality to be known. The scientific superiority of the standpoint of class (as against that of the individual) has become clear from the foregoing. Now we see the reason for this superiority: *only the class can actively penetrate the reality of society and transform it in its entirety*. For this reason, 'criticism' advanced from the standpoint of class is criticism from a total point of view and hence it provides the dialectical unity of theory and practice. In dialectical unity it is at once cause and effect, mirror and motor of the historical and dialectical process. The proletariat as the subject of thought in society destroys at one blow the dilemma of impotence: the dilemma created by the pure laws with their fatalism and by the ethics of pure intentions. . . .

Concrete analysis means then: the relation to society *as a whole*. For only when this relation is established does the consciousness of their exist-ence that men have at any given time emerge in all its essential charac-teristics. It appears, on the one hand, as something which is *subjectively* justified in the social and historical situation, as something which can and should be understood, i.e., as 'right'. At the same time, *objectively*, it bypasses the essence of the evolution of society and fails to pinpoint it and express it adequately. That is to say, objectively, it appears as a 'false consciousness'. On the other hand, we may see the same consciousness as something which fails *subjectively* to reach its self-appointed goals, while furthering and realizing the *objective* aims of society of which it is ignorant and which it did not choose.

This twofold dialectical determination of 'false consciousness' constitutes an analysis far removed from the naive description of what men *in fact* thought, felt and wanted at any moment in history and from any given point in the class structure. I do not wish to deny the great importance of this, but it remains after all merely the *material* of genuine historical analysis. The relation with concrete totality and the dialectical determinants arising from it transcend pure description and yield the category of objective possibility. By relating consciousness to the whole of society it becomes possible to infer the thoughts and feelings which men would have in a particular situation if they were *able* to assess both it and the interests arising from it in their impact on immmediate action and on the whole structure of society. That is to say, it would be possible to infer the thoughts and feelings appropriate to their objective situation. The number of such situations is not unlimited in any society. However much detailed researches are able to refine social typologies there will always be a number of clearly distinguished basic types whose characteristics are determined by the types of position available in the process of production. Now class-consciousness consists in fact of the appropriate and rational reactions 'imputed' [*zug-erechnet*] to a particular typical position in the process of production. This consciousness is, therefore, neither the sum nor the average of what is thought or felt by the single individuals who make up the class. And yet the historically significant actions of the class as a whole are determined in the last resort by this consciousness and not by the thought of the individual—and these actions can be understood only by reference to this consciousness.

This analysis establishes right from the start the distance that separates class-consciousness from the empirically given, and from the psychologically describable and explicable ideas which men form about their situation in life. But it is not enough just to state that this distance exists or even to define its implications in a formal and general way. We must discover, firstly, whether it is a phenomenon that differs according to the manner in which the various classes are related to society as a whole and whether the differences are so great as to produce *qualitative distinctions*. And we must discover, secondly, the *practical* significance of these different possible relations between the objective economic totality, the imputed class-consciousness, and the real, psychological thoughts of men about their lives. We must discover, in short, the *practical, historical function* of class-consciousness.

Only after such preparatory formulations can we begin to exploit the category of objective possibility systematically. The first question we must ask is how far is it *in fact* possible to discern the whole economy of a society from inside it? It is essential to transcend the limitations of particular individuals caught up in their own narrow prejudices. But it is no less vital not to overstep the frontier fixed for them by the economic structure of society and establishing their position in it. Regarded abstractly and formally, then, class-consciousness implies a class-conditioned *unconsciousness* of one's own socio-historical and economic condition. This condition is given as a definite structural relation, a definite formal nexus which appears to govern the whole of life. The 'falseness', the illusion implicit in this situation, is in no sense arbitrary; it is simply the intellectual reflex of the objective economic structure. Thus, for example, 'the value or price of labour-power takes on the appearance of the price or value of labour itself . . .' and 'the illusion is created that the totality is paid labour. . . . In contrast to that, under slavery even that portion of labour which is paid for appears unpaid for.' Now it requires the most painstaking historical analysis to use the category of objective possibility so as to isolate the conditions in which this illusion can be exposed and a real connection with the totality established. For if from the vantage point of a particular class the totality of existing society is not visible; if a class thinks the thoughts imputable to it and which bear upon its interests right through to their logical conclusion and yet fails to strike at the heart of that totality, then such a class is doomed to play only a subordinate role. It can never influence the course of history in either a conservative or progressive direction. Such classes are normally condemned to passivity, to an unstable oscillation between the ruling and the revolutionary classes, and if perchance they do erupt then such explosions are purely elemental and aimless. They may win a few battles but they are doomed to ultimate defeat.

For a class to be ripe for hegemony means that its interests and consciousness enable it to organize the whole of society in accordance with those interests. The crucial question in every class struggle is this: which class possesses this capacity and this consciousness at the decisive moment? This does not preclude the use of force. It does not mean that the class-interests destined to prevail and thus to uphold the interests of society as a whole can be guaranteed an automatic victory. On the contrary, such a transfer of power can often only be brought about by the most ruthless use of force (as e.g. the primitive accumulation of capital). But it often turns out that questions of class-consciousness prove to be decisive in just those situations where force is unavoidable and where classes are locked in a life-and-death struggle. Thus the noted Hungarian Marxist Erwin Szabó is mistaken in criticizing Engels for maintaining that the Great Peasant War (of 1525) was essentially a reactionary movement. Szabó argues that the peasants' revolt was suppressed *only* by the ruthless use of force and that its defeat was not grounded in socio-economic factors and in the class-consciousness of the peasants. He overlooks the fact that the deepest reason for the weakness of the peasantry and the superior strength of the princes is to be sought in class-consciousness. Even the most cursory student of the military aspects of the Peasants' War can easily convince himself of this.

It must not be thought, however, that all classes ripe for hegemony have a class-consciousness with the same inner structure. Everything hinges on the extent to which they can become conscious of the actions they need to perform in order to obtain and organize power. The question then becomes: how far does the class concerned perform the actions history has imposed on it 'consciously' or 'unconsciously'? And is that consciousness 'true' or 'false'. These distinctions are by no means academic. Quite apart from problems of culture where such fissures and dissonances are crucial, in all practical matters too the fate of a class depends on its ability to elucidate and solve the problems with which history confronts it. And here it becomes transparently obvious that class-consciousness is concerned neither with the thoughts of individuals, however advanced, nor with the state of scientific knowledge. For example, it is quite clear that ancient society was broken economically by the limitations of a system built on slavery. But it is equally clear that neither the ruling classes nor the classes that rebelled against them in the name of revolution or reform could perceive this. In consequence the practical emergence of these problems meant that the society was necessarily and irremediably doomed....

To say that class-consciousness has no psychological reality does not imply that it is a mere fiction. Its reality is vouched for by its ability to explain the infinitely painful path of the proletarian revolution, with its many reverses, its constant return to its starting-point, and the incessant self-criticism of which Marx speaks in the celebrated passage in *The Eighteenth Brumaire.*

Only the consciousness of the proletariat can point to the way that leads out of the impasse of capitalism. As long as this consciousness is lacking, the crisis remains permanent, it goes back to its starting-point, repeats the cycle, until after infinite sufferings and terrible detours the school of history completes the education of the proletariat and confers upon it the leadership of mankind. But the proleteriat is not given any choice. As Marx says, it must become a class not only 'as against capital' but also 'for itself'; that is to say, the class struggle must be raised from the level of economic necessity to the level of conscious aim and effective class-consciousness. The pacifists and humanitarians of the class struggle whose efforts tend whether they will or no to retard this lengthy, painful, and crisis-ridden process would be horrified if they could but see what sufferings they inflict on the proletariat by extending this course of education. But the proletariat cannot abdicate its mission. The only question at issue is how much it has to suffer before it achieves ideological maturity, before it acquires a true understanding of its class situation and a true class-consciousness.

Of course this uncertainty and lack of clarity are themselves the symptoms of the crisis in bourgeois society. As the product of capitalism the proletariat must necessarily be subject to the modes of existence of its creator. This mode of existence is inhumanity and reification. No doubt the very existence of the proletariat implies criticism and the negation of this form of life. But until the objective crisis of capitalism has matured and until the proletariat has achieved true class-consciousness, and the ability to understand the crisis fully, it cannot go beyond the criticism of reification and so it is only negatively superior to its antagonist. Indeed, if it can do no more than negate some aspects of capitalism, if it cannot at least aspire to a critique of the whole, then it will not even achieve a negative superiority. This applies to the petty-bourgeois attitudes of most trade unionists. Such criticism from the standpoint of capitalism can be seen more strikingly in the separation of the various theatres of war. The bare fact of separation itself indicates that the consciousness of the proletariat is still fettered by reification. And if the proletariat finds the economic inhumanity to which it is subject easier to understand than the political, and the political easier than the cultural, then all these separations point to the extent of the still unconquered power of capitalist forms of life in the proletariat itself....

Thus even the individual object which man confronts directly, either as producer or consumer, is distorted in its objectivity by its commodity character. If that can happen then it is evident that this process will be intensified in proportion as the relations which man establishes with objects as objects of the life process are mediated in the course of his social activity. It is obviously not possible here to give an analysis of the whole economic structure of capitalism. It must suffice to point out that modern capitalism does not content itself with transforming the relations of production in accordance with its own needs. It also integrates into its own system those forms of primitive capitalism that led an isolated existence in pre-capitalist times, divorced from production; it converts them into members of the henceforth unified process of radical capitalism. (Cf. merchant capital, the role of money as a hoard or as finance capital, etc.)

These forms of capital are objectively subordinated, it is true, to the real life-process of capitalism, the extraction of surplus value in the course of production. They are, therefore, only to be explained in terms of the nature of industrial capitalism itself. But in the minds of people in bourgeois society they constitute the pure, authentic, unadulterated forms of capital. In them the relations between men that lie hidden in the immediate commodity relation, as well as the relations between men and the objects that should really gratify their needs, have faded to the point where they can be neither recognized nor even perceived.

For that very reason the reified mind has come to regard them as the true representatives of his societal existence....

The divorce of the phenomena of reification from their economic bases and from the vantage-point from which alone they can be understood is facilitated by the fact that the [capitalist] process of transformation must embrace every manifestation of the life of society if the preconditions for the complete self-realization of capitalist production are to be fulfilled....

To give a detailed analysis of the various forms taken by the refusal to understand reality as a whole and as existence would be to go well beyond the framework of this study. Our aim here was to locate the point at which there appears in the thought of bourgeois society the double tendency characteristic of its evolution. On the one hand, it acquires increasing control over the details of its social existence, subjecting them to its needs. On the other hand it loses—likewise progressively—the possibility of gaining intellectual control of society as a whole and with that it loses its own qualifications for leadership.

Classical German philosophy marks a unique transitional stage in this process. It arises at a point of development where matters have progressed so far that these problems can be raised to the level of consciousness. At the same time this takes place in a milieu where the problems can only appear on an intellectual and philosophical plane. This has the drawback that the concrete problems of society and the concrete solutions to them cannot be seen. Nevertheless, classical philosophy is able to think the deepest and most fundamental problems of the development of bourgeois society through to the very end—on the plane of philosophy. It is able—in thought—to complete the evolution of class. And—in thought—it is able to take all the paradoxes of its position to the point where the necessity of going beyond this historical stage in mankind's development can at least be seen as a problem. . . .

Of course, the history of the dialectical method reaches back deep into the history of rationalistic thought. But the turn it now takes distinguishes it qualitatively from all earlier approaches. (Hegel himself underestimates the importance of this distinction, e.g. in his treatment of Plato.) In all earlier attempts to use dialectics in order to break out of the limits imposed by rationalism there was a failure to connect the dissolution of rigid concepts clearly and firmly to the problem of the logic of the content, to the problem of irrationality.

Hegel in his *Phenomenology* and *Logic* was the first to set about the task of consciously recasting all problems of logic by grounding them in the qualitative material nature of their content, in matter in the logical and philosophical sense of the word. This resulted in the establishment of a completely new logic of the *concrete concept*, the logic of totality—admittedly in a very problematic form which was not seriously continued after him.

Even more original is the fact that the subject is neither the unchanged observer of the objective dialectic of being and concept (as was true of the Eleatic philosophers and even of Plato), nor the practical manipulator of its purely mental possibilities (as with the Greek sophists): the dialectical process, the ending of a rigid confrontation of rigid forms, is enacted essentially *between the subject and the object*. No doubt, a few isolated earlier dialecticians were not wholly unaware of the different levels of subjectivity that arise in the dialectical process (consider for example the distinction between 'ratio' and 'intellectus' in the thought of Nicholas of Cusa). But this relativizing process only refers to the possibility of different subject-object relations existing simultaneously or with one subordinated to the other, or at best developing dialectically from each other; they do not involve the relativizing or the interpenetration of the subject and the object themselves. But only if that were the case, only if 'the true [were understood] not only as substance but also as subject', only if the subject (consciousness, thought) were both producer and product of the dialectical process, only if, as a result the subject moved in a self-created world of which it is the conscious form and only if the world imposed itself upon it in full objectivity, only then can the problem of dialectics, and with it the abolition of the antitheses of subject and object, thought and existence, freedom and necessity, be held to be solved. . . .

In addition to the mere contradiction—the automatic product of capitalism—a *new* element is required: the consciousness of the proletariat must become deed. But as the mere contradiction is raised to a consciously dialectical contradiction, as the act of becoming conscious turns into *a point of transition in practice*, we see once more in greater concreteness the character of proletarian dialectics as we have often described it: namely, since consciousness here is not the knowledge of an opposed object but is the self-consciousness of the object *the act of consciousness overthrows the objective form of its object*. . . .

But it now becomes quite clear that the social development and its intellectual reflex that was led to form 'facts' from a reality that had been undivided (originally, in its autochthonous state) did indeed make it possible to subject nature to the will of man. At the same time, however, they served to conceal the socio-historical grounding of these facts in relations between men 'so as to raise strange, phantom powers against them'. For the ossifying quality of reified thought with its tendency to oust the process is exemplified even more clearly in the 'facts' than in the 'laws' that would order them. In the latter it is still possible to detect a trace of human activity even though it often appears in a reified and false subjectivity. But in the 'facts' we find the crystallization of the essence of capitalist development into an ossified, impenetrable thing alienated from man. And the form assumed by the ossification and this alienation converts it into a foundation of reality and of philosophy that is perfectly self-evident and immune from every doubt. When confronted by the rigidity of these 'facts' every movement seems like a movement *impinging on* them, while every tendency to change them appears to be a merely subjective principle (a wish, a value judgement, an ought). Thus only when the theoretical primacy of the 'facts' has been broken, only when *every phenomenon is recognized to be a process*, will it be understood that what we are wont to call 'facts' consists of processes. Only then will it be understood that the facts are nothing but the parts, the *aspects* of the total process that have been broken off, artificially isolated,

and ossified. This also explains why the total process which is uncon-
taminated by any trace of reification and which allows the process-like
essence to prevail *in all its purity* should represent the authentic, higher
reality. Of course, it also becomes clear why in the reified thought of the
bourgeoisie the 'facts' have to play the part of its highest fetish in both
theory and practice. This petrified factuality in which everything is frozen
into a 'fixed magnitude', in which the reality that just happens to exist
persists in a totally senseless, unchanging way, precludes any theory that
could throw light on even this immediate reality.

This takes reification to its ultimate extreme: it no longer points dialectically to anything beyond itself: its dialectic is mediated only by the
reification of the immediate forms of production. But with that a climax is
reached in the conflict between existence in its immediacy together with
the abstract categories that constitute its thought, on the one hand, and a
vital societal reality on the other. For these forms (e.g. interest) appear to
capitalist thinkers as the fundamental ones that determine all the others
and serve as paradigms for them. And likewise, every decisive turn of events
in the production process must more or less reveal that the true categorical
structure of capitalism has been turned completely upside down.

Thus bourgeois thought remains fixated on these forms which it believes
to be immediate and original and from there it attempts to seek an understanding of economics, blithely unaware that the only phenomenon that
has been formulated is its own inability to comprehend its own social
foundations. Whereas for the proletariat the way is opened to a complete
penetration of the forms of reification. It achieves this by starting with what
is dialectically the clearest form (the immediate relation of capital and
labour). It then relates this to those forms that are more remote from the
production process and so includes and comprehends them, too, in the
dialectical totality. . . .

Among the factors that determine the direction to be taken, *the proletariat's
correct understanding of its own historical position is of the very first
importance*. The course of the Russian Revolution in 1917 is a classic
illustration of this. For we see there how at a crucial moment, the slogans of
peace, self-determination, and the radical solution to the agrarian problem
welded together an army that could be deployed for revolution whilst
completely disorganizing the whole power apparatus of counter-revolution
and rendering it impotent. It is not enough to object that the agrarian
revolution and the peace movement of the masses would have carried the
day without or even against the Communist Party. In the first place this is
absolutely unprovable: as counter-evidence we may point e.g. to Hungary

where a no less spontaneous agrarian uprising was defeated in October 1918. And even in Russia it might have been possible to crush the agrarian movement or allow it to dissipate itself, by achieving a 'coalition' (namely a counter-revolutionary coalition) of all the 'influential' 'workers' parties'. In the second place, if the 'same' agrarian movement had prevailed against the urban proletariat it would have become counter-revolutionary in character in the context of the social revolution.

This example alone shows the folly of applying mechanical and fatalistic criteria to the constellation of social forces in acute crisis-situations during a social revolution. It highlights the fact that the proletariat's correct insight and correct decision is *all-important*; it shows the extent to which the resolution of the crisis *depends upon the proletariat itself*. We should add that in comparison to the Western nations the situation in Russia was relatively simple. Mass movements there were more purely spontaneous and the opposing forces possessed no organization deeply rooted in tradition. It can be maintained without exaggeration, therefore, that our analysis would have an *even greater validity* for Western nations. All the more as the undeveloped character of Russia, the absence of a long tradition of a legal workers' movement—if we ignore for the moment the existence of a fully constituted Communist Party—gave the Russian proletariat the chance to resolve the ideological crisis with greater dispatch.

Thus the economic development of capitalism places the fate of society in the hands of the proletariat. Engels describes the transition accomplished by mankind *after* the revolution has been carried out as 'the leap from the realm of necessity into the realm of freedom'. For the dialectical materialist it is self-evident that despite the fact that this leap is a leap, or just because of it, it must represent in essence a process. Does not Engels himself say in the passage referred to that the changes that lead in this direction take place 'at a constantly increasing rate'? The only problem is to determine the *starting-point* of the process. It would, of course, be easiest to take Engels literally and to regard the realm of freedom simply as a *state* which will come into being after the completion of the social revolution. This would be simply to deny that the question had any immediate relevance. The only problem then would be to ask whether the question would really be exhausted by this formulation, which admittedly does correspond to Engels' literal statement. The question is whether a situation is even conceivable, let alone capable of being made social reality, if it has not been prepared by a lengthy *process* which has contained and developed the elements of that situation, albeit in a form that is inadequate in many ways and in great need of being subjected to a series of dialectical reversals. If we separate the 'realm of freedom' sharply from the process which is destined to call it into being, if we thus preclude all dialectical transitions, do we not thereby lapse into a Utopian outlook similar to that which has already been analysed in the case of the separation of final goal and the movement towards it?

If, however, the 'realm of freedom' is considered in the context of the process that leads up to it, then it cannot be doubted that even the earliest appearance of the proletariat on the stage of history indicated an aspiration towards that end—admittedly in a wholly unconscious way. However little the final goal of the proletariat is able, even in theory, to influence the initial stages of the early part of the process directly, it is a principle, a synthesizing factor and so can never be completely absent from any aspect of that process. It must not be forgotten, however, that the difference between the period in which the decisive battles are fought and the foregoing period does not lie in the extent and the intensity of the battles themselves. These quantitative changes are merely symptomatic of the fundamental differences in quality which distinguish these struggles from earlier ones. At an earlier stage, in the words of the *Communist Manifesto*, even 'the massive solidarity of the workers was not yet the consequence of their own unification but merely a consequence of the unification of the bourgeoisie'. Now, however, the process by which the proletariat becomes independent and 'organizes itself into a class' is repeated and intensified until the time when the final crisis of capitalism has been reached, the time when the decision comes more and more within the grasp of the proletariat.

This state of affairs should not be taken to imply that the objective economic 'laws' cease to operate. On the contrary, they will remain in effect until long *after the victory* of the proletariat and they will only wither away—like the State—when the classless society wholly in the control of mankind comes into being. What is novel in the present situation is merely—merely!!—that the blind forces of capitalist economics are driving society towards the abyss. The bourgeoisie no longer has the power to help society, after a few false starts, to break the 'deadlock' brought about by its economic laws. And the proletariat has the *opportunity* to turn events *in another direction* by the conscious exploitation of existing trends. This other direction is the conscious regulation of the productive forces of society. To desire this *consciously* is to desire the 'realm of freedom' and to take the *first conscious step* towards its realization.

This step follows 'necessarily' from the class situation of the proletariat. However, this necessity has itself the character of a leap. The *practical* relationship to the whole, the real unity of theory and practice which hitherto appeared only unconsciously, so to speak, in the actions of the proletariat, now emerges clearly and consciously. At earlier stages, too, the actions of the proletariat were driven to a climax in a series of leaps whose continuity with the previous development could only subsequently become conscious and be understood as the necessary consequence of that development. (An instance of this is the political form of the Commune of 1871.) In this case, however, the proletariat must take this step *consciously*. It is no wonder, therefore, that all those who remain imprisoned within the confines of capitalist thought recoil from taking this step and with all the mental energy at their disposal they hold fast to necessity which they see as a law of nature, as a 'law of the repetition' of phenomena. Hence, too, they reject as impossible the emergence of anything that is radically new of which we can have no 'experience'. It was Trotsky in his polemics against Kautsky who brought out this distinction most clearly, although it had been touched upon in the debates on the war: 'For the fundamental Bolshevist prejudice consists precisely in the idea that one can only learn to ride when one is sitting firmly on a horse.' But Kautsky and his like are only significant as symptoms of the state of affairs: they symbolize the ideological crisis of the working class, they embody that moment of its development when it 'once again recoils before the inchoate enormity of its own aims', and when it jibs at a task which it must take upon itself. Unless the proletariat wishes to share the fate of the bourgeoisie and perish wretchedly and ignominiously in the death-throes of capitalism, it must accomplish this task *in full consciousness*.

Source: History and Class Consciousness (London, 1971), pp. xvii ff., 1, 38 f., 50 ff., 75 ff., 93, 95, 120 f., 142, 178, 184 f., 311 ff.

FURTHER NOTES

WORKS

LUKACS, GEORG, *Conversations with Lukacs*, ed. T. Pinkus (London, 1973).
—— *History and Class Consciousness* (London, 1971).
—— *Lenin: A Study of the Unity of His Thought* (London, 1970).
—— *Marxism and Human Liberation*, ed. E. San Juan (New York, 1973).

COMMENTARIES

LICHTHEIM, G., *Lukacs* (London, 1970).
LOWY, M., *Georg Lukacs: From Idealism to Bolshevism* (London, 1979).
MESZAROS, I. (ed.), *Aspects of History and Class Consciousness* (London, 1971).
PARKINSON, G., *Georg Lukacs* (London, 1977).
—— (ed.), *Georg Lukacs: The Man, His Work and His Ideas* (London, 1970).
WATNICK, MORRIS, 'Relativism and Class Consciousness: Georg Lukacs', in L. Labedz (ed.), *Revisionism; Essays on the History of Marxist Ideas* (London, 1962).

11

ANTONIO GRAMSCI

It is even possible to affirm that present-day Marxism in its essential
trait is precisely the historical political concept of hegemony.

Lettere dal Carcere (Turin, 1965), p. 616.

ANTONIO GRAMSCI (1891–1937) *helped to found the Italian Communist
Party in 1921 after several years working as a radical journalist in the big
industrial city of Turin. Gramsci was active in Comintern politics and took
over the leadership of the Italian Communist Party in 1924 but was arrested
by the Fascist government in 1926 and spent the rest of his life in prison
where he composed his famous notebooks. These contain a radical attempt
to rethink Marxism in terms of its applicability to capitalist societies.
Gramsci had three aims in mind—all of them exemplified in the extracts
below. Firstly he discusses the important role of intellectuals in modern
society and the way they help organize a network of beliefs and socio-
political relationships which ensure the consent of the governed, a process
he referred to as 'hegemony'. Secondly, he drew the conclusion that a
Marxist strategy for the West would have to be very different from that
adopted by Lenin in Russia since in capitalist societies the bourgeoisie
exercised a hegemony that would have to be undermined before a frontal
assault on State power could be successful. Thirdly, Gramsci presented
Marxism as a historically relative movement which could and should
incorporate the best of European cultural history.*

1. Intellectuals and Hegemony

EVERY 'essential' social group which emerges into history out of the
preceding economic structure, and as an expression of a development of
this structure, has found (at least in all of history up to the present)
categories of intellectuals already in existence and which seemed indeed to
represent an historical continuity uninterrupted even by the most complicated and radical changes in political and social forms.

The most typical of these categories of intellectuals is that of the ecclesiastics, who for a long time (for a whole phase of history, which is partly
characterized by this very monopoly) held a monopoly of a number of
important services: religious ideology, that is the philosophy and science of

the age, together with schools, education, morality, justice, charity, good works, etc. The category of ecclesiastics can be considered the category of intellectuals organically bound to the landed aristocracy. It had equal status juridically with the aristocracy, with which it shared the exercise of feudal ownership of land, and the use of State privileges connected with property. But the monopoly held by the ecclesiastics in the superstructural field was not exercised without a struggle or without limitations, and hence there took place the birth, in various forms (to be gone into and studied concretely), of other categories, favoured and enabled to expand by the growing strength of the central power of the monarch, right up to absolutism. Thus we find the formation of the *noblesse de robe*, with its own privileges, a stratum of administrators, etc., scholars and scientists, theorists, non-ecclesiastical philosophers, etc.

Since these various categories of traditional intellectuals experience through an 'esprit de corps' their uninterrupted historical continuity and their special qualification, they thus put themselves forward as autonomous and independent of the dominant social group. This self-assessment is not without consequences in the ideological and political field, consequences of wide-ranging import. The whole of idealist philosophy can easily be connected with this position assumed by the social complex of intellectuals and can be defined as the expression of that social Utopia by which the intellectuals think of themselves as 'independent', autonomous, endowed with a character of their own, etc.

One should note however that if the pope and the leading hierarchy of the church consider themselves more linked to Christ and to the apostles than they are to senators Agnelli and Benni, the same does not hold for Gentile and Croce, for example: Croce in particular feels himself closely linked to Aristotle and Plato, but he does not conceal, on the other hand, his links with senators Agnelli and Benni, and it is precisely here that one can discern the most significant character of Croce's philosophy.

What are the 'maximum' limits of acceptance of the term 'intellectual'? Can one find a unitary criterion to characterize equally all the diverse and disparate activities of intellectuals and to distinguish these at the same time and in an essential way from the activities of other social groupings? The most widespread error of method seems to me that of having looked for this criterion of distinction in the intrinsic nature of intellectual activities, rather than in the ensemble of the system of relations in which these activities (and therefore the intellectual groups who personify them) have their place within the general complex of social relations. Indeed the worker or proletarian, for example, is not specifically characterized by his manual or instrumental work, but by performing this work in specific conditions and in specific social relations (apart from the consideration that purely physical labour does not exist and that even Taylor's phrase of 'trained gorilla' is a metaphor to indicate a limit in a certain direction: in any physical work, even the most degraded and mechanical, there exists a minimum of technical qualification, that is, a minimum of creative intellectual activity.) And we have already observed that the entrepreneur, by virtue of his very function, must have to some degree a certain number of qualifications of an intellectual nature although his part in society is determined not by these, but by the general social relations which specifically characterize the position of the entrepreneur within industry.

All men are intellectuals, one could therefore say: but not all men have in society the function of intellectuals.

When one distinguishes between intellectuals and non-intellectuals, one is referring in reality only to the immediate social function of the professional category of the intellectuals, that is, one has in mind the direction in which their specific professional activity is weighted, whether towards intellectual elaboration or towards muscular-nervous effort. This means that, although one can speak of intellectuals, one cannot speak of non-intellectuals, because non-intellectuals do not exist. But even the relationship between efforts of intellectual-cerebral elaboration and muscular-nervous effort is not always the same, so that there are varying degrees of specific intellectual activity. There is no human activity from which every form of intellectual participation can be excluded: *homo faber* cannot be separated from *homo sapiens*. Each man, finally, outside his professional activity, carries on some form of intellectual activity, that is, he is a 'philosopher', an artist, a man of taste, he participates in a particular conception of the world, has a conscious line of moral conduct, and therefore contributes to sustain a conception of the world or to modify it, that is, to bring into being new modes of thought.

The problem of creating a new stratum of intellectuals consists therefore in the critical elaboration of the intellectual activity that exists in everyone at a certain degree of development, modifying its relationship with the muscular-nervous effort towards a new equilibrium, and ensuring that the muscular-nervous effort iself, in so far as it is an element of a general practical activity, which is perpetually innovating the physical and social world, becomes the foundation of a new and integral conception of the world. The traditional and vulgarized type of the intellectual is given by the man of letters, the philosopher, the artist. Therefore journalists, who claim to be men of letters, philosophers, artists, also regard themselves as the 'true' intellectuals. In the modern world, technical education, closely bound to industrial labour even at the most primitive and unqualified level, must form the basis of the new type of intellectual. . . .

The relationship between the intellectuals and the world of production is not as direct as it is with the fundamental social groups but is, in varying degrees, 'mediated' by the whole fabric of society and by the complex of superstructures, of which the intellectuals are, precisely, the 'functionaries'. It should be possible both to measure the 'organic quality' [*organicità*] of the various intellectual strata and their degree of connection with a fundamental social group, and to establish a gradation of their functions and of the superstructures from the bottom to the top (from the structural base upwards). What we can do, for the moment, is to fix two major superstructural 'levels': the one that can be called 'civil society', that is the ensemble of organisms commonly called 'private', and that of 'political society' or 'the State'. These two levels correspond on the one hand to the function of 'hegemony' which the dominant group exercises throughout society and on the other hand to that of 'direct domination' or command exercised through the State and 'juridical' government. The functions in question are precisely organizational and connective. The intellectuals are the dominant group's 'deputies' exercising the subaltern functions of social hegemony and political government. These comprise:

1. The 'spontaneous' consent given by the great masses of the population to the general direction imposed on social life by the dominant fundamental group this consent is 'historically' caused by the prestige (and consequent confidence) which the dominant group enjoys because of its position and function in the world of production.

2. The apparatus of State coercive power which 'legally' enforces discipline on those groups who do not 'consent' either actively or passively. This apparatus is, however, constituted for the whole of society in anticipation of moments of crisis of command and direction when spontaneous consent has failed.

This way of posing the problem has as a result a considerable extension of the concept of intellectual, but it is the only way which enables one to reach a concrete approximation of reality. It also clashes with pre-conceptions of caste. The function of organizing social hegemony and state domination certainly gives rise to a particular division of labour and therefore to a whole hierarchy of qualifications in some of which there is no apparent attribution of directive or organizational functions. For example, in the apparatus of social and State direction there exist a whole series of jobs of a manual and instrumental character (non-executive work, agents rather than officials or functionaries). It is obvious that such a distinction has to be made just as it is obvious that other distinctions have to be made as well. Indeed, intellectual activity must also be distinguished in terms of its intrinsic characteristics, according to levels which in moments of extreme opposition represent a real qualitative difference—at the highest level would be the creators of the various sciences, philosophy, art, etc., at the lowest the most humble 'administrators' and divulgators of pre-existing, traditional, accumulated intellectual wealth....

The methodological criterion on which our own study must be based is the following: that the supremacy of a social group manifests itself in two ways, as 'domination' and as 'intellectual and moral leadership'. A social group dominates antagonistic groups, which it tends to 'liquidate', or to subjugate perhaps even by armed force; it leads kindred and allied groups. A social group can, and indeed must, already exercise 'leadership' before winning governmental power (this indeed is one of the principal conditions for the winning of such power); it subsequently becomes dominant when it exercises power, but even if it holds it firmly in its grasp, it must continue to 'lead' as well....

... there does not exist any independent class of intellectuals, but every social group has its own stratum of intellectuals, or tends to form one; however, the intellectuals of the historically (and concretely) progressive class, in the given conditions, exercise such a power of attraction that, in the last analysis, they end up by subjugating the intellectuals of the other social groups; they thereby create a system of solidarity between all the intellectuals, with bonds of a psychological nature (vanity, etc.) and often of a caste character (technico-juridical, corporate, etc.). This phenomenon manifests itself 'spontaneously' in the historical periods in which the given social group is really progressive—i.e. really causes the whole society to move forward, not merely satisfying its own existential requirements, but continuously augmenting its cadres for the conquest of ever new spheres of economic and productive activity. As soon as the dominant social group has exhausted its function, the ideological bloc tends to crumble away; then spontaneity may be replaced by 'constraint' in ever less disguised and in direct forms, culminating in outright police measures and *coups d'état*....

2.   Revolution in the West

It is the problem of the relations between structure and superstructure which must be accurately posed and resolved if the forces which are active in the history of a particular period are to be correctly analysed, and the relation between them determined. Two principles must orientate the discussion: 1. That no society sets itself tasks for whose accomplishment the necessary and sufficient conditions do not either already exist or are not at least beginning to emerge and develop; 2. that no society breaks down and can be replaced until it has first developed all the forms of life which are implicit in its internal relations. From a reflection on these two principles, one can move on to develop a whole series of further principles of historical methodology. Meanwhile, in studying a structure, it is necessary to distinguish organic movements (relatively permanent) from movements which may be termed 'conjunctural' (and which appear as occasional, immediate, almost accidental). Conjunctural phenomena too depend on organic movements to be sure, but they do not have any very far-reaching historical significance; they give rise to political criticism of a minor, day-to-day character, which has as its subject top political leaders and personalities with direct governmental responsibilities. Organic phenomena on the other hand give rise to socio-historical criticism, whose subject is wider social groupings—beyond the public figures and beyond the top leaders. When an historical period comes to be studied, the great importance of this distinction becomes clear. A crisis occurs, sometimes lasting for decades. This exceptional duration means that incurable structual contradictions have revealed themselves (reached maturity), and that, despite this, the political forces which are struggling to conserve and defend the existing structure itself are making every effort to cure them, within certain limits, and to overcome them. These incessant and persistent efforts (since no social formation will ever admit that it has been superseded) form the terrain of the 'conjunctural', and it is upon this terrain that the forces of opposition organize. These forces seek to demonstrate that the necessary and sufficient conditions already exist to make possible, and hence imperative, the accomplishment of certain historical tasks (imperative, because any falling short before an historical duty increases the necessary disorder, and prepares more serious catastrophes). (The demonstration in the last analysis only succeeds and is 'true' if it becomes a new reality, if the forces of opposition triumph; in the immediate, it is developed in a series of ideological, religious, philosophical, political, and juridical polemics, whose concreteness can be estimated by the extent to which they are convincing, and shift the previously existing disposition of social forces.)

A common error in historico-political analysis consists in an inability to find the correct relation between what is organic and what is conjunctural. This leads to presenting causes as immediately operative which in fact only operate indirectly, or to asserting that the immediate causes are the only effective ones. In the first case there is an excess of 'economism', or doctrinaire pedantry; in the second, an excess of 'ideologism'. In the first case there is an overestimation of mechanical causes, in the second an exaggeration of the voluntarist and individual element. The distinction between oganic 'movements' and facts and 'conjunctural' or occasional ones must be applied to all types of situation; not only to those in which a regressive development or an acute crisis takes place, but also to those in which there is a progressive development or one towards prosperity, or in which the productive forces are stagnant. The dialectical nexus between the two categories of movement, and therefore of research, is hard to establish precisely. Moreover, if error is serious in historiography, it becomes still more serious in the art of politics, when it is not the reconstruction of past history but the construction of present and future history which is at stake. One's own baser and more immediate desires and passions are the cause of error, in that they take the place of an objective and impartial analysis—and this happens not as a conscious 'means' to stimulate to action, but as self-deception. In this case too the snake bites the snakecharmer—in other words the demagogue is the first victim of his own demagogy.

The same reduction must take place in the art and science of politics, at least in the case of the most advanced States, where 'civil society' has become a very complex structure and one which is resistant to the catastrophic 'incursions' of the immediate economic element (crises, depressions, etc.). The superstructures of civil society are like the trench-systems of modern warfare. In war it would sometimes happen that a fierce artillery attack seemed to have destroyed the enemy's entire defensive system, whereas in fact it had only destroyed the outer perimeter; and at the moment of their advance and attack the assailants would find themselves confronted by a line of defence which was still effective. The same thing happens in politics, during the great economic crises. A crisis cannot give the attacking forces the ability to organize with lightning speed in time and in space; still less can it endow them with fighting spirit. Similarly, the defenders are not demoralized, nor do they abandon their positions, even among the ruins, nor do they lose faith in their own strength or their own future. Of course, things do not remain exactly as they were; but it is certain that one will not find the element of speed, of accelerated time, of the definitive forward march expected by the strategists of political Cadornism.

The last occurrence of the kind in the history of politics was the events of 1917. They marked a decisive turning-point in the history of the art and science of politics. Hence it is a question of studying 'in depth' which elements of civil society correspond to the defensive systems in a war of position. The use of the phrase 'in depth' is intentional, because 1917 has been studied—but only either from superficial and banal viewpoints, as when certain social historians study the vagaries of women's fashions, or from a 'rationalistic' viewpoint—in other words, with the conviction that certain phenomena are destroyed as soon as they are 'realistically' explained, as if they were popular superstitions (which anyway are not destroyed either merely by being explained).

The question of the meagre success achieved by new tendencies in the trade-union movement should be related to this series of problems. One attempt to begin a revision of the current tactical methods was perhaps that outlined by L. Dav. Br. [Trotsky] at the fourth meeting, when he made a comparison between the Eastern and Western fronts. The former had fallen at once, but unprecedented struggles had then ensued; in the case of the latter, the struggles would take place 'beforehand'. The question, therefore, was whether civil society resists before or after the attempt to seize power; where the latter takes place, etc. However, the question was outlined only in a brilliant, literary form, without directives of a practical character.

It should be seen whether Bronstein's famous theory about the *permanent* character of the movement is not the political reflection of the theory of war of manœuvre (recall the observation of the cossack general Krasnov)—i.e. in the last analysis, a reflection of the general-economic-cultural-social conditions in a country in which the structures of national life are embryonic and loose, and incapable of becoming 'trench or fortress'. In this case one might say that Bronstein, apparently 'Western', was in fact a cosmopolitan—i.e. superficially national and superficially Western or European. Ilich [Lenin] on the other hand was profoundly national and profoundly European.

Ilich, however, did not have time to expand his formula—though it should be borne in mind that he could only have expanded it theoretically, whereas the fundamental task was a national one; that is to say it required a reconnaissance of the terrain and identification of the elements of trench and fortress represented by the element of civil society, etc. In Russia the State was everything, civil society was primordial and gelatinous; in the West, there was a proper relation between State and civil society, and when the State trembled a sturdy structure of civil society was at once revealed. The State was only an outer ditch, behind which there stood a powerful system of fortresses and earthworks: more or less numerous from one State to the next, it goes without saying—but this precisely necessitated an accurate reconnaissance of each individual country.

Bronstein's theory can be compared to that of certain French syndicalists on the General Strike, and to Rosa [Luxemburg]'s theory in the work translated by Alessandri. Rosa's book and theories anyway influenced the French syndicalists, as is clear from some of Rosmer's articles on Germany in *Vie ouvrière* (first series in pamphlet form). It partly depends too on the theory of spontaneity.

The Transition from the War of Manœuvre (Frontal Attack) to the War of Position—in the Political Field as well

This seems to me to be the most important question of political theory that the post-war period has posed, and the most difficult to solve correctly. It is related to the problems raised by Bronstein [Trotsky], who in one way or another can be considered the political theorist of frontal attack in a period in which it only leads to defeats. This transition in political science is only indirectly (mediately) related to that which took place in the military field, although certainly a relation exists and an essential one. The war of position demands enormous sacrifices by infinite masses of people. So an unprecedented concentration of hegemony is necessary, and hence a more 'interventionist' government, which will take the offensive more openly against the oppositionists and organize permanently the 'impossibility' of internal disintegration—with controls of every kind, political, administrative, etc., reinforcement of the hegemonic 'positions' of the dominant group, etc. All this indicates that we have entered a culminating phase in the political-historical situation, since in politics the 'war of position', once won, is decisive definitively. In politics, in other words, the war of manœuvre subsists so long as it is a question of winning positions which are not decisive, so that all the resources of the State's hegemony cannot be mobilized. But when, for one reason or another, these positions have lost their value and only the decisive positions are at stake, then one passes over to siege warfare; this is concentrated, difficult, and requires exceptional qualities of patience and inventiveness. In politics, the siege is a reciprocal one, despite all appearances, and the mere fact that the ruler has to muster all his resources demonstrates how seriously he takes his adversary.

3.   The Culture of Marxism

The position of the philosophy of praxis is the antithesis of the Catholic. The philosophy of praxis does not tend to leave the 'simple' in their primitive philosophy of common sense, but rather to lead them to a higher conception of life. If it affirms the need for contact between intellectuals and simple it is not in order to restrict scientific activity and preserve unity at the low level of the masses, but precisely in order to construct an intellectual-moral bloc which can make politically possible the intellectual progress of the mass and not only of small intellectual groups.

The active man-in-the-mass has a practical activity, but has no clear theoretical consciousness of his practical activity, which nonetheless involves understanding the world in so far as it transforms it. His theoretical consciousness can indeed be historically in opposition to his activity. One might almost say that he has two theoretical consciousnesses (or one contradictory consciousness): one which is implicit in his activity and which in reality unites him with all his fellow workers in the practical transformation of the real world; and one, superficially explicit or verbal, which he has inherited from the past and uncritically absorbed. But this verbal conception is not without consequences. It holds together a specific social group, it influences moral conduct and the direction of will, with varying efficacity but often powerfully enough to produce a situation in which the contradictory state of consciousness does not permit of any action, any decision, or any choice, and produces a condition of moral and political passivity. Critical understanding of self takes place therefore through a struggle of political 'hegemonies' and of opposing directions, first in the ethical field and then in that of politics proper, in order to arrive at the working out at a higher level of one's own conception of reality. Consciousness of being part of a particular hegemonic force (that is to say, political consciousness) is the first stage towards a further progressive self-consciousness in which theory and practice will finally be one. Thus the unity of theory and practice is not just a matter of mechanical fact, but a part of the historical process, whose elementary and primitive phase is to be found in the sense of being 'different' and 'apart', in an instinctive feeling of independence, and which progresses to the level of real possession of a single and coherent conception of the world. This is why it must be stressed that the political development of the concept of hegemony represents a great philosophical advance as well as a politico-practical one. For it necessarily supposes an intellectual unity and an ethic in conformity with a conception of reality that has gone beyond common sense, and has become, if only within narrow limits, a critical conception.

However, in the most recent developments of the philosophy of praxis the exploration and refinement of the concept of the unity of theory and practice is still only at an early stage. There still remain residues of mechanicism, since people speak about theory as a 'complement' or an 'accessory' of practice, or as the handmaid of practice. It would seem right for this question too to be considered historically, as an aspect of the political question of the intellectuals. Critical self-consciousness means, historically and politically, the creation of an élite of intellectuals. A human mass does not 'distinguish' itself, does not become independent in its own right without, in the widest sense, organizing itself; and there is no organization without intellectuals, that is without organizers and leaders, in other words, without the theoretical aspect of the theory-practice nexus being distinguished concretely by the existence of a group of people 'specialized' in conceptual and philosophical elaboration of ideas. But the process of creating intellectuals is long, difficult, full of contradictions, advances and retreats, dispersals and regroupings, in which the loyalty of the masses is often sorely tried. (And one must not forget that at this early stage loyalty and discipline are the ways in which the masses participate and collaborate in the development of the cultural movement as a whole.)

The process of development is tied to a dialectic between the intellectuals and the masses. The intellectual stratum develops both quantitatively and qualitatively, but every leap forward towards a new breadth and complexity of the intellectual stratum is tied to an analogous movement on the part of the mass of the 'simple', who raise themselves to higher levels of culture and at the same time extend their circle of influence towards the stratum of specialized intellectuals, producing outstanding individuals and groups of greater or less importance. In the process, however, there continually recur moments in which a gap develops between the mass and the intellectuals (at any rate between some of them, or a group of them), a loss of contact, and thus the impression that theory is an 'accessory', a 'complement', and something subordinate. Insistence on the practical element of the theory-practice nexus, after having not only distinguished but separated and split the two elements (an operation which in itself is merely mechanical and conventional), means that one is going through a relatively primitive historical phase, one which is still economic-corporate, in which the general 'structural' framework is being quantitatively transformed and the appropriate quality-superstructure is in the process of emerging, but is not yet organically formed. One should stress the importance and significance which, in the modern world, political parties have in the elaboration and diffusion of conceptions of the world, because essentially what they do is to work out the ethics and the politics corresponding to these conceptions and act as it were as their historical 'laboratory'. The parties recruit individuals out of the working mass, and the selection is made on practical and theoretical criteria at the same time. The relation between theory and practice becomes even closer the more the conception is vitally and radically innovatory and opposed to old ways of thinking. For this reason one can say that the parties are the elaborators of new integral and totalitarian intelligentsias and the crucibles where the unification of theory and practice, understood as a real historical process, takes place. It is clear from this that the parties should be formed by individual memberships and not on the pattern of the British Labour Party, because, if it is a question of providing an organic leadership for the entire economically active mass, this leadership should not follow old schemas but should innovate. But innovation cannot come from the mass, at least at the beginning, except through the mediation of an élite for whom the conception implicit in human activity has already become to a certain degree a coherent and systematic ever-present awareness and a precise and decisive will....

The Lutheran Reformation and Calvinism created a vast national-popular movement through which their influence spread: only in later periods did they create a higher culture. The Italian reformers were infertile of any major historical success. It is true that even the Reformation, in its higher phase, necessarily adopted the style of the Renaissance and as such spread even in non-protestant countries where the movement had not had a popular incubation. But the phase of popular development enabled the Protestant countries to resist the crusade of the Catholic armies tenaciously and victoriously. Thus there was born the German nation as one of the most vigorous in modern Europe. France was lacerated by the wars of religion leading to an apparent victory of Catholicism, but it experienced a great popular reformation in the eighteenth century with the Enlightenment, Voltairianism, and the Encyclopaedia. This reformation preceded and accompanied the Revolution of 1789. It really was a matter here of a great intellectual and moral reformation of the French people, more complete than the German Lutheran Reformation, because it also embraced the great peasant masses in the countryside and had a distinct secular basis and attempted to replace religion with a completely secular ideology represented by the national and patriotic bond. Not even this reformation had an immediate flowering of high culture, except in political science in the form of the positive science of right.

A conception of the philosophy of praxis as a modern popular reformation (since those people who expect a religious reformation in Italy, a new Italian edition of Calvinism, like Missiroli and Co., are living in Cloud cuckoo-land) was perhaps hinted at by Georges Sorel, but his vision was fragmentary and intellectualistic, because of his kind of Jansenist fury against the squalor of parliamentarism and political parties. Sorel has taken from Renan the concept of the necessity of an intellectual and moral reformation; he has affirmed (in a letter to Missiroli) that great historical movements are often represented by a modern culture, etc. It seems to me, though, that a conception of this kind is implicit in Sorel when he uses primitive Christianity as a touchstone, in a rather literary way it is true, but nevertheless with more than a grain of truth; with mechanical and often contrived references, but nevertheless with occasional flashes of profound intuition.

The philosophy of praxis presupposes all this cultural past: Renaissance and Reformation, German philosophy and the French Revolution, Calvinism and English classical economics, secular liberalism and this historicism which is at the root of the whole modern conception of life. The philosophy of praxis is the crowning point of this entire movement of intellectual and moral reformation, made dialectical in the contrast between popular culture and high culture. It corresponds to the nexus Protestant Reformation plus French Revolution: it is a philosophy which is also politics, and a politics which is also philosophy. It is still going through its populist phase: creating a group of independent intellectuals is not an easy thing; it requires a long process, with actions and reactions, coming together and drifting apart, and the growth of very numerous and complex new formations. It is the conception of a subaltern social group, deprived of historical initiative, incontinuous but disorganic expansion, unable to go beyond a certain qualitative level, which still remains below the level of the possession of the State and of the real exercise of hegemony over the whole of society which alone permits a certain organic equilibrium in the development of the intellectual group. The philosophy of praxis has itself become 'prejudice' and 'superstition'. As it stands, it is the popular aspect of modern historicism, but it contains in itself the principle through which this historicism can be superseded. In the history of culture, which is much broader than the history of philosophy, every time that there has been a flowering of popular culture because a revolutionary phase was being passed through and because the metal of a new class was being forged from the ore of the people, there has been a flowering of 'materialism'; conversely, at the same time the traditional classes clung to philosophies of the spirit. Hegel, half-way between the French Revolution and the Restoration, gave dialectical form to the two moments of the life of thought, materialism and spiritualism, but his synthesis was 'a man walking on his head'. Hegel's successors destroyed this unity and there was a return to materialist systems on the one side and spiritualist on the other. The philosophy of praxis, through its founder, relived all this experience of Hegelianism, Feuerbachianism, and French materialism, in order to reconstruct the synthesis of dialectical unity, 'the man walking on his feet'. The laceration which happened to Hegelianism has been repeated with the philosophy of praxis. That is to say, from dialectical unity there has been a regress to philosophical materialism on the one hand, while on the other hand modern idealist high culture has tried to incorporate that part of the philosophy of praxis which was needed in order for it to find a new elixir.

'Politically' the materialist conception is close to the people, to 'common sense'. It is closely linked to many beliefs and prejudices, to almost all popular superstitions (witchcraft, spirits, etc.). This can be seen in popular Catholicism, and, even more so, in Byzantine orthodoxy. Popular religion is crassly materialistic, and yet the official religion of the intellectuals attempts to impede the formation of two distinct religions, two separate strata, so as not to become officially, as well as in reality, an ideology of restricted groups. But from this point of view it is important not to confuse the attitude of the philosophy of praxis with that of Catholicism. Whereas the former maintains a dynamic contact and tends continually to raise new strata of the population to a higher cultural life, the latter tends to maintain a purely mechanical contact, an external unity based in particular on the liturgy and on a cult visually imposing to the crowd. Many heretical movements were manifestations of popular forces aiming to reform the church and bring it closer to the people by exalting them. The reaction of the church was often very violent: it has created the Society of Jesus; it has clothed itself in the protective armour of the Council of Trent; although it has organized a marvellous mechanism of 'democratic' selection of its intellectuals, they have been selected as single individuals and not as the representative expression of popular groups.

In the history of cultural developments, it is important to pay special attention to the organization of culture and the personnel through whom this organization takes concrete form. G. De Ruggiero's volume on Renaissance and Reformation brings out the attitude of very many intellectuals, with Erasmus at their head: they gave way in the face of persecution and the stake. The bearer of the Reformation was therefore the German people itself in its totality, as undifferentiated mass, not the intellectuals. It is precisely this desertion of the intellectuals in the face of the enemy which explains the 'sterility' of the Reformation in the immediate sphere of high culture, until, by a process of selection, the people, which remained faithful to the cause, produced a new group of intellectuals culminating in classical philosophy.

Something similar has happened up to now with the philosophy of praxis. The great intellectuals formed on the terrain of this philosophy, besides being few in number, were not linked with the people, they did not emerge from the people, but were the expression of traditional intermediary classes, to which they returned at the great 'turning-points' of history. Some remained, but rather to subject the new conception to a systematic revision than to advance its autonomous development. The affirmation that the philosophy is a new, independent, and original conception, even though it is also a moment of world historical development, is an affirmation of the independence and originality of a new culture in incubation, which will develop with the development of social relations. What exists at any given time is a variable combination of old and new, a momentary equilibrium of cultural relations corresponding to the equilibrium of social relations. Only after the creation of the new State does the cultural problem impose itself in all its complexity and tend towards a coherent solution. In any case the attitude to be taken up before the formation of the new State can only be critico-polemical, never dogmatic; it must be a romantic attitude, but of a romanticism which is consciously aspiring to its classical synthesis. . . .

That the philosophy of praxis thinks of itself in a historicist manner, that is, as a transitory phase of philosophical thought, is not only implicit in its entire system, but is made quite explicit in the well-known thesis that historical development will at a certain point be characterized by the passage from the reign of necessity to the reign of freedom. All hitherto existing philosophies (philosophical systems) have been manifestations of the intimate contradictions by which society is lacerated. But each philosophical system taken by itself has not been the conscious expression of these contradictions, since this expression could be provided only by the *ensemble* of systems in conflict with each other. Every philosopher is, and cannot but be, convinced that he expresses the unity of the human spirit, that is, the unity of history and nature. Indeed, if such a conviction did not exist, men would not act, they would not create new history, philosophies would not become ideologies and would not in practice assume the fanatical granite compactness of the 'popular beliefs' which assume the same energy as 'material forces'.

In the history of philosophical thought Hegel represents a chapter on his own, since in his system, in one way or another, even in the form of a 'philosophical romance', one manages to understand what reality is. That is to say, one finds, in a single system and in a single philosopher, that consciousness of contradictions which one previously acquired from the ensemble of systems and of philosophers in polemic and contradiction with each other.

In a sense, moreover, the philosophy of praxis is a reform and a development of Hegelianism; it is a philosophy that has been liberated (or is attempting to liberate itself) from any unilateral and fanatical ideological elements; it is consciousness full of contradictions, in which the philosopher himself, understood both individually and as an entire social group, not only grasps the contradictions, but posits himself as an element of the contradiction and elevates this element to a principle of knowledge and therefore of action. 'Man in general', in whatever form he presents himself, is denied and all dogmatically 'unitary' concepts are spurned and destroyed as expressions of the concept of 'man in general' or of 'human nature' immanent in every man.

But even the philosophy of praxis is an expression of historical contradictions, and indeed their most complete, because most conscious, expression; this means that it too is tied to 'necessity' and not to a 'freedom' which does not exist and, historically, cannot yet exist. If, therefore, it is demonstrated that contradictions will disappear, it is also demonstrated implicitly that the philosophy of praxis too will disappear, or be superseded. In the reign of 'freedom' thought and ideas can no longer be born on the terrain of contradictions and the necessity of struggle. At the present time the philosopher—the philosopher of praxis—can only make this generic affirmation and can go no further; he cannot escape from the present field of contradictions, he cannot affirm, other than generically, a world without contradictions, without immediately creating a Utopia.

This is not to say that Utopia cannot have a philosophical value, for it has a political value and every politics is implicitly a philosophy, even if disconnected and crudely sketched. In this sense religion is the most gigantic Utopia, that is the most gigantic 'metaphysics', that history has ever known, since it is the most grandiose attempt to reconcile, in mythological form, the real contradictions of historical life. It affirms, in fact, that mankind has the same 'nature', that man in general exists, in so far as created by God, son of God, therefore brother of other men, equal to other men, and free among and as other men; and that he can conceive of himself as such, mirrored in God, who is the 'self-consciousness' of humanity; but it also affirms that all this is not of this world, but of another (the Utopia). Thus do ideas of equality, fraternity, and liberty ferment among men, among those strata of mankind who do not see themselves as equals nor as brothers of other men, nor as free in relation to them. Thus it has come about that in every radical stirring of the multitude, in one way or another, with particular forms and particular ideologies, these demands have always been raised.

At this point one can insert an element proposed by Vilich [Lenin]. The April 1917 programme, in the section devoted to the common school, and more exactly in the explanatory note to that section (see the Geneva edition of 1918) refers to the chemist and educationalist Lavoisier, guillotined under the Terror, who had put forward the concept of the common school, and had done so in accord with the popular sentiments of his age, which saw in the democratic movement of 1789 a developing reality and not just an ideology used as an instrument of government and which drew from this concrete egalitarian consequences. In Lavoisier this was still a Utopian element (an element which crops up more or less in all cultural currents that presuppose the singleness of human 'nature'), whereas for Vilich it had the demonstrative-theoretical significance of a political principle.

If the philosophy of praxis affirms theoretically that every 'truth' believed to be eternal and absolute has had practical origins and has represented a 'provisional' value (historicity of every conception of the world and of life), it is still very difficult to make people grasp 'practically' that such an interpretation is valid also for the philosophy of praxis itself, without in so doing shaking the convictions that are necessary for action. This is, moreover, a difficulty that recurs for every historicist philosophy; it is taken advantage of by cheap polemicists (particularly Catholics) in order to contrast within the same individual the 'scientist' and the 'demagogue', the philosopher and the man of action, and to deduce that historicism leads necessarily to moral scepticism and depravity. From this difficulty arise many dramas of conscience in little men, and in great men the 'Olympian' attitude à la Goethe. This is the reason why the proposition about the passage from the reign of necessity to that of freedom must be analysed and elaborated with subtlety and delicacy.

As a result even the philosophy of praxis tends to become an ideology in the worst sense of the word, that is to say a dogmatic system of eternal and absolute truths. This is particularly true when, as happens in the 'Popular Manual', it is confused with vulgar materialism, with its metaphysics of 'matter' which is necessarily eternal and absolute.

It is also worth saying that the passage from necessity to freedom takes place through the society of men and not through nature (although it may have effects on our intuition of nature, on scientific opinions, etc.). One can go so far as to affirm that, whereas the whole system of the philosophy of praxis may fall away in a unified world, many idealist conceptions, or at least certain aspects of them which are Utopian during the reign of necessity, could become 'truth' after the passage. One cannot talk of the 'spirit' when society is divided into groups without necessarily concluding that this 'spirit' is just *esprit de corps*'! (This fact is implicitly recognized when it is said, as is done by Gentile in his book on modernism, following Schopenhauer, that religion is the philosphy of the multitude, whereas philosophy is the religion of the elect, that is of the great intellectuals.) But it will be possible to talk in these terms after the unification has taken place (etc.).

The claim, presented as an essential postulate of historical materialism, that every fluctuation of politics and ideology can be presented and expounded as an immediate expression of the structure, must be contested in theory as primitive infantilism, and combated in practice with the authentic testimony of Marx, the author of concrete political and historical works. Particularly important from this point of view are *The Eighteenth Brumaire* and the writings on the Eastern Question, but also other writings (*Revolution and Counter-revolution in Germany, The Civil War in France*, and lesser works). An analysis of these works allows one to establish better the Marxist historical methodology, integrating, illuminating, and interpreting the theoretical affirmations scattered throughout his works.

One will be able to see from this the real precautions introduced by Marx into his concrete researches, precautions which could have no place in his general works. Among these precautions the following examples can be enumerated:

1. The difficulty of identifying at any given time, statically (like an instantaneous photographic image), the structure. Politics in fact is at any given time the reflection of the tendencies of development in the structure, but it is not necessarily the case that these tendencies must be realized. A structural phase can be concretely studied and analysed only after it has gone through its whole process of development, and not during the process itself, except hypothetically and with the explicit proviso that one is dealing with hypotheses.

2. From this it can be deduced that a particular political act may have been an error of calculation on the part of the leaders [*dirigenti*] of the dominant classes, an error which historical development, through the parliamentary and governmental 'crises' of the directive [*dirigenti*] classes, then corrects and goes beyond. Mechanical historical materialism does not allow for the possibility of error, but assumes that every political act is determined, immediately, by the structure, and therefore as a real and permanent (in the sense of achieved) modification of the structure. The principle of 'error' is a complex one: one may be dealing with an individual impulse based on mistaken calculations or equally it may be a manifestation of the attempts of specific groups or sects to take over hegemony within the directive grouping, attempts which may well be unsuccessful.

3. It is not sufficiently borne in mind that many political acts are due to internal necessities of an organizational character, that is they are tied to the need to give coherence to a party, a group, a society. This is made clear for example in the history of the Catholic church. If, for every ideological struggle within the Church, one wanted to find an immediate primary explanation in the structure, one would really be caught napping: all sorts of politico-economic romances have been written for this reason. It is evident on the contrary that the majority of these discussions are connected with sectarian and organizational necessities. In the discussion between Rome and Byzantium on the Procession of the Holy Spirit, it would be

ridiculous to look in the structure of the European East for the claim that
it proceeds only from the Father, and in that of the West for the claim that
it proceeds from the Father and the Son. The two Churches, whose existence
and whose conflict is dependent on the structure and on the whole of
history, posed questions which are principles of distinction and internal
cohesion for each side, but it could have happened that either of the
Churches could have argued what in fact was argued by the other. The
principle of distinction and conflict would have been upheld all the same,
and it is this problem of distinction and conflict that constitutes the historical
problem, and not the banner that happened to be hoisted by one side or
the other. . . .

Source: A Gramsci, *Selections from the Prison Notebooks*, ed. G. Nowell Smith
and Q. Hoare (London, 1971), pp. 6 ff., 12 f., 57 f., 60 f., 177 ff., 332 ff., 394 ff.,
404 ff.

FURTHER NOTES

WORKS

GRAMSCI, ANTONIO, *Letters from Prison* (New York, 1973).
——*Selections from Political Writings: 1910–1920* (London, 1977).
——*Selections from Political Writings: 1921–1926* (London, 1978).
——*Selections from the Prison Notebooks*, ed. Q. Hoare and G. Nowell Smith
(London, 1971).

COMMENTARIES

BOGGS, C., *Gramsci's Marxism* (London, 1976).
CAMMETT, J. M., *Antonio Gramsci and the Origins of Italian Communism*
(Stanford, 1969).
DAVIDSON, A. B., *Antonio Gramsci: Towards an Intellectual Biography*
(London, 1977).
FEMIA, J., *Gramsci's Political Thought* (Oxford, 1981).
JOLL, J., *Gramsci* (London, 1977).
SASSOON, A., *Gramsci's Politics* (New York, 1980).
WILLIAMS, G. A., *Proletarian Order: Antonio Gramsci, Factory Councils and
the Origins of Communism in Italy* (London, 1975).

PART II
CONTEMPORARY MARXISM

12

JOSEF STALIN: SOVIET MARXISM

The dictatorship of the proletariat cannot be contrasted to the leadership (the 'dictatorship') of the party. This is inadmissible because the leadership of the party is the principal thing in the dictatorship of the proletariat.

Leninism p. 142.

JOSEF VISSARIONOVICH DJUGASHVILI (1879-1953), *son of a poor Georgian cobbler, assumed the name of Stalin on joining Lenin and the Bolsheviks in 1904. He was elected to the Bolshevik Central Committee in 1912, was active in the 1917 revolution, and became general secretary of the party in 1922. Following Lenin's death in 1924, Stalin successively ousted his rivals Trotsky, Zinoviev, and Bukharin and became undisputed leader of the party by 1929. Although no great theoretician, Stalin expressed a lucid and rigidly dogmatic version of Leninism which was widely influential in its time as the unquestionable orthodoxy of the world-wide Communist movement. In the first two extracts, from* The Foundations of Leninism, *written in 1924, Stalin gives his own interpretation of Lenin's theory of revolution and of the dictatorship of the proletariat. The next two extracts represent Stalin's novel doctrines: the first, from* Problems of Leninism *of 1926, stressed the possibility of building Socialism in one country: the second, from a discussion of* Marxism and Linguistics *of 1951, emphasizes the active role of the superstructure and the concept of 'revolution from above'. In the final extract, from* Economic Problems of the USSR *(1952), Stalin discusses the inevitability of war among capitalist nations.*

1. Lenin's Theory of Revolution

FROM this theme I take three questions: (*a*) the importance of theory for the proletarian movement; (*b*) criticism of the 'theory' of spontaneity; (*c*) the theory of the proletarian revolution.

  1. *The importance of theory.* Some think that Leninism is the precedence of practice over theory in the sense that its main point is the translation of the Marxian theses into deeds, their 'execution'; as for theory, it is alleged that Leninism is rather unconcerned about it. We know that Plekhanov

occasionally chaffed Lenin about his 'unconcern' for theory, and particularly for philosophy. We also know that theory is not held in great favour by many present-day Leninist practical workers, particularly in view of the overwhelming amount of practical work imposed upon them by present circumstances. I must declare that this more than odd opinion about Lenin and Leninism is quite wrong and bears no relation whatever to the truth; that the attempt of practical workers to brush theory aside runs counter to the whole spirit of Leninism and is fraught with serious dangers to the cause.

Theory is the experience of the working-class movement in all countries taken in its general aspect. Of course, theory becomes aimless if it is not connected with revolutionary practice, just as practice gropes in the dark if its path is not illumined by revolutionary theory. But theory can become a tremendous force in the working-class movement if it is built up in indissoluble connection with revolutionary practice; for it, and it alone, can give the movement confidence, the power of orientation, and an understanding of the inherent connection between surrounding events; for it, and it alone, can help practice to discern not only how and in which direction classes are moving at the present time, but also how and in which direction they will move in the near future. None other than Lenin uttered and repeated scores of times the well-known thesis that: '*Without a revolutionary theory there can be no revolutionary movement.*' (Lenin, *Selected Works, ii.* 47.)

Lenin, better than anyone else, understood the great importance of theory, particularly for a party such as ours, in view of the role of vanguard fighter of the international proletariat which has fallen to its lot, and in view of the complicated internal and international situation in which it finds itself. Foreseeing this special role of our party as far back as 1902, he thought it necessary even then to point out that: '. . . *the role of vanguard can be fulfilled only by a party that is guided by the most advanced theory*'. (*Ibid.*, p. 48)

It need hardly be proved that now, when Lenin's prediction about the role of our party has come true, this thesis of Lenin's acquires particular force and particular importance. Perhaps the most striking expression of the great importance which Lenin attached to theory is the fact that none other than Lenin undertook the very serious task of generalizing, in line with the materialist philosophy, the most important achievements of science from the time of Engels down to his own time, as well as of subjecting to comprehensive criticism the anti-materialistic trends along Marxists. Engels said that 'materialism must assume a new aspect with every new great discovery'. It is well known that none other than Lenin accomplished this task for his own time in his remarkable work *Materialism and Empirio-criticism*. It is well known that Plekhanov, who loved to chaff Lenin about his 'unconcern' for philosophy, did not even dare to make a serious attempt to undertake such a task.

2. *Criticism of the 'theory' of spontaneity, of the role of the vanguard in the movement.* The 'theory' of spontaneity is a theory of opportunism, a theory of worshipping the spontaneity of the labour movement, a theory which actually repudiates the leading role of the vanguard of the working class, of the party of the working class.

The theory of worshipping spontaneity is decidedly opposed to the revolutionary character of the working-class movement; it is opposed to the movement taking the line of struggle against the foundations of capitalism; it stands for the idea of the movement proceeding exclusively along the line of 'realizable' demands, of demands 'acceptable' to capitalism; it stands entirely for the 'line of least resistance'. The theory of spontaneity is the ideology of trade unionism.

The theory of worshipping spontaneity is decidedly opposed to lending the spontaneous movement consciousness and system. It is opposed to the idea of the party marching at the head of the working class, of the party raising the masses to the level of class-consciousness, of the party leading the movement; it stands for the idea that the class-conscious elements of the movement must not hinder the movement from taking its own course; it stands for the idea that the party is only to heed the spontaneous movement and follow in its tail. The theory of spontaneity is the theory of belittling the role of the conscious element in the movement, the ideology of '*khvostism*'—the logical basis of *all* opportunism.

In practice this theory, which appeared on the scene even before the first revolution in Russia, led its adherents, the so-called 'economists', to deny the need for an independent workers' party in Russia, to oppose the revolutionary struggle of the working class for the overthrow of tsardom, to preach a purely trade-unionist policy in the movement, and, in general, to surrender the labour movement to the hegemony of the liberal bourgeoisie.

The fight of the old *Iskra* and the brilliant criticism of the theory of '*khvostism*' in Lenin's pamphlet *What is to be done?* not only smashed so-called 'economism', but also created the theoretical foundations for a truly revolutionary movement of the Russian working class.

Without this fight it would have been quite useless even to think of creating an independent workers' party in Russia and of its playing a leading part in the revolution.

But the theory of worshipping spontaneity is not peculiar to Russia. It is extremely widespread—in a somewhat different form, it is true—in all the parties of the Second International, without exception. I have in mind the so-called 'productive forces' theory, vulgarized by the leaders of the Second International—a theory that justifies everything and conciliates everybody, that states facts and explains them only after everyone has become sick and tired of them, and, having stated them, rests content with that. Marx said that the materialist theory could not confine itself to explaining the world, that it must also change it. But Kautsky and Co. are not concerned with this; they prefer to rest content with the first part of Marx's formula. Here is one of the numerous examples of the application of this 'theory': It is said that before the imperialist war the parties of the Second International threatened to declare 'war against war' if the imperialists should start a war. It is said that on the very eve of the war these parties pigeonholed the 'war against war' slogan and applied an opposite slogan, viz., 'war for the imperialist fatherland'. It is said that as a result of this change of slogans millions of workers were sent to their death. But it would be a mistake to think that there must have been people who were to blame for this, that someone was unfaithful to the working class or betrayed it. Not at all! Everything happened as it should have happened. Firstly, because the International is 'an instrument of peace', and not of war. Secondly, because, in view of the 'level of the productive forces' which then prevailed, there was nothing else that could be done. The 'productive forces' are 'to blame'. This is the precise explanation vouchsafed to 'us' by Mr Kautsky's 'productive forces theory'. And whoever does not believe in this 'theory' is not a Marxist. The role of the parties? Their part in the movement? But what can a party do against so decisive a factor as the 'level of the productive forces'? . . .

One could cite a host of similar examples of the falsification of Marxism.

It is hardly necessary to prove that this spurious Marxism, designed to hide the nakedness of opportunism, is merely a European variety of the selfsame theory of 'khvostism' which Lenin fought even before the first Russian revolution.

It is hardly necessary to prove that the demolition of this theoretical falsification is a prerequisite for the creation of truly revolutionary parties in the West.

3. *The theory of the proletarian revolution.* The Leninist theory of the proletarian revolution proceeds from three fundamental theses.

First Thesis: The domination of finance capital in the advanced capitalist countries; the issue of stocks and bonds as the principal operation of finance capital; the export of capital to the sources of raw materials, which is one of the foundations of imperialism; the omnipotence of a financial oligarchy, which is the result of the domination of finance capital—all this reveals the grossly parasitic character of monopolist capitalism, makes the yoke of the capitalist trusts and syndicates a hundred times more burdensome, quickens the revolt of the working class against the foundations of capitalism, and brings the masses to the proletarian revolution as their only salvation. (Cf. Lenin, *Imperialism, the Highest Stage of Capitalism*.)

Hence the first conclusion: intensification of the revolutionary crisis within the capitalist countries and growth of the elements of an explosion on the internal, proletarian front in the 'mother countries'.

Second Thesis: The increase in the export of capital to the colonies and dependent countries; the extension of 'spheres of influence' and colonial possessions until they cover the whole globe; the transformation of capitalism into a *world system* of financial enslavement and colonial oppression of the vast majority of the population of the earth by a handful of 'advanced' countries—all this has, on the one hand, converted the separate national economies and national territories into links in a single chain called world economy and, on the other hand, split the population of the globe into two camps: a handful of 'advanced' capitalist countries which exploit and oppress vast colonies and dependencies, and the vast majority of colonial and dependent countries which are compelled to fight for their liberation from the imperialist yoke. (Cf. Lenin, *Imperialism*.)

Hence the second conclusion: intensification of the revolutionary crisis in the colonial countries and growth of the elements of revolt against imperialism on the external, colonial front.

Third Thesis: The monopolistic possession of 'spheres of influence' and colonies; the uneven development of the different capitalist countries, leading to a frenzied struggle for the redivision of the world between the countries which have already seized territories and those claiming their 'share'; imperialist wars as the only method of restoring the disturbed 'equilibrium'—all this leads to the aggravation of the third front, the inter-capitalist front, which weakens imperialism: the front of the revolutionary proletariat and the front of colonial emancipation. (Cf. *Imperialism*.)

Hence the third conclusion: that under imperialism wars cannot be averted, and that a coalition between the proletarian revolution in Europe and the colonial revolution in the East in a united world front of revolution against the world front of imperialism is inevitable.

Lenin combines all these conclusions into one general conclusion that 'imperialism is the eve of the Socialist revolution'. (Lenin, *Selected Works*, v. 5).

The very approach to the question of the proletarian revolution, of the character of the revolution, of its scope, of its depth, the scheme of the revolution in general, changes accordingly.

Formerly, the analysis of the conditions for the proletarian revolution was usually approached from the point of view of the economic state of individual countries. Now, this aproach is no longer adequate. Now the matter must be approached from the point of view of the economic state of all or the majority of countries, from the point of view of the state of world economy; for individual countries and individual national economies have ceased to be self-sufficient units, have become links in a single chain called world economy; for the old 'cultured' capitalism has evolved into imperialism, and imperialism is a world system of financial enslavement and colonial oppression of the vast majority of the population of the earth by a handful of 'advanced' countries.

Formerly it was the accepted thing to speak of the existence or absence of objective conditions for the proletarian revolution in individual countries, or, to be more precise, in one or another developed country. Now this point of view is no longer adequate. Now we must speak of the existence of objective conditions for the revolution in the entire system of world imperialist economy as an integral unit; the existence within this system of some countries that are not sufficiently developed industrially cannot serve as an insurmountable obstacle to the revolution, *if* the system as a whole, or, more correctly, *because* the system as a whole is already ripe for revolution.

Formerly it was the accepted thing to speak of the proletarian revolution in one or another developed country as of something separate and self-sufficient, facing a separate national front of capital as its opposite. Now this point of view is no longer adequate. Now we must speak of the world proletarian revolution; for the separate national fronts of capital have become links in a single chain called the world front of imperialism, which must be opposed by a common front of the revolutionary movement in all countries.

Formerly the proletarian revolution was regarded exclusively as the result of the internal development of a given country. Now, this point of view is no longer adequate. Now the proletarian revolution must be regarded primarily as the result of the development of the contradictions within the world system of imperialism, as the result of the snapping of the chain of the imperialist world front in one country or another.

Where will the revolution begin? Where, in what country, can the front of capital be pierced first?

Where industry is more developed, where the proletariat constitutes the majority, where there is more culture, where there is more democracy—that was the reply usually given formerly.

No, objects the Leninist theory of revolution; *not necessarily where industry is more developed*, and so forth. The front of capital will be pierced where the chain of imperialism is weakest, for the proletarian revolution is the result of the breaking of the chain of the world imperialist front at its weakest link; and it may turn out that the country which has started the revolution, which has made a breach in the front of capital, is less developed in a capitalist sense than other, more developed, countries, which have, however, remained within the framework of capitalism.

In 1917 the chain of the imperialist world front proved to be weaker in Russia than in the other countries. It was there that the chain gave way and provided an outlet for the proletarian revolution. Why? Because in Russia a great popular revolution was unfolding, and at its head marched the revolutionary proletariat, which had such an important ally as the vast mass of the peasantry who were oppressed and exploited by the landlords. Because the revolution there was opposed by such a hideous representative of imperialism as tsarism, which lacked all moral prestige and was deservedly hated by the whole population. The chain proved to be weaker in Russia, although that country was less developed in a capitalist sense than, say, France or Germany, England or America.

Where will the chain break in the near future? Again, where it is weakest. It is not precluded that the chain may break, say in India. Why? Because that country has a young, militant, revolutionary proletariat, which has such an ally as the national liberation movement—an undoubtedly powerful and undoubtedly important ally. Because there the revolution is opposed by such a well-known foe as foreign imperialism, which lacks all moral credit and is deservedly hated by the oppressed and exploited masses of India.

It is also quite possible that the chain will break in Germany. Why? Because the factors which are operating, say, in India are beginning to operate in Germany as well; but, of course, the enormous difference in the level of development between India and Germany cannot but leave its impress on the progress and outcome of a revolution in Germany.

That is why Lenin said that:

The West-European capitalist countries ... are accomplishing their development towards Socialism ... not by the even 'ripening' of Socialism, but by the exploitation of some countries by others, by the exploitation of the first of the countries to be vanquished in the imperialist war combined with the exploitation of the whole of the East. On the other hand, precisely as a result of the first imperialist war, the East has been finally drawn into the revolutionary movement, has been drawn into the common maelstrom of the world revolutionary movement. (Lenin, *Selected Works*, ix. 399.)

Briefly, the chain of the imperialist front must, as a rule, give way where the links are weaker and, at all events, not necessarily where capitalism is more developed, where there is such and such a percentage of proletarians and such and such a percentage of peasants, and so on.

This is why in deciding the question of proletarian revolution statistical calculations of the percentage of the proletarian population in a given country lose the exceptional importance so eagerly attached to them by the pedants of the Second International, who have not understood imperialism and who fear revolution like the plague.

To proceed: the heroes of the Second International asserted (and continue to assert) that between the bourgeois-democratic revolution and the pro-letarian revolution there is a chasm, or at any rate a Chinese wall, separating one from the other by a more or less protracted interval of time, during which the bourgeoisie, having come into power, develops capitalism, while the proletariat accumulates strength and prepares for the 'decisive struggle' against capitalism. This interval is usually calculated to extend over many decades, if not longer. It is hardly necessary to prove that this Chinese wall 'theory' is totally devoid of scientific meaning under the conditions of imperialism, that it is and can be only a means of concealing and camou-flaging the counter-revolutionary aspirations of the bourgeoisie. It is hardly necessary to prove that under the conditions of imperialism, which is pregnant with collisions and wars; under the conditions of the 'eve of the Socialist revolution' when 'flourishing' capitalism is becoming 'moribund' capitalism (Lenin) and the revolutionary movement is growing in all coun-tries of the world; when imperialism is allying itself with all reactionary forces without exception, down to and including tsarism amd serfdom, thus making imperative the coalition of all revolutionary forces, from the proletarian movement of the West to the national liberation movement of the East; when the overthrow of the survivals of the regime of feudal serfdom becomes impossible without a revolutionary struggle against imperialism—it is hardly necessary to prove that the bourgeois-democratic revolution, in a more or less developed country, must under such cir-cumstances verge upon the proletarian revolution, that the former must pass into the latter. The history of the revolution in Russia has provided palpable proof that this thesis is correct and incontrovertible. It was not without reason that Lenin, as far back as 1905, on the eve of the first Russian revolution, in his pamphlet *Two Tactics*, depicted the bourgeois-democratic revolution and the Socialist revolution as two links in the same chain, as a single and integral picture of the sweep of the Russian revolution.

The proletariat must carry to completion the democratic revolution, by allying to itself the mass of the peasantry in order to crush by force the resistance of the autocracy and to paralyse the instability of the bourgeoisie. The proletariat must accomplish the Socialist revolution by allying to itself the mass of the semi-proletarian elements of the population in order to crush by force the resistance of the bourgeoisie and to paralyse the instability of the peasantry and the petty bourgeoisie. Such are the tasks of the proletariat, which the new *Iskra*-ists always present so narrowly in their arguments and resolutions about the scope of the revolution. (Lenin, *Selected Works*, iii. 110–11.)

I do not even mention other, later works of Lenin's, in which the idea of the bourgeois revolution passing into the proletarian revolution stands out in greater relief than in *Two Tactics* as one of the corner-stones of the Leninist theory of revolution.

It transpires that certain people believe that Lenin arrived at this idea only in 1916, that up to that time he had thought that the revolution in Russia would remain within the bourgeois framework, that power, consequently, would pass from the hands of the organ of the dictatorship of the proletariat and the peasantry to the hands of the bourgeoisie and not of the proletariat. It is said that this assertion has even penetrated into our Communist Press. I must say that this assertion is absolutely wrong, that it is totally at variance with the facts.

I might refer to Lenin's well-known speech at the Third Congress of the Party (1905), in which he described the dictatorship of the proletariat and the peasantry, i.e., the victory of the democratic revolution, not as the 'organization of order' but as the 'organization of war'. (Cf. Lenin, *Collected Works*, Russian edition, vii. 264.)

Further, I might refer to Lenin's well-known articles *On the Provisional Government* (1905), where, depicting the pospects of the unfolding Russian revolution, he assigns to the party the task of 'striving to make the Russian revolution not a movement of a few months, but a movement of many years, so that it may lead, not merely to slight concessions on the part of the powers that be, but to the complete overthrow of those powers'; where, enlarging further on these prospects and linking them with the revolution in Europe, he goes on to say:

And if we succeed in doing that, then . . . the revolutionary conflagration will spread all over Europe; the European worker, languishing under bourgeois reaction, will rise in his turn and will show us 'how it is done'; then the revolutionary wave in Europe will sweep back into Russia and will convert an epoch of a few revolutionary years into an epoch of several revolutionary decades. . . . (Lenin, *Selected Works*, iii. 31.)

I might also refer to a well-known article by Lenin published in November 1915, in which he writes;

The proletariat is fighting and will fight valiantly, to capture power, for a republic, for the confiscation of the land ... for the participation of the 'non-proletarian masses of the people' in freeing *bourgeois* Russia from *military-feudal* 'imperialism' (= tsarism). And the proletariat will *immediately* take advantage of this liberation of bourgeois Russia from tsarism, from the agrarian power of the landlords, not to aid the rich peasants in their struggle against the rural worker, but to bring about the Socialist revolution in alliance with the proletarians of Europe. (Lenin, *Selected Works*, v. 163.)

Finally, I might refer to the well-known passage in Lenin's pamphlet *The Proletarian Revolution and the Renegade Kautsky*, where, referring to the above-quoted passage in *Two Tactics* on the scope of the Russian revolution, he arrives at the following conclusion:

Things have turned out just as we said they would. The course taken by the revolution has confirmed the correctness of our reasoning. *First*, with the 'whole' of the peasantry against the monarchy, against the landlords, against the medieval regime (and to that extent the revolution remains bourgeois, bourgeois-democratic). *Then*, with the poorest peasants, with the semi-proletarians, with all the exploited, *against capitalism*, including the rural rich, the kulaks, the profiteers, and to that extent the revolution becomes a *Socialist* one. To attempt to raise an artificial Chinese wall between the first and second, to separate them by *anything else than* the degree of preparedness of the proletariat and the degree of its unity with the poor peasants, means monstrously to distort Marxism, to vulgarize it, to substitute liberalism in its place. (Lenin, *Selected Works*, vii. 191.)

Enough, I think.

Very well, we may be told: but if this be the case, why did Lenin combat the idea of 'permanent (uninterrupted) revolution'?

Because Lenin proposed that the revolutionary capacities of the peasantry be utilized 'to the utmost' and that the fullest use be made of their revolutionary energy for the complete liquidation of tsarism and for the transition to the proletarian revolution, whereas the adherents of 'permanent revolution' did not understand the important role of the peasantry in the Russian revolution, underestimated the strength of the revolutionary energy of the peasantry, underestimated the strength and capacity of the Russian proletariat to lead the peasantry, and thereby hampered the work of emancipating the peasantry from the influence of the bourgeoisie, the work of rallying the peasantry around the proletariat.

Because Lenin proposed that the work of the revolution *be crowned* with the transfer of power to the proletariat, whereas the adherents of 'permanent' revolution wanted to *begin* at once with the establishment of the power of the proletariat, failing to realize that in so doing they were closing their eyes to such a 'minor detail' as the survivals of serfdom and were leaving out of account so important a force as the Russian peasantry, failing to understand that such a policy could only retard the winning of the peasantry to the side of the proletariat.

Consequently, Lenin fought the adherents of 'permanent' revolution, not over the question of 'uninterruptedness', for he himself maintained the point of view of uninterrupted revolution, but because they underestimated the role of the peasantry, which is an enormous reserve force for the proletariat, because they failed to understand the idea of the hegemony of the proletariat.

The idea of 'permanent' revolution is not a new idea. It was first advanced by Marx at the end of the forties in his well-known *Address to the Communist League* (1850). It is from this document that our 'permanentists' took the idea of uninterrupted revolution. It should be noted, however, that in taking it from Marx our 'permanentists' altered it somewhat, and in altering it spoilt it and made it unfit for practical use. The experienced hand of Lenin was needed to rectify this mistake, to take Marx's idea of uninterrupted revolution in its pure form and make it a corner-stone of his theory of revolution.

Here is what Marx, in his *Address*, after enumerating a number of revolutionary-democratic demands which he calls upon the Communists to win, says about uninterrupted revolution:

While the democratic petty bourgeois wish to bring the revolution to a conclusion as quickly as possible, and with the achievement, at most, of the above demands, it is our interest and our task to make the revolution permanent, until all more or less possessing classes have been displaced from domination, until the proletariat has conquered State power, and the association of proletarians, not only in one country but in all the dominant countries of the world, has advanced so far that competition among the proletarians of these countries has ceased and that at least the decisive productive forces are concentrated in the hands of the proletarians.

In other words:

(*a*) Marx did not propose to *begin* the revolution in the Germany of the fifties with the immediate establishment of the proletarian power—*contrary* to the plans of our Russian 'permanentists'.

(*b*) Marx proposed only that the work of the revolution *be crowned* with the establishment of proletarian State power, by hurling, step by step, one section of the bourgeoisie after another from the heights of power, in order, after the attainment of power by the proletariat, to kindle the fire of revolution in every country—*fully in line* with everything that Lenin taught and carried out in the course of our revolution in pursuit of this theory of the proletarian revolution under the conditions of imperialism.

It follows, then, that our Russian 'permanentists' have not only under-estimated the role of the peasantry in the Russian revolution and the importance of the idea of the hegemony of the proletariat, but have altered (for the worse) Marx's idea of 'permanent' revolution, making it unfit for practical use.

That is why Lenin ridiculed the theory of our 'permanentists', calling it 'original' and 'fine', and accusing them of refusing to 'stop to think why, for ten whole years, life has passed by this fine theory'. (Lenin's article was written in 1915, ten years after the appearance of the theory of the 'permanentists' in Russia.) (Lenin, *Selected Works*, v. 162.)

That is why Lenin regarded this theory as a semi-Menshevik theory and said that it 'borrows from the Bolsheviks their call for a decisive revolutionary struggle and the conquest of political power by the proletariat, and from the Mensheviks the "repudiation" of the role of the peasantry'. (Ibid.)

This, then, is the position in regard to Lenin's idea of the bourgeois-democratic revolution passing into the proletarian revolution, of utilizing the bourgeois revolution for the 'immediate' transition to the proletarian revolution.

To proceed. Formerly, the victory of the revolution in one country was considered impossible, on the assumption that it would require the combined action of the proletarians of all or at least of a majority of the advanced countries to achieve victory over the bourgeoisie. Now this point of view no longer accords with the facts. Now we must proceed from the possibility of such a victory, for the uneven and spasmodic character of the development of the various capitalist countries under the conditions of imperialism, the development, within imperialism, of catastrophic con-tradictions leading to inevitable wars, the growth of the revolutionary movement in all countries of the world—all this leads, not only to the possibility, but also to the necessity of the victory of the proletariat in individual countries. The history of the Russian Revolution is direct proof of this. At the same time, however, it must be borne in mind that the overthrow of the bourgeoisie can be successfully accomplished only when certain absolutely necessary conditions exist, in the absence of which there can be even no question of the proletariat taking power.

Here is what Lenin says about these conditions in his pamphlet '*Left-wing' Communism, an Infantile Disorder*:

The fundamental law of revolution, which has been confirmed by all revolutions, and particularly by all three Russian revolutions in the twentieth century, consists in the following: it is not enough for revolution that the exploited and oppressed masses should understand the impossibility of living in the old way and demand changes; for revolution it is necessary that the exploiters should not be able to live and rule in the old way. Only when the *'lower classes' do not want* the old way, and when the 'upper classes' *cannot carry on in the old way*—only then can revolution triumph. This truth may be expressed in other words: *Revolution is impossible without a nation-wide crisis (affecting both the exploited and the exploiters)*. It follows that for revolution it is essential, first, that a majority of the workers (or at least a majority of the class-conscious, thinking, politically active workers) should fully understand the necessity for revolution and be ready to sacrifice their lives for it: secondly, that the ruling classes should be passing through a governmental crisis which would draw even the most backward masses into politics ... weaken the government and make it possible for the revolutionaries to overthrow it rapidly. (Lenin, *Selected Works*. x. 127.)

But the overthrow of the power of the bourgeoisie and establishment of the power of the proletariat in one country still does not mean that the complete victory of Socialism has been ensured. After consolidating its power and taking the peasantry in tow, the proletariat of the victorious country can and must build up a Socialist society. But does this mean that it will thereby achieve the complete and final victory of Socialism, i.e., does it mean that with the forces of only one country it can finally consolidate Socialism and fully guarantee that country against intervention and, consequently, also against restoration? No, it does not. For this the victory of the revolution in at least several countries is needed. Therefore, the development and support of revolution in other countries is an essential task of the victorious revolution. Therefore, the revolution in the victorious country must regard itself not as a self-sufficient entity but as an aid, as a means of hastening the victory of the proletariat in other countries.

Lenin expressed this thought in a nutshell when he said that the task of the victorious revolution is to do 'the utmost possible in one country *for* the development, support and awakening of the revolution *in all countries*'. (Lenin, *Selected Works*, vii. 182.)

These, in general, are the characteristic features of Lenin's theory of proletarian revolution.

2.   The Dictatorship of the Proletariat

From this theme I take the three main questions: (*a*) the dictatorship of the proletariat as the instrument of the proletarian revolution; (*b*) the dictatorship of the proletariat as the domination of the proletariat over the bourgeoisie; (*c*) the Soviet power as the state form of the dictatorship of the proletariat.

1. *The dictatorship of the proletariat as the instrument of the proletarian revolution.* The question of the proletarian dictatorship is above all a

question of the main content of the proletarian revolution. The proletarian revolution, its movement, its scope, and its achievements acquire flesh and blood only through the dictatorship of the proletariat. The dictatorship of the proletariat is the instrument of the proletarian revolution, its organ, its most important mainstay, brought into being for the purpose of, firstly, crushing the resistance of the overthrown exploiters and consolidating the achievements of the proletarian revolution, and, secondly, carrying the proletarian revolution to its completion, carrying the revolution to the complete victory of Socialism. The revolution can vanquish the bourgeoisie, can overthrow its power, without the dictatorship of the proletariat. But the revolution will be unable to crush to resistance of the bourgeoisie, to maintain its victory, and push forward to the final victory of Socialism unless at a certain stage in its development, it creates a special organ in the form of the dictatorship of the proletariat as its principal mainstay.

'The fundamental question of revolution is the question of power.' (*Lenin.*) Does this mean that all that is required is to assume power, to seize it? No, it does not mean that. The seizure of power is only the beginning. For many reasons the bourgeoisie that is overthrown in one country remains for a long time stronger than the proletariat which has overthrown it. Therefore, the whole point is to retain power, to consolidate it, to make it invincible. What is needed to attain this? To attain this it is necessary to carry out at least the three main tasks that confront the dictatorship of the proletariat 'on the morrow' of victory:

(*a*) to break the resistance of the landlords and capitalists who have been overthrown and expropriated by the revolution, to liquidate every attempt on their part to restore the power of capital;

(*b*) to organize construction in such a way as to rally all the labouring people around the proletariat, and to carry on this work along the lines of preparing for the liquidation, the abolition of classes;

(*c*) to arm the revolution, to organize the army of the revolution for the struggle against foreign enemies, for the struggle against imperialism.

The dictatorship of the proletariat is needed to carry out, to fulfil these tasks.

The transition from capitalism to Communism [says Lenin] represents an entire historical epoch. Until this epoch has terminated, the exploiters will inevitably cherish the hope of restoration, and this *hope* will be converted into *attempts* at restoration. And after their first serious defeat, the overthrown exploiters—who had not expected their overthrow, never believed it possible, never conceded the thought of it—will throw themselves with tenfold energy, with furious passion and hatred grown a hundredfold, into the battle for the recovery of their lost 'paradise', on behalf of their families, who had been leading such a sweet and easy life and whom now the 'common herd' is condemning to ruin and destitution (or to 'common' work). . . . In the train of the capitalist exploiters will be found the broad masses of the petty bourgeoisie, with regard to whom the historical experience of every country for decades testifies that they vacillate and hesitate, one day marching behind the proletariat and the next day taking fright at the difficulties of the revolution; that they become panic-stricken at the first defeat or semi-defeat of the workers, grow nervous, run about aimlessly, snivel and rush from one camp to the other. (Lenin, *Selected Works*. vii. 140–41.)

And the bourgeoisie has its grounds for making attempts at restoration, because for a long time after its overthrow it remains stronger than the proletariat which has overthrown it.

If the exploiters are defeated in one country only [says Lenin] and this, of course, is typical, since a simultaneous revolution in a number of countries is a rare exception, they *still* remain *stronger* than the exploited. (Ibid, p. 140.)

Wherein lies the strength of the overthrown bourgeoisie?

Firstly, 'in the strength of international capital, in the strength and durability of the international connections of the bourgeoisie.' (Lenin, *Selected Works*, x. 60.)

Secondly, in the fact that

for a long time after the revolution the exploiters inevitably continue to enjoy a number of great practical advantages: they still have money (since it is impossible to abolish money all at once), some movable property—often fairly considerable: they still have various connections, habits of organization and management, know-ledge of all the 'secrets' (customs, methods, means, and possibilities) of manage-ment, superior education, close connection with the higher technical personnel (who live and think like the bourgeoisie), incomparably greater experience in the art of war (this is very important), and so on, and so forth. (Lenin, *Selected Works*, vii. 140.)

Thirdly,

in the *force of habit*, in the strength of *small-scale production*. For unfortunately, there is still very, very much of small-scale production left in the world, and small-scale production *engenders* capitalism and the bourgeoisie continuously, daily, hourly, spontaneously, and on a mass scale; . . . [for] the abolition of classes means not only driving out the landlords and capitalists—that we accomplished with comparative ease: it means also *getting rid of the small commodity producers*, and they *cannot be driven out*, they cannot be crushed, we must live *in harmony* with them; they can (and must) be remoulded and re-educated only by very prolonged, slow, cautious organization work. (Lenin, *Selected Works*, x. 60, 83.)

That is why Lenin says:

The dictatorship of the proletariat is a most determined and most ruthless war waged by the new class against a *more powerful* enemy, the bourgeoisie, whose resistance is increased *tenfold* by its overthrow; ... [that] the dictatorship of the proletariat is a persistent struggle—sanguinary and bloodless, violent and peaceful, military and economic, educational and administrative—against the forces and traditions of the old society. (Lenin, *Selected Works*, x. 60, 84.)

It need hardly be proved that there is not the slightest possibility of carrying out these tasks in a short period, of doing all this in a few years. Therefore, the dictatorship of the proletariat, the transition from capitalism to Communism, must not be regarded as a fleeting period of 'super-revolutionary' acts and decrees, but as an entire historical era, replete with civil wars and external conflicts, with persistent organizational work and economic construction, with advances and retreats, victories and defeats. This historical era is needed not only to create the economic and cultural prerequisites for the complete victory of Socialism, but also to enable the proletariat, first, to educate itself and become steeled as a force capable of governing the country, and, secondly, to re-educate and remould the petty-bourgeois strata along such lines as will assure the organization of Socialist production.

Marx said to the workers:

You will have to go through fifteen, twenty, or fifty years of civil wars and international conflicts, not only to change existing conditions, but also to change yourselves and to make yourselves capable of wielding political power.

Continuing and developing Marx's idea still further, Lenin wrote that it will be necessary under the dictatorship of the proletariat to re-educate 'millions of peasants and small proprietors and hundreds of thousands of office employees, officials and bourgeois intellectuals', to subordinate 'all these to the proletarian state and to proletarian leadership', to overcome 'their bourgeois habits and traditions ... just as it will be necessary '... to re-educate—in a protracted struggle, on the basis of the dictatorship of the proletariat—the proletarians themselves, who do not abandon their petty-bourgeois prejudices at one stroke, by a miracle, at the behest of the Virgin Mary, at the behest of a slogan, resolution or decree, but only in the course of a long and difficult mass struggle against mass petty-bourgeois influences'. (Lenin, *Selected Works*, x. 157, 156.)

2. *The dictatorship of the proletariat as the domination of the proletariat over the bourgeoisie.* From the foregoing it is evident that the dictatorship of the proletariat is not a mere change of personalities in the government, a change of 'cabinet', etc., leaving the old economic and political order intact. The Mensheviks and opportunists of all countries, who fear dictatorship like fire and in their fright substitute the concept 'conquest of power' for the concept 'dictatorship of the proletariat', usually reduce the

meaning of 'conquest of power' to a change of 'cabinet', to the accession to power of a new ministry made up of people like Scheidemann and Noske, MacDonald and Henderson. It is hardly necessary to explain that these and similar cabinet changes have nothing in common with the dictatorship of the proletariat, with the conquest of real power by the real proletariat. The MacDonalds and Scheidemanns in power, while the old bourgeois order is allowed to remain, their so-called governments cannot be anything else than an apparatus serving the bourgeoisie, a screen to hide the ulcers of imperialism, a weapon in the hands of the bourgeoisie against the revolutionary movement of the oppressed and exploited masses. Capital needs such governments as a screen when it finds it inconvenient, unprofitable, difficult to oppress and exploit the masses without the aid of a screen. Of course, the appearance of such governments is a symptom that 'over there' (i.e., in the capitalist camp) 'all is not quiet at the Shipka Pass'; nevertheless, governments of this kind necessarily remain governments of capital in disguise. The government of a MacDonald or a Scheidemann is as far removed from the conquest of power by the proletariat as the sky from the earth. The dictatorship of the proletariat is not a mere change of government, but a new State, with new organs of power, both central and local; it is the State of the proletariat, which has arisen on the ruins of the old State, the State of the bourgeoisie.

The dictatorship of the proletariat arises not on the basis of the bourgeois order, but in the process of the breaking up of this order after the overthrow of the bourgeoisie, in the process of the expropriation of the landlords and capitalists, in the process of the socialization of the principal instruments and means of production, in the process of a violent proletarian revolution. the dictatorship of the proletariat is a revolutionary power based on the use of force against the bourgeoisie.

The State is a machine in the hands of the ruling class for suppressing the resistance of its class enemies. *In this respect* the dictatorship of the proletariat does not differ essentially from the dictatorship of any other class, for the proletarian State is a machine for the suppression of the bourgeoisie. But there is one *substantial* difference. This difference consists in the fact that all hitherto existing class States have been dictatorships of an exploiting minority over the exploited majority, whereas the dictatorship of the proletariat is the dictatorship of the exploited majority over the exploiting minority.

Briefly: *the dictatorship of the proletariat is the rule—unrestricted by law and based on force—of the proletariat over the bourgeoisie, a rule enjoying the sympathy and support of the labouring and exploited masses.* (*The State and Revolution.*)

From this follow two main conclusions:

First conclusion: the dictatorship of the proletariat cannot be 'complete' democracy, democracy for *all*, for the rich as well as for the poor; the dictatorship of the proletariat 'must be a State that is democratic *in a new way—for* the proletarians and the propertyless in general—and dictatorial *in a new way—against* the bourgeoisie....' (Lenin, *Selected Works*, vii. 34.) The talk of Kautsky and Co, about universal equality, about 'pure' democracy, about 'perfect' democracy, and the like, is but a bourgeois screen to conceal the indubitable fact that equality between exploited and exploiters is impossible. The theory of 'pure' democracy is the theory of the upper stratum of the working class, which has been broken in and is being fed by the imperialist robbers. It was brought into being for the purpose of concealing the ulcers of capitalism, of touching up imperialism and lending it moral strength in the struggle against the exploited masses. Under capitalism there are no real 'liberties' for the exploited, nor can there be, if for no other reason than that the premises, printing plants, paper supplies, etc., indispensable for the actual enjoyment of 'liberties' are the privilege of the exploiters. Under capitalism the exploited masses do not, nor can they, really participate in the administration of the country, if for no other reason than that, even under the most democratic regime, governments, under the conditions of capitalism, are not set up by the people but by the Rothschilds and Stinneses, the Rockefellers and Morgans. Democracy under capitalism is *capitalist* democracy, the democracy of the exploiting minority, based on the restriction of the rights of the exploited majority, and directed against this majority. Only under the dictatorship of the proletariat are real 'liberties' for the exploited and real participation in the administration of the country by the proletarians and peasants possible. Under the dictatorship of the proletariat, democracy is *proletarian* democracy, the democracy of the exploited majority, based upon the restriction of the rights of the exploiting minority and directed against this minority.

Second conclusion: The dictatorship of the proletariat cannot arise as the result of the peaceful development of bourgeois society and of bourgeois democracy; it can arise only as the result of the smashing of the bourgeois State machine, the bourgeois army, the bourgeois bureaucratic machine, the bourgeois police.

In a preface to *The Communist Manifesto* Marx and Engels wrote: 'The working class cannot simply lay hold of the ready-made State machine and wield it for its own purposes.'

In a letter to Kugelmann (1871) Marx wrote that the task of the proletarian revolution is 'no longer, as before, to transfer the bureaucratic military machine from one hand to another, but to *smash* it, and that is a preliminary condition for every real people's revolution on the Continent'.

Marx's qualifying phrase about the Continent gave the opportunists and Mensheviks of all countries a pretext for proclaiming that Marx had thus conceded the possibility of the peaceful evolution of bourgeois democracy into a proletarian democracy, at least in certain countries outside the European continent (England, America). Marx did in fact concede that possibility, and he had good grounds for conceding it in regard to England and America in the seventies of the last century, when monopoly capitalism and imperialism did not yet exist, and when these countries, owing to the special conditions of their development, had as yet no developed militarism and bureaucracy. That was the situation before the appearance of developed imperialism. But later, after a lapse of thirty or forty years, when the situation in these countries had radically changed, when imperialism had developed and had embraced all capitalist countries without exception, when militarism and bureaucracy had appeared in England and America also, when the special conditions for peaceful development in England and the United States had disappeared—then the qualification in regard to these countries necessarily could no longer hold good.

Today [said Lenin] in 1917, in the epoch of the first great imperialist war, this qualification made by Marx is no longer valid. Both England and America, the greatest and the last representatives—in the whole world—of Anglo-Saxon 'liberty', in the sense that militarism and bureaucracy were absent, have slid down entirely into the all-European, filthy, bloody morass of military-bureaucratic institutions to which everything is subordinated and which trample everything underfoot. Today, both in England and in America, the 'preliminary condition for every real people's revolution' is the smashing, the *destruction* of the 'ready-made State machine' (brought in those countries, between 1914 and 1917, to general 'European' imperialist perfection). (Lenin, *Selected Works*, vii. 37.)

In other words, the law of violent proletarian revolution, the law of the smashing of the bourgeois State machine as a preliminary condition for such a revolution, is an inevitable law of the revolutionary movement in the imperialist countries of the world.

Of course, in the remote future, if the proletariat is victorious in the most important capitalist countries, and if the present capitalist encirclement is replaced by a Socialist encirclement, a 'peaceful' path of development is quite possible for certain capitalist countries, whose capitalists, in view of the 'unfavourable' international situation, will consider it expedient 'voluntarily' to make substantial concessions to the proletariat. But this supposition applies only to a remote and possible future. With regard to the immediate future, there is no ground whatsoever for this supposition.

Therefore, Lenin is right in saying: 'The proletarian revolution is impossible without the forcible destruction of the bourgeois State machine and the substitution for it of a *new one.* . . . (Lenin, *Selected Works*, vii. 124.)

3. *The Soviet power as the State form of the dictatorship of the proletariat.* The victory of the dictatorship of the proletariat signifies the suppression of the bourgeoisie, the smashing of the bourgeois State machine, and the substitution of proletarian democracy for bourgeois democracy. That is clear. But by means of what organizations can this colossal task be carried out? The old forms of organization of the proletariat, which grew up on the basis of bourgeois parliamentarism, are inadequate for this task—of that there can hardly be any doubt. What then, are the new forms of organization of the proletariat that are capable of serving as the grave-diggers of the bourgeois State machine, that are capable not only of smashing this machine, not only of substituting proletarian democracy for bourgeois democracy, but also of becoming the foundation of the pro-letarian State power?

This new form of organization of the proletariat is the Soviets.

Wherein lies the strength of the Soviets as compared with the old forms of organization?

In that the Soviets are the most *all-embracing* mass organizations of the proletariat, for they and they alone embrace all workers without exception.

In that the Soviets are the *only* mass organizations which embrace all the oppressed and exploited, workers and peasants, soldiers and sailors, and in which the vanguard of the masses, the proletariat, can, for this reason, most easily and most completely exercise its political leadership of the mass struggle.

In that the Soviets are the *most powerful organs* of the revolutionary struggle of the masses, of the political actions of the masses, of the insur-rection of the masses—organs capable of breaking the omnipotence of finance capital and of its political appendages.

In that the Soviets are the *immediate* organizations of the masses them-selves, i.e., they are *the most democratic* and therefore the most author-itative organizations of the masses, which facilitate to the utmost their participation in the work of building up the new state and in its admin-istration, and which bring into full play the revolutionary energy, initiative, and creative abilities of the masses in the struggle for the destruction of the old order, in the struggle for the new, proletarian order.

The Soviet power is the amalgamation and formation of the local Soviets into one common State organization, into the State organization of the proletariat as the vanguard of the oppressed and exploited masses and as the ruling class—their amalgamation into the republic of Soviets.

The essence of the Soviet power is contained in the fact that these organizations of a most pronounced mass character, these most revolutionary organizations of precisely those classes that were oppressed by the capitalists and landlords, are now the '*permanent and sole* basis of the whole power of the State, of the whole State apparatus': that 'precisely those masses which even in the most democratic bourgeois republics while being equal in law, have in fact been prevented by thousands of tricks and devices from taking part in political life and from enjoying democratic rights and liberties, are now drawn unfailingly into *constant* and, moreover, *decisive* participation in the democratic administration of the State.' (Lenin, *Selected Works*, vii. 231.)

This is why the Soviet power is a *new form* of State organization, different in principle from the old bourgeois-democratic amd parliamentary form, a *new type* of State, adapted not to the task of exploiting and oppressing the labouring masses, but to the task of completely emancipating them from all oppression and exploitation, to the tasks facing the dictatorship of the proletariat.

Lenin rightly says that with the appearance of the Soviet power 'the era of bourgeois-democratic parliamentarism has come to an end, and a new chapter in world history—the era of proletarian dictatorship—has commenced'.

What are the characteristic features of the Soviet power?

The Soviet power has a most pronounced mass character and is the most democratic State organization of all possible State organizations while classes continue to exist; for, being the arena of the bond and collaboration between the workers and the exploited peasants in their struggle against the exploiters, and basing itself in its work on this bond and on this collaboration, it represents, by virtue of this, the power of the majority of the population over the minority, it is the state of the majority, the expression of its dictatorship.

The Soviet power is the most internationalist of all State organizations in class society, for, since it destroys every kind of national oppression and rests on the collaboration of the labouring masses of the various nationalities, it facilitates, by virtue of this, the amalgamation of these masses into a single State union.

The Soviet power, by its very structure, facilitates the task of leading the oppressed and exploited masses for the vanguard of these masses—for the proletariat, as the most consolidated and most class-conscious core of the Soviets.

'The experience of all revolutions and of all movements of the oppressed classes, the experience of the world Socialist movement teaches', says Lenin, 'that the proletariat alone is able to unite and lead the scattered and backward strata of the toiling and exploited population.' (Lenin, *Selected Works*, vii. 232.) The structure of the Soviet power facilitates the practical application of the lessons drawn from this experience.

The Soviet power, by combining the legislative and executive functions in a single State body and replacing territorial electoral constituencies by industrial units, factories, and mills thereby directly links the workers and the labouring masses in general with the apparatus of State administration, teaches them how to administer the country.

The Soviet power alone is capable of releasing the army from its subordination to bourgeois command and of converting it from the instrument of oppression of the people which it is under the bourgeois order into an instrument for the liberation of the people from the yoke of the bourgeoisie, both native and foreign.

'The Soviet organization of the State alone is capable of immediately and effectively smashing and finally destroying the old, i.e., the bourgeois, bureaucratic and judicial apparatus.' (Ibid.)

The Soviet form of State alone, by drawing the mass organizations of the toilers and exploited into constant and unrestricted participation in State administration, is capable of preparing the ground for the withering away of the State, which is one of the basic elements of the future Stateless Communist society.

The republic of Soviets is thus the political form, so long sought and finally discovered, within the framework of which the economic emancipation of the proletariat, the complete victory of Socialism, is to be accomplished.

The Paris Commune was the embryo of this form; the Soviet power is its development and culmination.

Hence the three main aspects of the dictatorship of the proletariat:

1. The utilization of the power of the proletariat for the suppression of the exploiters, for the defence of the country, for the consolidation of the ties with the proletarians of other lands, and for the development and the victory of the revolution in all countries.

2. The utilization of the power of the proletariat in order to detach the toiling and exploited masses once and for all from the bourgeoisie, to consolidate the alliance of the proletariat with these masses, to enlist these masses for the work of Socialist construction, and to ensure the State leadership of these masses by the proletariat.

3. The utilization of the power of the proletariat for the organization of Socialism, for the abolition of classes, for the transition to a society without classes, to a society without a State.

The proletarian dictatorship is a combination of all three aspects. Not one of these three aspects can be advanced as the *sole* characteristic feature of the dictatorship of the proletariat. On the other hand, it is sufficient, under the conditions of capitalist encirclement, that even one of these three features be lacking for the dictatorship of the proletariat to cease being a dictatorship. Therefore, not one of these three aspects can be omitted without running the risk of distorting the concept of the dictatorship of the proletariat. Only all these three aspects taken together give us a complete and finished concept of the dictatorship of the proletariat.

The dictatorship of the proletariat has its periods, its special forms, diverse methods, of work. During the period of civil war, the violent side of the dictatorship is most conspicuous. But it by no means follows from this that no constructive work is carried on during the period of civil war. Without constructive work it is impossible to wage civil war. During the period of Socialist construction on the other hand, the peaceful organizational, and cultural work of the dictatorship, revolutionary law, etc., are most conspicuous. But here, again, it by no means follows that the violent side of the dictatorship has fallen away, or can fall away, in the period of construction. The organs of suppression, the army and other organizations, are as necessary now, in the period of construction, as they were during the period of civil war. Without these organs, constructive work by the dictatorship with any degree of security would be impossible. It should not be forgotten that for the time being the revolution has been victorious in only one country. It should not be forgotten that as long as the capitalist encirclement exists the danger of intervention, with all the consequences resulting from this danger, will exist.

3. Socialism in One Country

What do we mean by the *possibility* of the victory of Socialism in one country?

We mean the possibility of solving the contradictions between the proletariat and the peasantry with the aid of the internal forces of our country, the possibility of the proletariat assuming power and using that power to build a complete Socialist society in our country, with the sympathy and the support of the proletarians of other countries, but without the preliminary victory of the proletarian revolution in other countries.

Without such a possibility, the building of Socialism is building without prospects, building without being sure that Socialism will be built. It is no use building Socialism without being sure that we can build it, without being sure that the technical backwardness of our country is not an *insuperable* obstacle to the building of a complete Socialist society. To deny such a possibility is to display lack of faith in the cause of building Socialism, to abandon Leninism.

What do we mean by the *impossibility* of the complete, final victory of Socialism in one country without the victory of the revolution in other countries?

We mean the impossibility of having full guarantees against intervention, and consequently against the restoration of the bourgeois order, without the victory of the revolution in at least a number of countries. To deny this indisputable thesis is to abandon internationalism, to abandon Leninism. . . .

For what else is our country, 'the country that is building Socialism', if not the base of the world revolution? But can it be a real base of the world revolution if it is incapable of building Socialist society? Can it remain the mighty centre of attraction for the workers of all countries that it undoubtedly is now, if it is incapable of achieving victory over the capitalist elements in its economy, the victory of Socialist construction? I think not. But does it not follow from this that scepticism regarding the victory of Socialist construction, the dissemination of this scepticism, will lead to our country being discredited as the base of the world revolution? And if our country is discredited the world revolutionary movement will be weakened. How did Messrs the Social Democrats try to scare the workers away from us? By preaching that 'the Russians will get nowhere'. Wherewith do we beat the Social Democrats now, when we attract numerous workers' delegations to our country and thereby strengthen the position of Communism all over the world? By our successes in building Socialism. Is it not obvious, then, that whoever disseminates scepticism regarding our successes in building Socialism thereby indirectly helps the Social Democrats, reduces the sweep of the international revolutionary movement, and inevitably departs from internationalism. . . ?

4. Revolution from Above

Marxism holds that the transition of a language from an old quality to a new does not take place by way of an explosion, by the destruction of an existing language and the creation of a new one, but by the gradual accumulation of the elements of the new quality, and, hence, by the gradual dying away of the elements of the old quality.

It should be said in general for the benefit of comrades who have an infatuation for such explosions that the law of transition from an old quality to a new by means of an explosion is inapplicable not only to the history of the development of languages; it is not always applicable to some other social phenomena of a basal or superstructural character. It is compulsory for a society divided into hostile classes. But it is not at all compulsory for a society which has no hostile classes. In a period of eight to ten years we effected a transition in the agriculture of our country from the bourgeois individual-peasant system to the Socialist, collective-farm system. This was a revolution which eliminated the old bourgeois economic system in the countryside and created a new Socialist system. But this revolution did not take place by means of an explosion, that is, by the overthrow of the existing power and the creation of a new power, but by a gradual transition from the old bourgeois system of the countryside to a new system. And we succeeded in doing this because it was a revolution from above, because the revolution was accomplished on the initiative of the existing power with the support of the overwhelming mass of the peasantry....

Further, the superstructure is a product of the base; but this does not mean that it merely reflects the base, that it is passive, neutral, indifferent to the fate of its base, to the fate of the classes, to the character of the system. On the contrary, no sooner does it arise than it becomes an exceedingly active force, actively assisting its base to take shape and consolidate itself, and doing everything it can to help the new system finish off and eliminate the old base and the old classes.

It cannot be otherwise. The base creates the superstructure precisely in order that it may serve it, that it may actively help it to take shape and consolidate itself, that it may actively strive for the elimination of the old, moribund base and its old superstructure. The superstructure has only to renounce its role of auxiliary, it has only to pass from a postion of active defense of its base to one of indifference toward it, to adopt the same attitude to all classes, and it loses its virtue and ceases to be a superstructure.

5. Wars Between Capitalist Nations

Some comrades hold that, owing to the development of new international conditions since the Second World War, wars between capitalist countries have ceased to be inevitable. They consider that the contradictions between the Socialist camp and the capitalist camp are more acute than the contradictions among the capitalist countries; that the USA has brought the other capitalist countries sufficiently under its sway to be able to prevent them going to war among themselves and weakening one another; that the foremost capitalist minds have been sufficiently taught by the two world wars and the severe damage they caused to the whole capitalist world not to venture to involve the capitalist countries in war with one another again—and that, because of all this, wars between capitalist countries are no longer inevitable.

These comrades are mistaken. They see the outward phenomena that come and go on the surface, but they do not see those profound forces which, although they are so far operating imperceptibly, will nevertheless determine the course of developments.

Outwardly, everything would seem to be 'going well': the USA has put Western Europe, Japan, and other capitalist countries on rations; Germany (Western), Britain, France, Italy, and Japan have fallen into the clutches of the USA and are meekly obeying its commands. But it would be mistaken to think that things can continue to 'go well' for 'all eternity', that these countries will tolerate the domination and oppression of the United States endlessly, that they will not endeavour to tear loose from American bondage and take the path of independent development.

Take, first of all, Britain and France. Undoubtedly, they are imperialist countries. Undoubtedly, cheap raw materials and secure markets are of paramount importance to them. Can it be assumed that they will endlessly tolerate the present situation, in which, under the guise of 'Marshall Plan aid', Americans are penetrating into the economies of Britain and France and trying to convert them into adjuncts of the Unitrd States economy, and Americal capital is seizing raw materials and markets in the British and French colonies and thereby plotting disaster for the high profits of the British and French capitalists? Would it not be truer to say that capitalist Britain and, after her, capitalist France will be compelled in the end to break from the embrace of the USA and enter into conflict with it in order to secure an independent position and, of course, high profits?

Let us pass to the major vanquished countries, Germany (Western) and Japan. These countries are now languishing in misery under the jackboot of American imperialism. Their industry and agriculture, their trade, their foreign and home policies, and their whole life are fettered by the American occupation 'regime'. Yet only yesterday these countries were great imperialist powers and were shaking the foundations of the domination of Britain, the USA and France in Europe and Asia, To think that these countries will not try to get on their feet again, will not try to smash the US 'regime', and force their way to independent development, is to believe in miracles.

It is said that the contradictions between capitalism and socialism are stronger than the contradictions among the capitalist countries. Theoretically, of course, that is true. It is not only true now, today; it was true before the Second World War. And it was more or less realized by the leaders of the capitalist countries. Yet the Second World War began not as a war with the USSR, but as a war between capitalist countries. Why?

Firstly, because war with the USSR, as a socialist land, is more dangerous to capitalism than war between capitalist countries; for whereas war between capitalist countries puts in question only the supremacy of certain capitalist countries over others, war with the USSR must certainly put in question the existence of capitalism itself. Secondly, because the capitalists, although they clamour, for 'propaganda' purposes, about the aggressiveness of the Soviet Union, do not themselves believe that it is aggressive, because they are aware of the Soviet Union's peaceful policy and know that it will not itself attack capitalist countries.

After the First World War it was similarly believed that Germany had been definitely put out of action, just as certain comrades now believe that Japan and Germany have been definitely put out of action. Then, too, it was said and clamoured in the press that the United States had put Europe on rations; that Germany would never rise to her feet again, and that there would be no more wars between capitalist countries. In spite of this, Germany rose to her feet again as a great power within the space of some fifteen or twenty years after her defeat, having broken out of bondage and taken the path of independent development. And it is significant that it was none other than Britain and the United States that helped Germany to recover economically and enhance her economic war potential. Of course, when the United States and Britain assisted Germany's economic recovery, they did so with a view to setting a recovered Germany against the Soviet Union, to utilizing her against the land of Socialism. But Germany directed her forces in the first place against the Anglo-French-American bloc. And when Hitler Germany declared war on the Soviet Union, the Anglo-French-Americal bloc, far from joining with Hitler Germany, was compelled to enter into a coalition with the USSR against Hitler Germany.

Consequently, the struggle of the capitalist countries for markets and their desire to crush their competitors proved in practice to be stronger than the contradictions between the capitalist camp and the Socialist camp.

What guarantee is there, then, that Germany and Japan will not rise to their feet again, will not attempt to break out of American bondage and live their own independent lives? I think there is no such guarantee,

But it follows from this that the inevitability of wars between capitalist countries remains in force.

It is said that Lenin's thesis that imperialism inevitably generates war must now be regarded as obsolete, since powerful popular forces have come forward today in defence of peace and against another world war. That is not true.

The object of the present-day peace movement is to rouse the masses of the people to fight for the preservation of peace and for the prevention of another world war. Consequently, the aim of this movement is not to overthrow capitalism and establish Socialism—it confines itself to the democratic aim of preserving peace. In this respect, the present-day peace movement differs from the movement of the time of the First World War for the conversion of the imperialist war into Civil war, since the latter movement went farther and pursued Socialist aims.

It is possible that in a definite conjuncture of circumstances the fight for peace will develop here or there into a fight for Socialism. But then it will no longer be the present-day peace movement; it will be a movement for the overthrow of capitalism.

What is most likely is that the present-day peace movement, as a movement for the preservation of peace, will, if it succeeds, result in preventing a *particular* war, in its temporary postponement, in the temporary preservation of a *particular* peace, in the resignation of a bellicose government and its supersession by another that is prepared temporarily to keep the peace. That, of course, will be good. Even very good. But, all the same, it will not be enough to eliminate the inevitability of wars between capitalist countries generally. It will not be enough, because, for all the successes of the peace movement, imperialism will remain, continue in force—and, consequently, the inevitability of wars will also continue in force.

To eliminate the inevitability of war, it is necessary to abolish imperialism.

Sources: J. Stalin, *Leninism* (Moscow, 1940), pp. 14 ff., 28 ff., 156, 162. *The Essential Stalin*, ed. B. Franklin (New York, 1972), pp. 408 f., 425 f., 469 ff.

FURTHER NOTES

WORKS

STALIN, JOSEPH, *Works*, 13 vols. (Moscow, 1952–5).
—— *The Essential Stalin: Major Theoretical Writings, 1905–1952*, ed. Bruce Franklin (New York, 1972).
MCNEAL, ROBERT H. (ed.), *Stalin's Works: An Annotated Bibliography* (Stanford, 1967).

COMMENTARIES

CARR, H., *Socialism in One Country: 1924–1926*, 3 vols. (London, 1958–64).
CONQUEST, R., *The Great Terror: Stalin's Purge of the Thirties* (New York, 1968).
MEDVEDEV, R. A., *Let History Judge: The Origins and Consequences of Stalinism* (New York, 1971).
NOVE, A., *Economic Rationality and Soviet Politics: or, Was Stalin Really Necessary?* (New York, 1964).

RIGBY, THOMAS H. (ed.), *Stalin* (Englewood Cliffs, NJ, 1966).
SOUVARINE, BORIS, *Stalin: A Critical Survey of Bolshevism* (London, 1939).
TUCKER, ROBERT C., *Stalin as Revolutionary, 1879–1929: A Study in History and Personality* (New York, 1973).
—— (ed.), *Stalinism: Essays in Historical Interpretation* (New York, 1977).
ULAM, ADAM B., *Stalin: The Man and His Era* (London, 1974).

13

MAO ZEDONG: CHINESE MARXISM

Contemporary China has grown out of the China of the past; we are
Marxist in our historical approach and must not lop off our history ...
Being Marxists, Communists are internationalists, but we can put
Marxism into practice only when it is integrated with the specific
characteristics of our country and acquires a definite national form.

'The Role of the Chinese Communist Party in the National War',
Selected Works (Peking, 1965), ii. 209.

MAO ZEDONG (1893–1976) *came from a peasant family in South China
and was one of the thirteen founding members of the Chinese Communist
Party in 1921. He worked out his adaptation of Marxism to Chinese
conditions in the course of long armed struggles, first against his former
allies, the nationalist Kuomintang led by Chiang Kai-shek, then against the
Japanese invaders. The first extract below, from a report commissioned
the party in 1926, shows Mao's early enthusiasm for the peasantry as a
basis for revolution. This was to become the distinctive feature of Chinese
Marxism. The second extract gives a succinct summary of Mao's tactics in
guerrilla warfare. The third extract, from Mao's 1932 essay* On Contradiction, *goes to the heart of his philosophy. Possibly influenced by China's
Taoist and Buddhist heritage, he emphasizes the constant flux and imper-
manence of the universe with contradiction as the essential and continuous
principle in the development of all things. But he was also concerned to
sort out which, among the many contradictions in society, was the principal
contradiction. For example, during the previous ten years the principal
contradiction for the Communists had been between themselves and the
Kuomintang, but with the outbreak of the Sino-Japanese war the principal
contradiction had become that between the Chinese as a whole and the
Japanese invaders. In the fourth extract, from* On the Correct Handling of
Contradictions Among the People, *written in 1957, Mao discusses the
changing nature of contradiction in emergent Chinese Socialist society.
Finally, the fifth extract, a talk given in 1966, gives a glimpse of the
practical, rather anarchic flavour of the vast upheaval known as the Cultural
Revolution, unleashed by Mao in the same year.*

1. The Peasantry as a Revolutionary Force

FOR the present upsurge of the peasant movement is a colossal event. In a very short time, in China's central, southern, and northern provinces, several hundred million peasants will rise like a mighty storm, like a hurricane, a force so swift and violent that no power, however great, will be able to hold it back. They will smash all the trammels that bind them and rush forward along the road to liberation. They will sweep all the imperialists, warlords, corrupt officials, local tyrants, and evil gentry into their graves. Every revolutionary party and every revolutionary comrade will be put to the test, to be accepted or rejected as they decide. There are three alternatives. To march at their head and lead them? To trail behind them, gesticulating and criticizing? Or to stand in their way and oppose them? Every Chinese is free to choose, but events will force you to make the choice quickly....

The poor peasants have always been the main force in the bitter fight in the countryside. They have fought militantly through the two periods of underground work and of open activity. They are the most responsive to Communist Party leadership. They are deadly enemies of the camp of the local tyrants and evil gentry and attack it without the slightest hesitation. 'We joined the peasant association long ago,' they say to the rich peasants, 'why are you still hesitating?' The rich peasants answer mockingly, 'What is there to keep you from joining? You people have neither a tile over your heads nor a speck of land under your feet!' It is true the poor peasants are not afraid of losing anything. Many of them really have 'neither a tile over their heads nor a speck of land under their feet'. What, indeed, is there to keep them from joining the associations? According to the survey of Changsha County, the poor peasants comprise 70 per cent, the middle peasants 20 per cent, and the landlords and the rich peasants 10 per cent of the population in the rural areas. The 70 per cent, the poor peasants, may be subdivided into two categories, the utterly destitute and the less destitute. The utterly destitute, comprising 20 per cent, are the completely dispossessed, that is, people who have neither land nor money, are without any means of livelihood, and are forced to leave home and become mercenaries or hired labourers or wandering beggars. The less destitute, the other 50 per cent, are the partially dispossessed, that is, people with just a little land or a little money who eat up more than they earn and live in toil and distress the year round, such as the handicraftsmen, the tenant-peasants (not including the rich tenant-peasants), and the semi-owner-peasants. This great mass of poor peasants, or altogether 70 per cent of the rural population, are the backbone of the peasant associations, the vanguard in the overthrow

of the feudal forces, and the heroes who have performed the great revolutionary task which for long years was left undone. Without the poor
peasant class (the 'riff-raff', as the gentry call them), it would have been
impossible to bring about the present revolutionary situation in the countryside, or to overthrow the local tyrants and evil gentry and complete the
democratic revolution. The poor peasants, being the most revolutionary
group, have gained the leadership of the peasant associations. In both the
first and second periods almost all the chairmen and committee members
in the peasant associations at the lowest level were poor peasants (of
the officials in the township associations in Hengshan County the utterly
destitute comprise 50 per cent, the less destitute 40 per cent, and poverty-
stricken intellectuals 10 per cent). Leadership by the poor peasants is
absolutely necessary. Without the poor peasants there would be no revolution. To deny their role is to deny the revolution. To attack them is to
attack the revolution. They have never been wrong on the general direction
of the revolution. They have discredited the local tyrants and evil gentry.
They have beaten down the local tyrants and evil gentry, big and small,
and kept them underfoot. Many of their deeds in the period of revolutionary
action, which were labelled as 'going too far', were in fact the very things
the revolution required. Some county governments, county headquarters of
the Kuomintang, and county peasant associations in Hunan have already
made a number of mistakes; some have even sent soldiers to arrest officials
of the lower-level associations at the landlords' request. A good many
chairmen and committee members of township associations in Hengshan
and Hsianghsiang Counties have been thrown in jail. This mistake is very
serious and feeds the arrogance of the reactionaries. To judge whether or
not it is a mistake, you have only to see how joyful the lawless landlords
become and how reactionary sentiments grow, wherever the chairmen or
committee members of local peasant associations are arrested. We must
combat the counter-revolutionary talk of the 'movement of riff-raff' and a
'movement of lazy peasants' and must be especially careful not to commit
the error of helping the local tyrants and evil gentry in their attacks on the
poor peasant class. Though a few of the poor peasant leaders undoubtedly
did have shortcomings, most of them have changed by now. They themselves are energetically prohibiting gambling and suppressing banditry.
Where the peasant association is powerful, gambling has stopped altogether
and banditry has vanished. In some places it is literally true that people do
not take any articles left by the wayside and that doors are not bolted at
night. According to the Hengshan survey, 85 per cent of the poor peasant
leaders have made great progress and have proved themselves capable and
hard-working. Only 15 per cent retain some bad habits. The most one can

call these is 'an unhealthy minority', and we must not echo the local tyrants and evil gentry in undiscriminatingly condemning them as 'riff-raff'. This problem of the 'unhealthy minority' can be tackled only under the peasant associations' own slogan of 'strengthen discipline', by carrying on propaganda among the masses, by educating the 'unhealthy minority', and by tightening the associations' discipline; in no circumstances should soldiers be arbitrarily sent to make such arrests as would damage the prestige of the poor peasants and feed the arrogance of the local tyrants and evil gentry. This point requires particular attention....

2. On Guerrilla Warfare

Communists do not fight for personal military power (they must in no circumstances do that, and let no one ever again follow the example of Chang Kuo-tao), but they must fight for military power for the party, for military power for the people. As a national war of resistance is going on, we must also fight for military power for the nation. Where there is naïvety on the question of military power, nothing whatsoever can be achieved. It is very difficult for the labouring people, who have been deceived and intimidated by the reactionary ruling classes for thousands of years, to awaken to the importance of having guns in their own hands. Now that Japanese imperialist oppression and the nation-wide resistance to it have pushed our labouring people into the arena of war, Communists should prove themselves the most politically conscious leaders in this war. Every Communist must grasp the truth, 'Political power grows out of the barrel of a gun.' Our principle is that the party commands the gun, and the gun must never be allowed to command the party. Yet, having guns, we can create party organizations, as witness the powerful party organizations which the Eighth Route Army has created in northern China. We can also create cadres, create schools, create culture, create mass movements. Everything in Yenan has been created by having guns. All things grow out of the barrel of a gun. According to the Marxist theory of the State, the army is the chief component of State power. Whoever wants to seize and retain State power must have a strong army. Some people ridicule us as advocates of the 'omnipotence of war'. Yes, we are advocates of the omnipotence of revolutionary war; that is good, not bad, it is Marxist. The guns of the Russian Communist Party created Socialism. We shall create a democratic republic. Experience in the class struggle in the era of imperialism teaches us that it is only by the power of the gun that the working class and the labouring masses can defeat the armed bourgeoisie and landlords; in this sense we may say that only with guns can the whole world be transformed. We are advocates of the abolition of war, we do not want war; but war can only be abolished through war, and in order to get rid of the gun it is necessary to take up the gun. . . .

3.   On Contradiction

Throughout the history of human knowledge, there have been two conceptions concerning the law of development of the universe, the metaphysical conception and the dialectical conception, which form two opposing world outlooks. Lenin said:

The two basic (or two possible? or two historically observable?) conceptions of development (evolution) are: development as decrease and increase, as repetition *and* development as a unity of opposites (the division of a unity into mutually exclusive opposites and their reciprocal relation).

Here Lenin was referring to these two different world outlooks.

In China another name for metaphysics is *hsuan-hsueh*. For a long period in history whether in China or in Europe, this way of thinking, which is part and parcel of the idealist world outlook, occupied a dominant position in human thought. In Europe, the materialism of the bourgeoisie in its early days was also metaphysical. As the social economy of many European countries advanced to the stage of highly developed capitalism, as the forces of production, the class struggle, and the sciences developed to a level unprecedented in history, and as the industrial proletariat became the greatest motive force in historical development, there arose the Marxist world outlook of materialist dialectics. Then, in addition to open and barefaced reactionary idealism, vulgar evolutionism emerged among the bourgeoisie to oppose materialist dialectics.

The metaphysical or vulgar evolutionist world outlook sees things as isolated, static, and one-sided. It regards all things in the universe, their forms and their species, as eternally isolated from one another and immutable. Such change as there is can only be an increase or decrease in quantity or a change of place. Moreover, the cause of such an increase or decrease or change of place is not inside things but outside them, that is, the motive force is external. Metaphysicians hold that all the different kinds of things in the universe and all their characteristics have been the same ever since they first came into being. All subsequent changes have simply been increases or decreases in quantity. They contend that a thing can only keep on repeating itself as the same kind of thing and cannot change into anything different. In their opinion, capitalist exploitation, capitalist competition, the individualist ideology of capitalist society, and so on, can all be found in ancient slave society, or even in primitive society, and will exist for ever unchanged.

They ascribe the causes of social development to factors external to society, such as geography and climate. They search in an over-simplified way outside a thing for the causes of its development, and they deny the theory of materialist dialectics which holds that development arises from the contradictions inside a thing. Consequently they can explain neither the qualitative diversity of things, nor the phenomenon of one quality changing into another. In Europe, this mode of thinking existed as mechanical materialism in the seventeenth and eighteenth centuries and as vulgar evolutionism at the end of the nineteenth and the beginning of the twentieth centuries. In China, there was the metaphysical thinking exemplified in the saying 'Heaven changeth not, likewise the Tao changeth not', and it was supported by the decadent feudal ruling classes for a long time. Mechanical materialism and vulgar evolutionism, which were imported from Europe in the last hundred years, are supported by the bourgeoisie.

As opposed to the metaphysical world outlook, the world outlook of materialist dialectics holds that in order to understand the development of a thing we should study it internally and in its relations with other things; in other words, the development of things should be seen as their internal and necessary self-movement, while each thing in its movement is interrelated with and interacts on the things around it. The fundamental cause of the development of a thing is not external but internal; it lies in the contradictoriness within the thing. There is internal contradiction in every single thing, hence its motion and development. Contradictoriness within a thing is the fundamental cause of its development, while its interrelations and interactions with other things are secondary causes. Thus materialist dialectics effectively combats the theory of external causes, or of an external motive force, advanced by metaphysical mechanical materialism and vulgar evolutionism....

There are many contradictions in the process of development of a complex thing, and one of them is necessarily the principal contradiction whose existence and development determine or influence the existence and development of the other contradictions.

For instance, in capitalist society the two forces in contradiction, the proletariat and the bourgeoisie, form the principal contradiction. The other contradictions, such as those between the remnant feudal class and the bourgeoisie, between the peasant petty bourgeoisie and the bourgeoisie, between the proletariat and the peasant petty bourgeoisie, between the non-monopoly capitalists and the monopoly capitalists, between bourgeois democracy and bourgeois Fascism, among the capitalist countries, and between imperialism and the colonies, are all determined or influenced by this principal contradiction.

In a semi-colonial country such as China, the relationship between the principal contradiction and the non-principal contradictions presents a complicated picture.

When imperialism launches a war of aggression against such a country, all its various classes, except for some traitors, can temporarily unite in a national war against imperialism. At such a time, the contradiction between imperialism and the country concerned becomes the principal contradiction, while all the contradictions among the various classes within the country (including what was the principal contradiction, between the feudal system and the great masses of the people) are temporarily relegated to a secondary and subordinate position. So it was in China in the Opium War of 1840, the Sino-Japanese War of 1894, and the Yi Ho Tuan War of 1900, and so it is now in the present Sino-Japanese War.

But in another situation, the contradictions change position. When imperialism carries on its oppression not by war, but by milder means—political, economic, and cultural—the ruling classes in semi-colonial countries capitulate to imperialism, and the two form an alliance for the joint oppression of the masses of the people. At such a time, the masses often resort to civil war against the alliance of imperialism and the feudal classes, while imperialism often employs indirect methods rather than direct action in helping the reactionaries in the semi-colonial countries to oppress the people, and thus the internal contradictions become particularly sharp. This is what happened in China in the Revolutionary War of 1911, the Revolutionary War of 1924–7, and the ten years of Agrarian Revolutionary War after 1927. Wars among the various reactionary ruling groups in the semi-colonial countries, e.g., the wars among the warlords in China, fall into the same category.

When a revolutionary civil war develops to the point of threatening the very existence of imperialism and its running dogs, the domestic reactionaries, imperialism often adopts other methods in order to maintain its rule; it either tries to split the revolutionary front from within or sends armed forces to help the domestic reactionaries directly. At such a time, foreign imperialism and domestic reaction stand quite openly at one pole while the masses of the people stand at the other pole, thus forming the principal contradiction which determines or influences the development of the other contradictions. The assistance given by various capitalist countries to the Russian reactionaries after the October revolution is an example of armed intervention. Chiang Kai-shek's betrayal in 1927 is an example of splitting the revolutionary front.

But whatever happens, there is no doubt at all that at every stage in the development of a process, there is only one principal contradiction which plays the leading role.

Hence, if in any process there are a number of contradictions, one of them must be the principal contradiction playing the leading and decisive role, while the rest occupy a secondary and subordinate position. Therefore, in studying any complex process in which there are two or more contradictions, we must devote every effort to finding its principal contradiction. Once this principal contradiction is grasped, all problems can be readily solved. This is the method Marx taught us in his study of capitalist society. Likewise Lenin and Stalin taught us this method when they studied imperialism and the general crisis of capitalism and when they studied the Soviet economy. There are thousands of scholars and men of action who do not understand it, and the result is that, lost in a fog, they are unable to get to the heart of a problem and naturally cannot find a way to resolve its contradictions.

As we have said, one must not treat all the contradictions in a process as being equal but must distinguish between the principal and the secondary contradictions, and pay special attention to grasping the principal one. But, in any given contradiction, whether principal or secondary, should the two contradictory aspects be treated as equal? Again, no. In any contradiction the development of the contradictory aspects is uneven. Sometimes they seem to be in equilibrium, which is however only temporary and relative, while unevenness is basic. Of the two contradictory aspects, one must be principal and the other secondary. The principal aspect is the one playing the leading role in the contradiction. The nature of a thing is determined mainly by the principal aspect of a contradiction, the aspect which has gained the dominant position.

But this situation is not static; the principal and the non-principal aspects of a contradiction transform themselves into each other and the nature of the thing changes accordingly. In a given process or at a given stage in the development of a contradiction, A is the principal aspect and B is the non-principal aspect; at another stage or in another process the roles are reversed—a change determined by the extent of the increase or decrease in the force of each aspect in its struggle against the other in the course of the development of a thing.

We often speak of 'the new superseding the old'. The supersession of the old by the new is a general, eternal, and inviolable law of the universe. The transformation of one thing into another, through leaps of different forms in accordance with its essence and external conditions—this is the process of the new superseding the old. In each thing there is contradiction between its new and its old aspects, and this gives rise to a series of struggles with many twists and turns. As a result of these struggles, the new aspect changes from being minor to being major and rises to predominance, while the old aspect changes from being major to being minor and gradually dies out. And the moment the new aspect gains dominance over the old, the old thing changes qualitatively into a new thing. It can thus be seen that the nature of a thing is mainly determined by the principal aspect of the contradiction, the aspect which has gained predominance. When the principal aspect which has gained predominance changes, the nature of a thing changes accordingly.

In capitalist society, capitalism has changed its position from being a subordinate force in the old feudal era to being the dominant force, and the nature of society has accordingly changed from feudal to capitalist. In the new, capitalist era, the feudal forces changed from their former dominant position to a subordinate one, gradually dying out. Such was the case, for example, in Britain and France. With the development of the productive forces, the bourgeoisie changes from being a new class playing a progressive role to being an old class playing a reactionary role, until it is finally overthrown by the proletariat and becomes a class deprived of privately owned means of production and stripped of power, when it, too, gradually dies out. The proletariat, which is much more numerous than the bourgeoisie and grows simultaneously with it but under its rule, is a new force which, initially subordinate to the bourgeoisie, gradually gains strength, becomes an independent class playing the leading role in history, and finally seizes political power and becomes the ruling class. Thereupon the nature of society changes and the old capitalist society becomes the new Socialist society. This is the path already taken by the Soviet Union, a path that all other countries will inevitably take.

Look at China, for instance. Imperialism occupies the principal position in the contradiction in which China has been reduced to a semi-colony, it oppresses the Chinese people, and China has been changed from an independent country into a semi-colonial one. But this state of affairs will inevitably change; in the struggle between the two sides, the power of the Chinese people which is growing under the leadership of the proletariat will inevitably change China from a semi-colony into an independent country, whereas imperialism will be overthrown and old China will inevitably change into New China.

The change of old China into New China also involves a change in the relation between the old feudal forces and the new popular forces within the country. The old feudal landlord class will be overthrown, and from being the ruler it will change into being the ruled; and this class, too, will gradually die out. From being the ruled the people, led by the proletariat, will become the rulers. Thereupon, the nature of Chinese society will change and the old, semi-colonial and semi-feudal society will change into a new democratic society.

Instances of such reciprocal transformation are found in our past experience. The Ching Dynasty which ruled China for nearly three hundred years was overthrown in the revolution of 1911, and the revolutionary *Tung Meng Hui* under Sun Yat-sen's leadership was victorious for a time. In the Revolutionary War of 1924–7, the revolutionary forces of the Communist-Kuomintang alliance in the south changed from being weak to being strong and won victory in the Northern Expedition, while the northern warlords who once ruled the roost were overthrown. In 1927, the people's forces led by the Communist Party were greatly reduced numerically under the attacks of Kuomintang reaction, but with the elimination of opportunism within their ranks they gradually grew again. In the revolutionary base areas under communist leadership, the peasants have been transformed from being the ruled to being the rulers, while the landlords have undergone a reverse transformation. It is always so in the world, the new displacing the old, the old being superseded by the new, the old being eliminated to make way for the new, and the new emerging out of the old.

At certain times in the revolutionary struggle, the difficulties outweigh the favourable conditions and so constitute the principal aspect of the contradiction and the favourable conditions constitute the secondary aspect. But through their efforts the revolutionaries can overcome the difficulties step by step and open up a favourable new situation; thus a difficult situation yields place to a favourable one. This is what happened after the failure of the revolution in China in 1927 and during the Long March of the Chinese Red Army. In the present Sino-Japanese War, China is again in a difficult position, but we can change this and fundamentally transform the situation as between China and Japan. Conversely, favourable conditions can be transformed into difficulty if the revolutionaries make mistakes. Thus the victory of the revolution of 1924–7 turned into defeat. The revolutionary base areas which grew up in the southern provinces after 1927 had all suffered defeat by 1934.

When we engage in study, the same holds good for the contradiction in the passage from ignorance to knowledge. At the very beginning of our study of Marxism, our ignorance of or scanty acquaintance with Marxism stands in contradiction to knowledge of Marxism. But by assiduous study, ignorance can be transformed into knowledge, scanty knowledge into substantial knowledge, and blindness in the application of Marxism into mastery of its application.

Some people think that this is not true of certain contradictions. For instance, in the contradiction between the productive forces and the relations of production, the productive forces are the principal aspect; in the contradiction between theory and practice, practice is the principal aspect; in the contradiction between the economic base and the superstructure, the economic base is the principal aspect; and there is no change in their respective positions. This is the mechanical materialist conception, not the dialectical materialist conception. True, the productive forces, practice, and the economic base generally play the principal and decisive role; whoever denies this is not a materialist. But it must also be admitted that, in certain conditions, such aspects as the relations of production, theory, and the superstructure in turn manifest themselves in the principal and decisive role. When it is impossible for the productive forces to develop without a change in the relations of production, then the change in the relations of production plays the principal and decisive role. The creation and advocacy of revolutionary theory plays the principal and decisive role in those times of which Lenin said, 'Without revolutionary theory there can be no revolutionary movement.' When a task, no matter which, has to be performed, but there is as yet no guiding line, method, plan, or policy, the principal and decisive thing is to decide on a guiding line, method, plan, or policy. When the superstructure (politics, culture, etc.) obstructs the development of the economic base, political and cultural changes become principal and decisive. Are we going against materialism when we say this? No. The reason is that while we recognize that in the general development of history the material determines the mental and social being determines social consciousness, we also—and indeed must—recognize the reaction of mental on material things, of social consciousness on social being, and of the superstructure on the economic base. This does not go against materialism; on the contrary, it avoids mechanical materialism and firmly upholds dialectical materialism.

In studying the particularity of contradiction, unless we examine these two facets—the principal and the non-principal contradictions in a process, and the principal and the non-principal aspects of a contradiction—that is, unless we examine the distinctive character of these two facets of contradiction, we shall get bogged down in abstractions, be unable to understand contradiction concretely, and consequently be unable to find the correct method of resolving it. The distinctive character or particularity of these two facets of contradiction represents the unevenness of the forces that are in contradiction. Nothing in this world develops absolutely evenly; we must oppose the theory of even development or the theory of equilibrium. Moreover, it is these concrete features of a contradiction and the changes in the principal and non-principal aspects of a contradiction in the course of its development that manifest the force of the new superseding the old. The study of the various states of unevenness in contradictions, of the principal and non-principal contradictions, and of the principal and the non-principal aspects of a contradiction constitutes an essential method by which a revolutionary political party correctly determines its strategic and tactical policies both in political and in military affairs. All Communists must give it attention.

4. On the Correct Handling of Contradictions Among the People

Never before has our country been as united as it is today. The victories of the bourgeois democratic revolution and of the Socialist revolution and our achievements in Socialist construction have rapidly changed the face of the old China. A still brighter future lies ahead for our motherland. The days of national disunity and chaos which the people detested are gone, never to return. Led by the working class and the Communist Party, our 600 million people, united as one, are engaged in the great task of building Socialism. The unification of our country, the unity of our people, and the unity of our various nationalities—these are the basic guarantees for the sure triumph of our cause. However, this does not mean that contradictions no longer exist in our society. To imagine that none exist is a naïve idea which is at variance with objective reality. We are confronted with two types of social contradictions—those between ourselves and the enemy and those among the people. The two are totally different in nature.

To understand these two different types of contradictions correctly, we must first be clear on what is meant by 'the people' and what is meant by 'the enemy'. The concept of 'the people' varies in content in different countries and in different periods of history in a given country. Take our own country for example. During the War of Resistance against Japan, all those classes, strata, and social groups opposing Japanese aggression came within the category of the people, while the Japanese imperialists, their Chinese collaborators, and the pro-Japanese elements were all enemies of the people. During the War of Liberation, the US imperialists and their running dogs—the bureaucrat-capitalists, the landlords, and the Kuomintang reactionaries who represented these two classes—were the enemies of the people, while the other classes, strata, and social groups, which opposed them, all came within the category of the people. At the present stage, the period of building Socialism, the classes, strata, and social groups which favour, support, and work for the cause of Socialist construction all come within the category of the people, while the social forces and groups which resist the Socialist revolution and are hostile to or sabotage Socialist construction are all enemies of the people.

The contradictions between ourselves and the enemy are antagonistic contradictions. Within the ranks of the people, the contradictions among the working people are non-antagonistic, while those between the exploited and the exploiting classes have a non-antagonistic as well as an antagonistic aspect. There have always been contradictions among the people, but they are different in content in each period of the revolution and in the period of building Socialism. In the conditions prevailing in China today, the contradictions among the people comprise the contradictions within the working class, the contradictions within the peasantry, the contradictions within the intelligentsia, the contradictions between the working class and the peasantry, the contradictions between the workers and peasants on the one hand and the intellectuals on the other, the contradictions between the working class and other sections of the working people on the one hand and the national bourgeoisie on the other, the contradictions within the national bourgeoisie, and so on. Our People's Government is one that genuinely represents the people's interests, it is a government that serves the people. Nevertheless, there are still certain contradictions between this government and the people. These include the contradictions between the interests of the State and the interests of the collective on the one hand and the interests of the individual on the other, between democracy and centralism, between the leadership and the led, and the contradictions arising from the bureaucratic style of work of some of the State personnel in their relations with the masses. All these are also contradictions among the people. Generally speaking, the fundamental identity of the people's interests underlies the contradictions among the people.

In our country, the contradiction between the working class and the national bourgeoisie comes under the category of contradictions among the people. By and large, the class struggle between the two is a class struggle within the ranks of the people, because the Chinese national bourgeoisie has a dual character. In the period of the bourgeois-democratic revolution, it had both a revolutionary and a conciliationist side to its character. In the period of the Socialist revolution, exploitation of the working class for profit constitutes one side of the character of the national bourgeoisie, while its support of the Constitution and its willingness to accept Socialist transformation constitute the other. The national bourgeoisie differs from the imperialists, the landlords, and the bureaucrat-capitalists. The contradiction between the national bourgeoisie and the working class is one between exploiter and exploited, and is by nature antagonistic. But in the concrete conditions of China, this antagonistic contradiction between the two classes, if properly handled, can be transformed into a non-antagonistic one and be resolved by peaceful methods. However, the contradiction between the working class and the national bourgeoisie will change into a contradiction between ourselves and the enemy if we do not handle it properly and do not follow the policy of uniting with, criticising, and educating the national bourgeoisie, or if the national bourgeoisie does not accept this policy of ours.

Since they are different in nature, the contradictions between ourselves and the enemy and the contradictions among the people must be resolved by different methods. To put it briefly, the former entail drawing a clear distinction between ourselves and the enemy, and the latter entail drawing a clear distinction between right and wrong. It is of course true that the distinction between ourselves and the enemy is also one of right and wrong. For example, the question of who is in the right, we or the domestic and foreign reactionaries, the imperialists, the feudalists, and bureaucrat-capitalists, is also one of right and wrong, but it is in a different category from questions of right and wrong among the people.

Our State is a people's democratic dictatorship led by the working class and based on the worker-peasant alliance. What is this dictatorship for? Its first function is internal, namely, to suppress the reactionary classes and elements and those exploiters who resist the Socialist revolution, to suppress those who try to wreck our Socialist construction, or, in other words, to resolve the contradictions between ourselves and the internal enemy. For instance, to arrest, try, and sentence certain counter-revolutionaries, and to deprive landlords and bureaucrat-capitalists of their right to vote and their freedom of speech for a certain period of time—all this comes within the scope of our dictatorship. To maintain public order and safeguard the interests of the people, it is necessary to exercise dictatorship as well over thieves, swindlers, murderers, arsonists, criminal gangs, and other scoundrels who seriously disrupt public order. The second function of this dictatorship is to protect our country from subversion and possible aggression by external enemies. In such contingencies, it is the task of this dictatorship to resolve the contradiction between ourselves and the external enemy. The aim of this dictatorship is to protect all our people so that they can devote themselves to peaceful labour and make China a Socialist country with modern industry, modern agriculture, and modern science and culture. Who is to exercise this dictatorship? Naturally, the working class and the entire people under its leadership. Dictatorship does not apply within the ranks of the people. The people cannot exercise dictatorship over themselves, nor must one section of the people oppress another. Law-breakers among the people will be punished according to law, but this is different in principle from the exercise of dictatorship to suppress enemies of the people. What applies among the people is democratic centralism. Our Constitution lays it down that citizens of the People's Republic of China enjoy freedom of speech, the press, assembly, association, procession, demonstration, religious belief, and so on. Our Constitution also provides that the organs of State must practise democratic centralism, that they must rely on the masses, and that their personnel must serve the people. Our Socialist democracy is the broadest kind of democracy, such as is not to be found in any bourgeois State. Our dictatorship is the people's democratic dictatorship led by the working class and based on the worker-peasant alliance. That is to say, democracy operates within the ranks of the people, while the working class, uniting with all others enjoying civil rights, and in the first place with the peasantry, enforces dictatorship over the reactionary classes and elements and all those who resist Socialist transformation and oppose Socialist construction. By civil rights, we mean, politically, the rights of freedom and democracy.

But this freedom is freedom with leadership and this democracy is democracy under centralized guidance, not anarchy. Anarchy does not accord with the interests or wishes of the people.

Certain people in our country were delighted by the Hungarian incident. They hoped that something similar would happen in China, that thousands upon thousands of people would take to the streets to demonstrate against the People's Government. Their hopes ran counter to the interests of the masses and therefore could not possibly win their support. Deceived by domestic and foreign counter-revolutionaries, a section of the people in Hungary made the mistake of resorting to violence against the people's government, with the result that both the State and the people suffered. The damage done to the country's economy in a few weeks of rioting will take a long time to repair. In our country there were some others who wavered on the question of the Hungarian incident because they were ignorant of the real state of affairs in the world. They think that there is too little freedom under our people's democracy and that there is more freedom under Western parliamentary democracy. They ask for a two-party system as in the West, with one party in office and the other in opposition. But this so-called two-party system is nothing but a device for maintaining the dictatorship of the bourgeoisie; it can never guarantee freedoms to the working people. As a matter of fact, freedom and democracy exist not in the abstract, but only in the concrete. In a society where class struggle exists, if there is freedom for the exploiting classes to exploit the working people, there is no freedom for the working people not to be exploited. If there is democracy for the bourgeoisie, there is no democracy for the proletariat and other working people. The legal existence of the Communist Party is tolerated in some capitalist countries, but only to the extent that it does not endanger the fundamental interests of the bourgeoisie; it is not tolerated beyond that. Those who demand freedom and democracy in the abstract regard democracy as an end and not as a means. Democracy as such sometimes seems to be an end, but it is in fact only a means. Marxism teaches us that democracy is part of the superstructure and belongs to the realm of politics. That is to say, in the last analysis, it serves the economic base. The same is true of freedom. Both democracy and freedom are relative, not absolute, and they come into being and develop in specific historical conditions. Within the ranks of the people, democracy is correlative with centralism and freedom with discipline. They are the two opposites of a single entity, contradictory as well as united, and we should not one-sidedly emphasize one to the exclusion of the other. Within the ranks of the people, we cannot do without freedom, nor can we do without discipline; we cannot do without democracy, nor can we do without centralism. This unity of democracy and centralism, of freedom and discipline, constitutes our democratic centralism. Under this system, the people enjoy broad democracy and freedom, but at the same time they have to keep within the bounds of Socialist discipline. All this is well understood by the masses.

In advocating freedom with leadership and democracy under centralized guidance, we in no way mean that coercive measures should be taken to settle ideological questions or questions involving the distinction between right and wrong among the people. All attempts to use administrative orders or coercive measures to settle ideological questions or questions of right and wrong are not only ineffective but harmful. We cannot abolish religion by administrative order or force people not to believe in it. We cannot compel people to give up idealism, any more than we can force them to embrace Marxism. The only way to settle questions of an ideological nature or controversial issues among the people is by the democratic method, the method of discussion, criticism, persuasion, and education, and not by the method of coercion or repression. To be able to carry on their production and studies effectively and to lead their lives in peace and order, the people want their government and those in charge of production and of cultural and educational organizations to issue appropriate administrative regulations of an obligatory nature. It is common sense that without them the maintenance of public order would be impossible. Administrative regulations and the method of persuasion and education complement each other in resolving contradictions among the people. In fact, administrative regulations for the maintenance of public order must be accompanied by persuasion and education, for in many cases regulations alone will not work.

This democratic method of resolving contradictions among the people was epitomized in 1942 in the formula 'unity–criticism–unity'. To elaborate, that means starting from the desire for unity, resolving contradictions through criticism or struggle, and arriving at a new unity on a new basis. In our experience this is the correct method of resolving contradictions among the people. In 1942 we used it to resolve contradictions inside the Communist Party, namely, the contradictions between the dogmatists and the great majority of the membership, and between dogmatism and Marxism. The 'left' dogmatists had resorted to the method of 'ruthless struggle and merciless blows' in inner-party struggle. It was the wrong method. In criticizing 'left' dogmatism, we did not use this old method but adopted a new one, that is, one of starting from the desire for unity, distinguishing between right and wrong through criticism or struggle, and arriving at a new unity on a new basis. This was the method used in the rectification movement of 1942. Within a few years, by the time the Chinese Communist Party held its Seventh National Congress in 1945, unity was achieved throughout the party as anticipated, and consequently the people's revolution triumphed. Here, the essential thing is to start from the desire for unity. For without this desire for unity, the struggle, once begun, is certain to throw things into confusion and get out of hand. Wouldn't this be the same as 'ruthless struggle and merciless blows'? And what party unity would there be left? It was precisely this experience that led us to the formula 'unity–criticism–unity'. Or, in other words, 'learn from past mistakes to avoid future ones and cure the sickness to save the patient'. We extended this method beyond our party. We applied it with great success in the anti-Japanese base areas in dealing with the relations between the leadership and the masses, between the army and the people, between officers and men, between the different units of the army, and between the different groups of cadres. The use of this method can be traced back to still earlier times in our party's history. Ever since 1927, when we built our revolutionary armed forces and base areas in the south, this method had been used to deal with the relations between the party and the masses, between the army and the people, between officers and men, and with other relations among the people. The only difference was that during the anti-Japanese war we employed this method much more consciously. And since the liberation of the whole country, we have employed this same method of 'unity–criticism–unity' in our relations with the democratic parties and with industrial and commercial circles. Our task now is to continue to extend and make still better use of this method throughout the ranks of the people; we want all our factories, co-operatives, shops, schools, offices, and people's organizations, in a word, all our 600 million people, to use it in resolving contradictions among themselves.

In ordinary circumstances, contradictions among the people are not antagonistic. But if they are not handled properly, or if we relax our vigilance and lower our guard, antagonism may arise. In a Socialist country, a development of this kind is usually only a localized and temporary phenomenon. The reason is that the system of exploitation of man by man has been abolished and the interests of the people are fundamentally identical. The antagonistic actions which took place on a fairly wide scale during the Hungarian incident were the result of the operations of both domestic and foreign counter-revolutionary elements. This was a particular as well as a temporary phenomenon. It was a case of the reactionaries inside a Socialist country, in league with the imperialists, attempting to achieve their conspiratorial aims by taking advantage of contradictions among the people to foment dissension and stir up disorder. The lesson of the Hungarian incident merits attention.

Many people seem to think that the use of the democratic method to resolve contradictions among the people is something new. Actually it is not. Marxists have always held that the cause of the proletariat must depend on the masses of the people and that Communists must use the democratic method of persuasion and education when working among the labouring people and must on no account resort to commandism or coercion. The Chinese Communist Party faithfully adheres to this Marxist-Leninist principle. It has been our consistent view that under the people's democratic dictatorship two different methods, one dictatorial and the other democratic, should be used to resolve the two types of contradictions which differ in nature—those between ourselves and the enemy and those among the people. This idea has been explained again and again in many party documents and in speeches by many leading comrades of our party. In my article 'On the People's Democratic Dictatorship', written in 1949, I said, 'The combination of these two aspects, democracy for the people and dictatorship over the reactionaries, is the people's democratic dictatorship.' I also pointed out that in order to settle problems within the ranks of the people 'the method we employ is democratic, the method of persuasion, not of compulsion'. Again, in addressing the Second Session of the First National Committee of the Political Consultative Conference in June 1950, I said:

The people's democratic dictatorship uses two methods. Towards the enemy, it uses the method of dictatorship, that is, for as long a period of time as is necessary it does not permit them to take part in political activity and compels them to obey the law of the People's Government, to engage in labour and, through such labour, be transformed into new men. Towards the people, on the contrary, it uses the method of democracy and not of compulsion, that is, it must necessarily let them take part in political activity and does not compel them to do this or that but uses the method of democracy to educate and persuade. Such education is self-education for the people, and its basic method is criticism and self-criticism.

Thus, on many occasions we have discussed the use of the democratic method for resolving contradictions among the people; furthermore, we have in the main applied it in our work, and many cadres and many other people are familiar with it in practice. Why then do some people now feel that it is a new issue? Because, in the past, the struggle between ourselves and the enemy, both internal and external, was most acute, and contradictions among the people therefore did not attract as much attention as they do today.

Quite a few people fail to make a clear distinction between these two different types of contradictions—those between ourselves and the enemy and those among the people—and are prone to confuse the two. It must be admitted that it is sometimes quite easy to do so. We have had instances of such confusion in our work in the past. In the course of cleaning out counter-revolutionaries good people were sometimes mistaken for bad, and such things still happen today. We are able to keep mistakes within bounds because it has been our policy to draw a sharp line between ourselves and the enemy and to rectify mistakes whenever discovered.

Marxist philosophy holds that the law of the unity of opposites is the fundamental law of the universe. This law operates universally, whether in the natural world, in human society, or in man's thinking. Between the opposites in a contradiction there is at once unity and struggle, and it is this that impels things to move and change. Contradictions exist everywhere, but their nature differs in accordance with the different nature of different things. In any given thing, the unity of opposites is conditional, temporary, and transitory, and hence relative, whereas the struggle of opposites is absolute. Lenin gave a very clear exposition of this law. It has come to be understood by a growing number of people in our country. But for many people it is one thing to accept this law and quite another to apply it in examining and dealing with problems. Many dare not openly admit that contradictions still exist among the people of our country, while it is precisely these contradictions that are pushing our society forward. Many do not admit that contradictions still exist in Socialist society, with the result that they become irresolute and passive when confronted with social contradictions; they do not understand that Socialist society grows more united and consolidated through the ceaseless process of correctly handling and resolving contradictions. For this reason, we need to explain things to our people, and to our cadres in the first place, in order to help them understand the contradictions in Socialist society and learn to use correct methods for handling them.

Contradictions in Socialist society are fundamentally different from those in the old societies, such as capitalist society. In capitalist society contradictions find expression in acute antagonisms and conflicts, in sharp class struggle; they cannot be resolved by the capitalist system itself and can only be resolved by Socialist revolution. The case is quite different with contradictions in Socialist society; on the contrary, they are not antagonitic and can be ceaselessly resolved by the Socialist system itself.

In Socialist society the basic contradictions are still those between the relations of production and the productive forces and between the superstructure and the economic base. However, they are fundamentally different in character and have different features from the contradictions between the relations of production and the productive forces and between the superstructure and the economic base in the old societies. The present social system of our country is far superior to that of the old days. If it were not so, the old system would not have been overthrown and the new system could not have been established. In saying that the Socialist relations of production correspond better to the character of the productive forces than did the old relations of production, we mean that they allow the productive forces to develop at a speed unattainable in the old society, so that production can expand steadily and increasingly meet the constantly growing needs of the people. Under the rule of imperialism, feudalism, and bureaucrat-capitalism, the productive forces of the old China grew very slowly. For more than fifty years before liberation, China produced only a few tens of thousands of tons of steel a year, not counting the output of the northeastern provinces. If these provinces are included, the peak annual steel output only amounted to a little over 900,000 tons. In 1949, the national steel output was a little over 100,000 tons. Yet now, a mere seven years after the liberation of our country, steel output already exceeds 4,000,000 tons. In the old China, there was hardly any machine-building industry, to say nothing of the automobile and aircraft industries; now we have all three. When the people overthrew the rule of imperialism, feudalism, and bureaucrat-capitalism, many were not clear as to which way China should head—towards capitalism or towards Socialism. Facts have now provided the answer: Only Socialism can save China. The Socialist system has promoted the rapid development of the productive forces of our country, a fact even our enemies abroad have had to acknowledge.

But our Socialist system has only just been set up; it is not yet fully established or fully consolidated. In joint State–private industrial and commercial enterprises, capitalists still get a fixed rate of interest on their capital, that is to say, exploitation still exists. So far as ownership is concerned, these enterprises are not yet completely Socialist in nature. A number of our agricultural and handicraft producers' co-operatives are still semi-Socialist, while even in the fully Socialist co-operatives certain specific problems of ownership remain to be solved. Relations between production and exchange in accordance with Socialist principles are being gradually established within and between all branches of our economy, and more and more appropriate forms are being sought. The problem of the proper relation of accumulation to consumption within each of the two sectors of the Socialist economy—the one where the means of production are owned by the whole people and the other where the means of production are owned by the collective—and the problem of the proper relation of accumulation to consumption between the two sectors themselves are complicated problems for which it is not easy to work out a perfectly rational solution all at once. To sum up, Socialist relations of production have been established and are in correspondence with the growth of the productive forces, but these relations are still far from perfect, and this imperfection stands in contradiction to the growth of the productive forces. Apart from correspondence as well as contradiction between the relations of production and the growth of the productive forces, there is correspondence as well as contradiction between the superstructure and the economic base. The superstructure, comprising the State system and laws of the people's democratic dictatorship and the socialist ideology guided by Marxism-Leninism, plays a positive role in facilitating the victory of Socialist transformation and the Socialist way of organizing labour; it is in correspondence with the Socialist economic base, that is, with Socialist relations of production. But the existence of bourgeois ideology, a certain bureaucratic style of work in our State organs, and defects in some of the links in our State institutions are in contradiction with the Socialist economic base. We must continue to resolve all such contradictions in the light of our specific conditions. Of course, new problems will emerge as these contradictions are resolved. And further efforts will be required to resolve the new contradictions. For instance, a constant process of readjustment through State planning is needed to deal with the contradiction between production and the needs of society, which will long remain an objective reality. Every year our country draws up an economic plan in order to establish a proper ratio between accumulation and consumption and achieve an equilibrium between production and needs. Equilibrium is nothing but a temporary, relative, unity of opposites. By the end of each year, this equilibrium, taken as a whole, is upset by the struggle of opposites; the unity undergoes a change, equilibrium becomes disequilibrium, unity becomes disunity, and once again it is necessary to work out an equilibrium and unity for the next year. Herein lies the superiority of our planned economy. As a matter of fact, this equilibrium, this unity, is partially upset every month or every quarter, and partial readjustments are called for. Sometimes, contradictions arise and the equilibrium is upset because our subjective arrangements do not conform to objective reality; this is what we call making a mistake. The ceaseless emergence and ceaseless resolution of contradictions constitute the dialectical law of the development of things.

5. Difficulties of the Cultural Revolution

I have just a few words to say about two matters.

For the past seventeen years there is one thing which in my opinion we haven't done well. Out of concern for State security and in view of the lessons of Stalin in the Soviet Union, we set up a first and second line. I have been in the second line, other comrades in the first line. Now we can see that wasn't so good; as a result our forces were dispersed. When we entered the cities we could not centralize our efforts, and there were quite a few independent kingdoms. Hence the Eleventh Plenum carried out changes. This is one matter. I am in the second line, I do not take charge of day-to-day work. Many things are left to other people so that other people's prestige is built up, and when I go to see God there won't be such a big upheaval in the State. Everybody was in agreement with this idea of mine. It seems that there are some things which the comrades in the first line have not managed too well. There are some things I should have kept a grip on which I did not. So I am responsible, we cannot just blame them. Why do I say that I bear some responsibility?

First, it was I who proposed that the Standing Committee be divided into two lines and that a secretariat be set up. Everyone agreed with this. Moreover I put too much trust in others. It was at the time of the Twenty-three Articles that my vigilance was aroused. I could do nothing in Peking; I could do nothing at the Centre. Last September and October I asked, if revisionism appeared at the Centre, what could the localities do? I felt that my ideas couldn't be carried out in Peking. Why was the criticism of Wu Han initiated not in Peking but in Shanghai? Because there was nobody to do it in Peking. Now the problem of Peking has been solved.

Second, the Great Cultural Revolution wreaked havoc after I approved Nieh Yüan-tzu's big-character poster in Peking University, and wrote a letter to Tsinghua University Middle School, as well as writing a big-character poster of my own entitled 'Bombard the Headquarters'. It all happened within a very short period, less than five months in June, July, August, September, and October. No wonder the comrades did not understand too much. The time was so short and the events so violent. I myself had not foreseen that as soon as the Peking University poster was broadcast, the whole country would be thrown into turmoil. Even before the letter to the Red Guards had gone out, Red Guards had mobilized throughout the country, and in one rush they swept you off your feet. Since it was I who caused the havoc, it is understandable if you have some bitter words for me. Last time we met I lacked confidence and I said that our decisions would not necessarily be carried out. Indeed all that time quite a few comrades still did not understand things fully, though now after a couple of months we have had some experience, and things are a bit better. This meeting has had two stages. In the first stage the speeches were not quite normal, but during the second stage, after speeches and the exchange of experience by comrades at the Centre, things went more smoothly and the ideas were understood a bit better. It has only been five months. Perhaps the movement may last another five months, or even longer.

Our democratic revolution went on for twenty-eight years, from 1921 to 1949. At first nobody knew how to conduct the revolution or how to carry on the struggle; only later did we acquire some experience. Our path gradually emerged in the course of practice. Did we not carry on for twenty-eight years, summarizing our experience as we went along? Have we not been carrying on the Socialist revolution for seventeen years, whereas the Cultural Revolution has been going on for only five months? Hence we cannot ask comrades to understand so well now. Many comrades did not read the articles criticizing Wu Han last year and did not pay much attention to them. The articles criticizing the film *The Life of Wu Hsün* and studies of the novel *Dream of the Red chamber* could not be grasped if taken separately, but only if taken as a whole. For this I am responsible. If you take them separately it is like treating only the head when you have a headache and treating only the feet when they hurt, the problem cannot be solved. During the first several months of this Great Cultural Revolution— in January, February, March, April, and May— articles were written and the Centre issued directives, but they did not arouse all that much attention. It was the big-character posters and onslaughts of the Red Guards which drew your attention, you could not avoid it because the revolution was right on top of you. You must quickly summarize your experience and properly carry out political and ideological work. Why are we meeting again after two months? It is to summarize our experience and carry out political and ideological work. You also have a great deal of political and ideological work to do after you go back. The Political Bureau, the provincial committees, the regional committees, and county committees must meet for ten days or more and thrash out the problems. But they mustn't think that everything can be cleared up. Some people have said, 'We understand the principles, but when we run up against concrete problems we cannot deal with them properly.' At first I could not understand why, if the principles were clear, the concrete problems could not be dealt with. I can see some reason for this: it may be that political and ideological work has not been done properly. When you went back after our last meeting some places did not find time to hold proper meetings. In Honan there were ten secretaries. Out of the ten there were seven or eight who were receiving people. The Red Guards rushed in and caused havoc. The students were angry, but they did not realize it and had not prepared themselves to answer questions. They thought that to make a welcoming speech lasting a quarter of an hour or so would do. But the students were thoroughly enraged. The fact that there were a number of questions which they could not immediately answer put the secretaries on the defensive. Yet this defensive attitude can be changed, can be transformed so that they take the initiative. Hence my confidence in this meeting has increased. I don't know what you think. If when you go back you do things according to the old system, maintaining the status quo, putting yourself in opposition to one group of Red Guards and letting another group hold sway, then I think things cannot change, the situation cannot improve. But I think things can change and things can improve. Of course we shouldn't expect too much. We can't be certain that the mass of central, provincial, regional, and county cadres should all be so enlightened. There will always be some who fail to understand, and there will be a minority on the opposite side. But I think it will be possible to make the majority understand.

I have talked about two matters. The first concerns history. For seventeen years the two lines have not been united. Others have some responsibility for this, so have I. The second issue is the five months of the Great Cultural Revolution, the fire of which I kindled. It has been going on only five months, not even half a year, a very brief span compared to the twenty-eight years of democratic revolution and the seventeen years of Socialist revolution. So one can see why it has not been thoroughly understood and there were obstacles. Why hasn't it been understood? In the past you have only been in charge of industry, agriculture, and communications and you have never carried out a Great Cultural Revolution. You in the Foreign Affairs Ministry and the Military Affairs Committee are the same. That which you never dreamed of has come to pass. What's come has come. I think that there are advantages in being assailed. For so many years you

SCHRAM, S., *Mao Tse-tung* (Harmondsworth, 1966).

SNOW, E., *Red Star over China*, rev. edn. (New York, 1972).

SOLOMON, R., *Mao's Revolution and the Chinese Political Culture* (Berkeley, 1971).

STARR, J., *Continuing the Revolution: The Political Thoughts of Mao* (Princeton, 1979).

SUYIN, H., *The Morning Deluge: Mao Tse-tung and the Chinese Revolution 1893–1953* (New York, 1972).

WOMACK, B., *The Foundations of Mao Zedong's Political Thought* (Honolulu, 1982).

14

HERBERT MARCUSE:
WESTERN MARXISM

What is denounced as 'Utopian' is no longer that which has 'no place' and cannot have any place in the historical universe, but rather that which is blocked from coming about by the power of established societies.

An Essay on Liberation (Harmondsworth, 1972), p. 13.

HERBERT MARCUSE (1898–1979) *was the most prominent theorist of the New Left in America in the 1960s and early 1970s. Born in Berlin, Marcuse joined the Frankfurt School in 1933 and retained more of a commitment to classical Marxism than its other leading members—Horkheimer and Adorno. It was through him that the Frankfurt School's criticism of the authoritarian trends implicit in contemporary culture became widely known. Marcuse was particularly interested in the changing nature of human labour, the possibility of a truly emancipated society, and the importance of retaining a 'Utopian' dimension to the human project. The extracts below are taken from* One-dimensional Man, *written in 1964, which became by far the most popular of Marcuse's works and was particularly influential at the time of the student unrest in the late 1960s. They contain striking claims about the forms of domination engendered by modern science and the 'affluent society' together with, in the last few pages, a pessimistic account of the possibilities of revolution.*

One-dimensional Society

ADVANCED industrial society confronts the critique with a situation which seems to deprive it of its very basis. Technical progress, extended to a whole system of domination and co-ordination, creates forms of life (and of power) which appear to reconcile the forces opposing the system and to defeat or refute all protest in the name of the historical prospects of freedom from toil and domination. Contemporary society seems to be capable of containing social change—qualitative change which would establish essentially different institutions, a new direction of the productive process, new modes of human existence. This containment of social change is perhaps the most singular achievement of advanced industrial society; the general acceptance of the National Purpose, bipartisan policy, the decline of pluralism, the collusion of Business and Labour within the strong State testify to the integration of opposites which is the result as well as the prerequisite of this achievement.

A brief comparison between the formative stage of the theory of industrial society and its present situation may help to show how the basis of the critique has been altered. At its origins in the first half of the nineteenth century, when it elaborated the first concepts of the alternatives, the critique of industrial society attained concreteness in a historical mediation between theory and practice, values and facts, needs and goals. This historical mediation occurred in the consciousness and in the political action of the two great classes which faced each other in the society: the bourgeoisie and the proletariat. In the capitalist world, they are still the basic classes. However, the capitalist development has altered the structure and function of these two classes in such a way that they no longer appear to be agents of historical transformation. An overriding interest in the preservation and improvement of the institutional status quo unites the former antagonists in the most advanced areas of contemporary society. And to the degree to which technical progress assures the growth and cohesion of Communist society, the very idea of qualitative change recedes before the realistic notions of a non-explosive evolution. In the absence of demonstrable agents and agencies of social change, the critique is thus thrown back to a high level of abstraction. There is no ground on which theory and practice, thought and action, meet. Even the most empirical analysis of historical alternatives appears to be unrealistic speculation, and commitment to them a matter of personal (or group) preference.

And yet: does this absence refute the theory? In the face of apparently contradictory facts, the critical analysis continues to insist that the need for qualitative change is as pressing as ever before. Needed by whom? the answer continues to be the same: by the society as a whole, for every one of its members. The union of growing productivity and growing destruction; the brinkmanship of annihilation; the surrender of thought, hope, and fear to the decisions of the powers that be; the preservation of misery in the face of unprecedented wealth, constitute the most impartial indictment—even if they are not the *raison d'être* of this society but only its by-product: its sweeping rationality, which propels efficiency and growth, is itself irrational.

The fact that the vast majority of the population accepts, and is made to accept, this society does not render it less irrational and less reprehensible. The distinction between true and false consciousness, real and immediate interest still is meaningful. But this distinction itself must be validated. Men must come to see it and to find their way from false to true consciousness, from their immediate to their real interest. They can do so only if they live in need of changing their way of life, of denying the positive, of refusing. It is precisely this need which the established society manages to repress to the degree to which it is capable of 'delivering the goods' on an increasingly large scale, and using the scientific conquest of nature for the scientific conquest of man.

Confronted with the total character of the achievements of advanced industrial society, critical theory is left without the rationale for transcending this society. The vacuum empties the theoretical structure itself, because the categories of a critical social theory were developed during the period in which the need for refusal and subversion was embodied in the action of effective social forces. These categories were essentially negative and oppositional concepts, defining the actual contradictions in nineteenth-century European society. The category 'society' itself expressed the acute conflict between the social and political sphere—society as antagonistic to the State. Similarly, 'individual', 'class,' 'private', 'family', denoted spheres and forces not yet integrated with the established conditions—spheres of tension and contradiction. With the growing integration of industrial society, these categories are losing their critical connotation, and tend to become descriptive, deceptive, or operational terms.

An attempt to recapture the critical intent of these categories, and to understand how the intent was cancelled by the social reality, appears from the outset to be regression from a theory joined with historical practice to abstract, speculative thought: from the critique of political economy to philosophy. This ideological character of the critique results from the fact that the analysis is forced to proceed from a position 'outside' the positive as well as negative, the productive as well as destructive tendencies in society. Modern industrial society is the pervasive identity of these opposites—it is the whole that is in question. At the same time, the position of theory cannot be one of mere speculation. It must be a historical position in the sense that it must be grounded on the capabilities of the given society.

This ambiguous situation involves a still more fundamental ambiguity. *One-dimensional Man* will vacillate throughout between two contradictory hypotheses: (1) that advanced industrial society is capable of containing qualitative change for the foreseeable future; (2) that forces and tendencies exist which may break this containment and explode the society. I do not think that a clear answer can be given. Both tendencies are there, side by side—and even the one in the other. The first tendency is dominant, and whatever pre-conditions for a reversal may exist are being used to prevent it. Perhaps an accident may alter the situation, but unless the recognition of what is being done and what is being prevented subverts the consciousness and the behaviour of man, not even a catastrophe will bring about the change.

A comfortable, smooth, reasonable, democratic unfreedom prevails in advanced industrial civilization, a token of technical progress. Indeed, what could be more rational than the suppression of individuality in the mechanization of socially necessary but painful performances; the concentration of individual enterprises in more effective, more productive corporations; the regulation of free competition among unequally equipped economic subjects; the curtailment of prerogatives and national sovereignties which impede the international organization of resources. That this technological order also involves a political and intellectual co-ordination may be a regrettable and yet promising development.

The rights and liberties which were such vital factors in the origins and earlier stages of industrial society yield to a higher stage of this society: they are losing their traditional rationale and content. Freedom of thought, speech, and conscience were—just as free enterprise, which they served to promote and protect—essentially *critical* ideas, designed to replace an obsolescent material and intellectual culture by a more productive and rational one. Once institutionalized, these rights and liberties shared the fate of the society of which they had become an integral part. The achievement cancels the premises.

To the degree to which freedom from want, the concrete substance of all freedom, is becoming a real possibility, the liberties which pertain to a state of lower productivity are losing their former content. Independence of thought, autonomy, and the right to political opposition are being deprived of their basic critical function in a society which seems increasingly capable of satisfying the needs of the individuals through the way in which it is organized. Such a society may justly demand acceptance of its principles and institutions, and reduce the opposition to the discussion and promotion of alternative policies *within* the status quo. In this respect, it seems to make little difference whether the increasing satisfaction of needs is accomplished by an authoritarian or a non-authoritarian system. Under the conditions of a rising standard of living, non-conformity with the system itself appears to be socially useless, and the more so when it entails tangible economic and political disadvantages and threatens the smooth operation of the whole. Indeed, at least in so far as the necessities of life are involved, there seems to be no good reason why the production and distribution of goods and services should proceed through the competitive concurrence of individual liberties.

Freedom of enterprise was from the beginning not altogether a blessing. As the liberty to work or to starve, it spelled toil, insecurity, and fear for the vast majority of the population. If the individual were no longer compelled to prove himself on the market, as a free economic subject, the disappearance of this kind of freedom would be one of the greatest achievements of civilization. The technological processes of mechanization and standardization might release individual energy into a yet uncharted realm of freedom beyond necessity. The very structure of human existence would be altered; the individual would be liberated from the work world's imposing upon him alien needs and alien possibilities. The individual would be free to exert autonomy over a life that would be his own. If the productive apparatus could be organized and directed toward the satisfaction of the vital needs, its control might well be centralized; such control would not prevent individual autonomy, but render it possible.

This is a goal within the capabilities of advanced industrial civilization, the 'end' of technological rationality. In actual fact, however, the contrary trend operates; the apparatus imposes its economic and political requirements for defence and expansion on labour time and free time, on the material and intellectual culture. By virtue of the way it has organized its technological base, contemporary industrial society tends to be totalitarian. For 'totalitarian' is not only a terroristic political co-ordination of society, but also a non-terroristic economic-technical co-ordination which operates through the manipulation of needs by vested interests. It thus precludes the emergence of an effective opposition against the whole. Not only a specific form of government or party rule makes for totalitarianism, but also a specific system of production and distribution which may well be compatible with a 'pluralism' of parties, newspapers, 'countervailing powers', etc.

Today political power asserts itself through its power over the machine process and over the technical organization of the apparatus. The government of advanced and advancing industrial societies can maintain and secure itself only when it succeeds in mobilizing, organizing, and exploiting the technical, scientific, and mechanical productivity available to industrial civilization. And this productivity mobilizes society as a whole, above and beyond any particular individual or group interests. The brute fact that the machine's physical (only physical?) power surpasses that of the individual, and of any particular group of individuals, makes the machine the most effective political instrument in any society whose basic organization is that of the machine process. But the political trend may be reversed; essentially the power of the machine is only the stored-up and projected power of man. To the extent to which the work world is conceived of as a machine and mechanized accordingly, it becomes the *potential* basis of a new freedom for man.

Contemporary industrial civilization demonstrates that it has reached the stage at which 'the free society' can no longer be adequately defined in the traditional terms of economic, political, and intellectual liberties, not because these liberties have become insignificant, but because they are too significant to be confined within the traditional forms. New modes of realization are needed, corresponding to the new capabilities of society.

Such new modes can be indicated only in negative terms because they would amount to the negation of the prevailing modes. Thus economic freedom would mean freedom *from* the economy—from being controlled by economic forces and relationships; freedom from the daily struggle for existence, from earning a living. Political freedom would mean liberation of the individuals *from* politics over which they have no effective control. Similarly, intellectual freedom would mean the restoration of individual thought now absorbed by mass communication and indoctrination, abolition of 'public opinion' together with its makers. The unrealistic sound of these propositions is indicative, not of their Utopian character, but of the strength of the forces which prevent their realization. The most effective and enduring form of warfare against liberation is the implanting of material and intellectual needs that perpetuate obsolete forms of the struggle for existence.

The intensity, the satisfaction and even the character of human needs, beyond the biological level, have always been preconditioned. Whether or not the possibility of doing or leaving, enjoying or destroying, possessing or rejecting something is seized as a *need* depends on whether or not it can be seen as desirable and necessary for the prevailing societal institutions and interests. In this sense, human needs are historical needs and, to the extent to which the society demands the repressive development of the individual, his needs themselves and their claim for satisfaction are subject to overriding critical standards.

We may distinguish both true and false needs. 'False' are those which are superimposed upon the individual by particular social interests in his repression: the needs which perpetuate toil, aggressiveness, misery, and injustice. Their satisfaction might be most gratifying to the individual, but this happiness is not a condition which has to be maintained and protected if it serves to arrest the development of the ability (his own and others) to recognize the disease of the whole and grasp the chances of curing the disease. The result then is euphoria in unhappiness. Most of the prevailing needs to relax, to have fun, to behave and consume in accordance with the advertisements, to love and hate what others love and hate, belong to this category of false needs.

Such needs have a societal content and function which are determined by external powers over which the individual has no control; the development and satisfaction of these needs is heteronomous. No matter how much such needs may have become the individual's own, reproduced and fortified by the conditions of his existence; no matter how much he identifies himself with them and finds himself in their satisfaction, they continue to be what they were from the beginning—products of a society whose dominant interest demands repression.

The prevalence of repressive needs is an accomplished fact, accepted in ignorance and defeat, but a fact that must be undone in the interest of the happy individual as well as all those whose misery is the price of his satisfaction. The only needs that have an unqualified claim for satisfaction are the vital ones—nourishment, clothing, lodging at the attainable level of culture. The satisfaction of these needs is the prerequisite for the realization of *all* needs, of the unsublimated as well as the sublimated ones.

For any consciousness and conscience, for any experience which does not accept the prevailing societal interest as the supreme law of thought and behaviour, the established universe of needs and satisfactions is a fact to be questioned—questioned in terms of truth and falsehood. These terms are historical throughout, and their objectivity is historical. The judgement of needs and their satisfaction, under the given conditions, involves standards of *priority*—standards which refer to the optimal development of the individual, of all individuals, under the optimal utilization of the material and intellctual resources available to man. The resources are calculable. 'Truth' and 'falsehood' of needs designate objective conditions to the extent to which the universal satisfaction of vital needs and, beyond it, the progressive alleviation of toil and poverty and are universally valid standards. But as historical standards, they do not only vary according to area and stage of development, they also can be defined only in (greater or lesser) *contradiction* to the prevailing ones. What tribunal can possibly claim the authority of decision?

In the last analysis, the question of what are true and false needs must be answered by the individuals themselves, but only in the last analysis; that is, if and when they are free to give their own answer. As long as they are kept incapable of being autonomous, as long as they are indoctrinated and manipulated (down to their very instincts), their answer to this question cannot be taken as their own. By the same token, however, no tribunal can justly arrogate to itself the right to decide which needs should be developed and satisfied. Any such tribunal is reprehensible, although our revulsion does not do away with the question: how can the people who have been the object of effective and productive domination by themselves create the conditions of freedom?

The more rational, productive, technical, and total the repressive admin-istration of society becomes, the more unimaginable the means and ways by which the administered individuals might break their servitude and seize their own liberation. To be sure, to impose Reason upon an entire society is a paradoxial and scandalous idea—although one might dispute the righteousness of a society which ridicules this idea while making its own population into objects of total administration. All liberation depends on the consciousness of servitude, and the emergence of this consciousness is always hampered by the predominance of needs and satisfactions which, to a great extent, have become the individual's own. The process always replaces one system of pre-conditioning by another; the optimal goal is the replacement of false needs by true ones, the abandonment of repressive satisfaction.

The distinguishing feature of advanced industrial society is its effective suffocation of those needs which demand liberation—liberation also from that which is tolerable and rewarding and comfortable—while it sustains and absolves the destructive power and repressive function of the affluent society. Here, the social controls exact the overwhelming need for the production and consumption of waste; the need for stupefying work where it is no longer a real necessity; the need for modes of relaxation which soothe and prolong this stupefication; the need for maintaining such deceptive liberties as free competition at administered prices, a free press which censors itself, free choice between brands and gadgets.

Under the rule of a repressive whole, liberty can be made into a powerful instrument of domination. The range of choice open to the individual is not the decisive factor in determining the degree of human freedom, but *what* can be chosen and what *is* chosen by the individual. The criterion for free choice can never be an absolute one, but neither is it entirely relative. Free election of masters does not abolish the masters or the slaves. Free choice among a wide variety of goods and services does not signify freedom if these goods and services sustain social controls over a life of toil and fear—that is, if they sustain alienation. And the spontaneous reproduction of superimposed needs by the individual does not establish autonomy; it only testifies to the efficacy of the controls.

Our insistence on the depth and efficacy of these controls is open to the objection that we overrate greatly the indoctrinating power of the 'media', and that by themselves the people would feel and satisfy the needs which are now imposed upon them. The objection misses the point. The pre-conditioning does not start with the mass production of radio and television and with the centralization of their control. The people enter this stage as pre-conditioned receptacles of longstanding; the decisive difference is in the flattening out of the contrast (or conflict) between the given and the possible, between the satisfied and the unsatisfied needs. Here, the so-called equalization of class distinctions reveals its ideological function. If the worker and his boss enjoy the same television programme and visit the same resort places, if the typist is as attractively made up as the daughter of her employer, if the Negro owns a Cadillac, if they all read the same newspaper, then this assimilation indicates not the disappearance of classes, but the extent to which the needs and satisfactions that serve the preservation of the Establishment are shared by the underlying population.

Indeed, in the most highly developed areas of contemporary society, the transplantation of social into individual needs is so effective that the difference between them seems to be purely theoretical. Can one really distinguish between the mass media as instruments of information and entertainment, and as agents of manipulation and indoctrination? Between the automobile as nuisance and as convenience? Between the horrors and the comforts of functional architecture? Between the work for national defence and the work for corporate gain? Between the private pleasure and the commercial and political utility involved in increasing the birth rate?

We are again confronted with one of the most vexing aspects of advanced industrial civilization: the rational character of its irrationality. Its productivity and efficiency, its capacity to increase and spread comforts, to turn waste into need, and destruction into construction, the extent to which this civilization transforms the object world into an extension of man's mind and body makes the very notion of alienation questionable. The people recognize themselves in their commodities; they find their soul in their automobile, hi-fi set, split-level home, kitchen equipment. The very mechanism which ties the individual to his society has changed, and social control is anchored in the new needs which it has produced.

Having discussed the political integration of advanced industrial society, an achievement rendered possible by growing technological productivity and the expanding conquest of man and nature, we will now turn to a corresponding integration in the realm of culture. In this chapter, certain key notions and images of literature and their fate will illustrate how the progress of technological rationality is liquidating the oppositional and transcending elements in the 'higher culture'. They succumb in fact to the process of *desublimation* which prevails in the advanced regions of contemporary society.

The achievements and the failures of this society invalidate its higher culture. The celebration of the autonomous personality, of humanism, of tragic and romantic love, appears to be the ideal of a backward stage of the development. What is happening now is not the deterioration of higher culture into mass culture but the refutation of this culture by the reality. The reality surpasses its culture. Man today can do *more* than the culture heroes and half-gods; he has solved many insoluble problems. But he has also betrayed the hope and destroyed the truth which were preserved in the sublimations of higher culture. To be sure, the higher culture was always in contradiction with social reality, and only a privileged minority enjoyed its blessings and represented its ideals. The two antagonistic spheres of society have always coexisted; the higher culture has always been accommodating, while the reality was rarely disturbed by its ideals and its truth.

Today's novel feature is the flattening out of the antagonism between culture and social reality through the obliteration of the oppositional, alien, and transcendent elements in the higher culture by virtue of which it constituted *another dimension* of reality. This liquidation of *two-dimensional* culture takes place not through the denial and rejection of the 'cultural values', but through their wholesale incorporation into the established order, through their reproduction and display on a massive scale.

In fact, they serve as instruments of social cohesion. The greatness of a free literature and art, the ideals of humanism, the sorrows and joys of the individual, the fulfilment of the personality are important items in the competitive struggle between East and West. They speak heavily against the present forms of Communism, and they are daily administered and sold. The fact that they contradict the society which sells them does not count. Just as people know or feel that advertisements and political platforms must not be necessarily true or right, and yet hear and read them and even let themselves be guided by them, so they accept the traditional values, and make them part of their mental equipment. If mass communications blend together harmoniously and often unnoticeably art, politics, religion, and philosophy with commercials, they bring these realms of culture to their common denominator—the commodity form. The music of the soul is also the music of salesmanship. Exchange value, not truth value, counts. On it centres the rationality of the status quo, and all alien rationality is bent to it.

As the great words of freedom and fulfilment are pronounced by campaigning leaders and politicians, on the screens and radios and stages, they turn into meaningless sounds which obtain meaning only in the context of propaganda, business, discipline, and relaxation. This assimilation of the ideal with reality testifies to the extent to which the ideal has been surpassed. It is brought down from the sublimated realm of the soul or the spirit or the inner man, and translated into operational terms and problems. Here are the progressive elements of mass culture. The perversion is indicative of the fact that advanced industrial society is confronted with the possibility of a materialization of ideals. The capabilities of this society are progressively reducing the sublimated realm in which the condition of man was represented, idealized, and indicted. Higher culture becomes part of the material culture. In this transformation, it loses the greater part of its truth.

The higher culture of the West—whose moral, aesthetic, and intellectual values industrial society will profess—was a pre-technological culture in a functional as well as chronological sense. Its validity was derived from the experience of a world which no longer exists and which cannot be recaptured because it is in a strict sense invalidated by technological society. Moreover, it remained to a large degree a feudal culture, even when the bourgeois period gave it some of its most lasting formulations. It was feudal not only because of its confinement to privileged minorities, not only because of its inherent romantic element (which will be discussed presently), but also because its authentic works expressed a conscious, methodical alienation from the entire sphere of business and industry, and from its calculable and profitable order.

While this bourgeois order found its rich—and even affirmative—representation in art and literature (as in the Dutch painters of the seventeenth century, in Goethe's *Wilhelm Meister*, in the English novel of the nineteenth century, in Thomas Mann), it remained an order which was overshadowed, broken, refuted by another dimension which was irreconcilably antagonistic to the order of business, indicting it and denying it. And in the literature, this other dimension is represented *not* by the religious, spiritual, moral heroes (who often sustain the established order), but rather by such disruptive characters as the artist, the prostitute, the adulteress, the great criminal and outcast, the warrior, the rebel-poet, the devil, the fool—those who don't earn a living, at least not in an orderly and normal way.

To be sure, these characters have not disappeared from the literature of advanced industrial society, but they survive essentially transformed. The vamp, the national hero, the beatnik, the neurotic housewife, the gangster, the star, the charismatic tycoon perform a function very different from and even contrary to that of their cultural predecessors. They are no longer images of another way of life but rather freaks or types of the same life, serving as an affirmation rather than negation of the established order.

Surely, the world of their predecessors was a backward, pre-technological world, a world with the good conscience of inequality and toil, in which labour was still a fated misfortune; but a world in which man and nature were not yet organized as things and instrumentalities. With its code of forms and manners, with the style and vocabulary of its literature and philosophy, this past culture expressed the rhythm and content of a universe in which valleys and forests, villages and inns, nobles and villains, salons and courts were a part of the experienced reality. In the verse and prose of this pre-technological culture is the rhythm of those who wander or ride in carriages, who have the time and the pleasure to think, contemplate, feel, and narrate.

It is outdated and surpassed culture, and only dreams and childlike regressions can recapture it. But this culture is, in some of its decisive elements, also a *post*-technological one. Its most advanced images and positions seem to survive their absorption into administered comforts and stimuli; they continue to haunt the consciousness with the possibility of their rebirth in the consummation of technical progress. They are the expression of that free and conscious alienation from the established forms of life with which literature and the arts opposed these forms even where they adorned them.

In contrast to the Marxian concept, which denotes man's relation to himself and to his work in capitalist society, the *artistic alienation* is the conscious transcendence of the alienated existence—a 'higher level' or mediated alienation. The conflict with the world of progress, the negation of the order of business, the anti-bourgeois elements in bourgeois literature and art are due neither to the aesthetic lowliness of this order nor to romantic reaction—nostalgic consecration of a disappearing stage of civilization. 'Romantic' is a term of condescending defamation which is easily applied to disparaging avant-garde positions, just as the term 'decadent' far more often denounces the genuinely progressive traits of a dying culture than the real factors of decay. The traditional images of artistic alienation are indeed romantic in as much as they are in aesthetic incompatibility with the developing society. This incompatibility is the token of their truth. What they recall and preserve in memory pertains to the future: images of a gratification that would dissolve the society which suppresses it. The great surrealist art and literature of the 'twenties and 'thirties has still recaptured them in their subversive and liberating function. Random examples from the basic literary vocabulary may indicate the range and the kinship of these images, and the dimension which they reveal: Soul and Spirit and Heart; *la recherche de l'absolu, Les Fleurs du mal, la femme-enfant*: the Kingdom by the Sea; *Le Bateau ivre* and the Long-legged Bait; *Ferne* and *Heimat*; but also demon rum, demon machine, and demon money; Don Juan and Romeo; the Master Builder and When We Dead Awake.

Their mere enumeration shows that they belong to a lost dimension.

They are invalidated not because of their literary obsolescence. Some of these images pertain to contemporary literature and survive in its most advanced creations. What has been invalidated is their subversive force, their destructive content—their truth. In this transformation, they find their home in everyday living. The alien and alienating *œuvres* of intellectual culture become familiar goods and services. Is their massive reproduction and consumption only a change in quantity, namely, growing appreciation and understanding, democratization of culture?

The truth of literature and art has always been granted (if it was granted at all) as one of a 'higher' order, which should not and indeed did not disturb the order of business. What has changed in the contemporary period is the difference between the two orders and their truths. The absorbent power of society depletes the artistic dimension by assimilating its antagonistic contents. In the realm of culture, the new totalitarianism manifests itself precisely in a harmonizing pluralism, where the most contradictory works and truths peacefully coexist in indifference.

For example, compare love-making in a meadow and in an automobile, on a lovers' walk outside the town walls and on a Manhattan street. In the former cases, the environment partakes of and invites libidinal cathexis and tends to be eroticized. Libido transcends beyond the immediate erotogenic zones—a process of non-repressive sublimation. In contrast, a mechanized environment seems to block such self-transcendence of libido. Impelled in the striving to extend the field of erotic gratification, libido becomes less 'polymorphous', less capable of eroticism beyond localized sexuality, and the *latter* is intensified.

Thus diminishing erotic and intensifying sexual energy, the technological reality *limits the scope of sublimation*. It also reduces the *need* for sublimation. In the mental apparatus, the tension between that which is desired and that which is permitted seems considerably lowered, and the Reality Principle no longer seems to require a sweeping and painful transformation of instinctual needs. The individual must adapt himself to a world which does not seem to demand the denial of his innermost needs—a world which is not essentially hostile.

The organism is thus being pre-conditioned for the spontaneous acceptance of what is offered. Inasmuch as the greater liberty involves a contraction rather than extension and development of instinctual needs, it works *for* rather than *against* the status quo of general repression—one might speak of 'institutionalized desublimation'. The latter appears to be a vital factor in the making of the authoritarian personality of our time.

It has often been noted that advanced industrial civilization operates with a greater degree of sexual freedom—'operates' in the sense that the latter becomes a market value and a factor of social mores. Without ceasing to be an instrument of labour, the body is allowed to exhibit its sexual features in the everyday work world and in work relations. This is one of the unique achievements of industrial society—rendered possible by the reduction of dirty and heavy physical labour; by the availability of cheap, attractive clothing, beauty culture, and physical hygiene; by the requirements of the advertising industry, etc. The sexy office and sales girls, the handsome, virile junior executive and floor walker, are highly-marketable commodities, and the possession of suitable mistresses—once the prerogative of kings, princes, and lords—facilitates the career of even the less exalted ranks in the business community.

Functionalism, going artistic, promotes this trend. Shops and offices open themselves through huge glass windows and expose their personnel; inside, high counters and non-transparent partitions are coming down. The corrosion of privacy in massive apartment houses and suburban homes breaks the barrier which formerly separated the individual from the public existence and exposes more easily the attractive qualities of other wives and other husbands.

This socialization is not contradictory but complementary to the de-erotization of the environment. Sex is integrated into work and public relations and is thus made more susceptible to (controlled) satisfaction. Technical progress and more comfortable living permit the systematic inclusion of libidinal components into the realm of commodity production and exchange. But no matter how controlled the mobilization of instinctual energy may be (it sometimes amounts to a scientific management of libido), no matter how much it may serve as a prop for the status quo—it is also gratifying to the managed individuals, just as racing the outboard motor, pushing the power lawn mower, and speeding the automobile are fun.

This mobilization and administration of libido may account for much of the voluntary compliance, the absence of terror, the pre-established harmony between individual needs and socially required desires, goals, and aspirations. The technological and political conquest of the transcending factors in human existence, so characteristic of advanced industrial civilization, here asserts itself in the instinctual sphere: satisfaction in a way which generates submission and weakens the rationality of protest.

The range of socially permissible and desirable satisfaction is greatly enlarged, but through this satisfaction, the Pleasure Principle is reduced—deprived of the claims which are irreconcilable with the established society. Pleasure, thus adjusted, generates submisson.

Does this mean that the critical theory of society abdicates and leaves the field to an empirical sociology which, freed from all theoretical guidance except a methodological one, succumbs to the fallacies of misplaced concreteness, thus performing an ideological service while proclaiming the elimination of value judgements? Or do the dialectical concepts once again testify to their truth—by comprehending their own situation as that of the society which they analyse? A response might suggest itself if one considers the critical theory precisely at the point of its greatest weakness—its inability to demonstrate the liberating tendencies *within* the established society.

The critical theory of society was, at the time of its origin, confronted with the presence of real forces (objective and subjective) *in* the established society which moved (or could be guided to move) toward more rational and freer institutions by abolishing the existing ones which had become obstacles to progress. These were the empirical grounds on which the theory was erected, and from these empirical grounds derived the idea of the liberation of *inherent* possibilities—the development, otherwise blocked and distorted, of material and intellectual productivity, faculties, and needs. Without the demonstration of such forces, the critique of society would still be valid and rational but it would be incapable of translating its rationality into terms of historical practice. The conclusion? 'Liberation of inherent possibilities' no longer adequately expresses the historical alternative.

The enchained possibilities of advanced industrial societies are: development of the productive forces on an enlarged scale, extension of the conquest of nature, growing satisfaction of needs for a growing number of people, creation of new needs and faculties. But these possibilities are gradually being realized through means and institutions which cancel their liberating potential, and this process affects not only the means but also the ends. The instruments of productivity and progress, organized into a totalitarian system, determine not only the actual but also the possible utilizations.

At its most advanced stage, domination functions as administration, and in the overdeveloped areas of mass consumption, the administered life becomes the good life of the whole, in the defence of which the opposites are united. This is the pure form of domination. Conversely, its negation appears to be the pure form of negation. All content seems reduced to the one abstract demand for the end of domination—the only truly revolutionary exigency, and the event that would validate the achievements of industrial civilization. In the face of its efficient denial by the established system, this negation appears in the politically impotent form of the 'absolute refusal'— a refusal which seems the more unreasonable the more the established system develops its productivity and alleviates the burden of life. In the words of Maurice Blanchot:

Ce que nous refusons n'est pas sans valeur ni sans importance. C'est bien à cause de cela que le refus est nécessaire. Il y a une raison que nous n'accepterons plus, il y a une apparence de sagesse qui nous fait horreur, il y a une offre d'accord et de conciliation que nous n'entendrons pas. Une rupture s'est produite. Nous avons été remenés à cette franchise qui ne tolére plus la complicité.[1]

But if the abstract character of the refusal is the result of total reification, then the concrete ground for refusal must still exist, for reification is an illusion. By the same token, the unification of opposites in the medium of technological rationality must be, *in all its reality*, an illusory unification, which elimates neither the contradiction between the growing productivity and its repressive use, nor the vital need for solving the contradiction.

But the struggle for the solution has outgrown the traditional forms. The totalitarian tendencies of the one-dimensional society render the traditional ways and means of protest ineffective—perhaps even dangerous because they preserve the illusion of popular sovereignty. This illusion contains some truth: 'the people', previously the ferment of social change, have 'moved up' to become the ferment of social cohesion. Here rather than in the redistribution of wealth and equalization of classes is the new stratification characteristic of advanced industrial society.

However, underneath the conservative popular base is the substratum of the outcasts and outsiders, the exploited and persecuted of other races and other colours, the unemployed and the unemployable. They exist outside the democratic process; their life is the most immediate and the most real need for ending intolerable conditions and institutions. Thus their opposition is revolutionary even if their consciousness is not. Their opposition hits the system from without and is therefore not deflected by the system; it is an elementary force which violates the rules of the game, and, in doing so, reveals it as a rigged game. When they get together and go out into the streets, without arms, without protection, in order to ask for the most primitive civil rights, they know that they face dogs, stones, and bombs, jail, concentration camps, even death. Their force is behind every political demonstration for the victims of law and order. The fact that they start refusing to play the game may be the fact which marks the beginning of the end of a period.

Nothing indicates that it will be a good end. The economic and technical capabilities of the established societies are sufficiently vast to allow for

[1] 'What we refuse is not without value or importance. Precisely because of that, the refusal is necessary. There is a reason which we no longer accept, there is an appearance of wisdom which horrifies us, there is a plea for agreement and conciliation which we will no longer heed. A break has occurred. We have been reduced to that frankness which no longer tolerates complicity.' 'Le Refus', in *Le 14 juillet*, no. 2, Paris, Oct. 1958.

adjustments and concessions to the underdog, and their armed forces sufficiently trained and equipped to take care of emergency situations. However, the spectre is there again, inside and outside the frontiers of the advanced societies. The facile historical parallel with the barbarians threatening the empire of civilization prejudges the issue; the second period of barbarism may well be the continued empire of civilization itself. But the chance is that, in this period, the historical extremes may meet again: the most advanced consciousness of humanity, and its most exploited force. It is nothing but a chance. The critical theory of society possesses no concepts which could bridge the gap between the present and its future; holding no promise and showing no success, it remains negative. Thus it wants to remain loyal to those who, without hope, have given and give their life to the Great Refusal.

At the beginning of the fascist era, Walter Benjamin wrote:

Nur um der Hoffnungslosen willen ist uns die Hoffnung gegeben.
It is only for the sake of those without hope that hope is given to us.

Source: H. Marcuse, *One-dimensional Man* (Boston, 1964), pp. 11 ff., 19 ff., 58 ff., 70 f., 199 ff.,

FURTHER NOTES

WORKS
MARCUSE, H., *Reason and Revolution: Hegel and the Rise of Social Theory* (New York, 1964).
—— *Eros and Civilization* (Boston, 1964).
—— *Societ Marxism* (Boston, 1958).
—— *One-dimensional Man* (Boston, 1964).
—— *An Essay on Liberation*, (Boston, 1972).

COMMENTARIES
BREINES, P. (ed.), *Critical Interruptions: New Left Perspectives on Herbert Marcuse* (New York, 1970).
FRY, J., *Marcuse: Dilemma on Liberation: A Critical Analysis* (Brighton, 1978).
GEOGHEGAN, V., *Reason and Eros: The Social Theory of Herbert Marcuse* (London, 1981).
KATZ, B., *Herbert Marcuse and the Art of Liberation* (London, 1982).
LEISS, W., *The Domination of Nature* (New York, 1974).
MACINTYRE, A., *Marcuse* (New York, 1970).
MATTICK, P., *Critique of Marcuse: One Dimensional Man in Class Society* (London, 1972).

15

EDVARD KARDELJ:
YUGOSLAV MARXISM

Tito paced up and down as though completely wrapped up in his own thoughts. Suddenly he exclaimed 'The factories belonging to the workers: something that has never been achieved!' And with these words the theories worked out by Kardelj and myself seemed to shed their complexity and to acquire a better prospect of being carried into effect.

M. Djilas, quoted in N. Beloff, *Tito's Flawed Legacy* (London, 1985), p. 219.

EDWARD KARDELJ (1910–79) *trained as a teacher, joined forces with Tito in 1934, and remained one of his closest associates. After the two other members of Tito's original triumvirate were disgraced—Djilas in 1954 and Rankovic in 1966—Kardelj emerged as second in importance only to Tito. As Foreign Minister of post-war Yugoslavia, Kardelj became a specialist in the exposition of the special Yugoslav interpretation of Marxism which dates from the political break with Moscow in 1948. In the extract below, taken from his book* Democracy and Socialism, *Kardelj defends Yugoslavia's deviation from the orthodox model. The most striking aspect is their emphasis on workers' self-management in the context of a considerably decentralized economy and a strongly market-orientated Socialism. Beginning with the need to find a method by which the partisans could govern liberated areas during the war, Yugoslav self-management has become increasingly pragmatic in the pursuit of economic growth and technical development based upon legitimate self-interest and strong material incentives. Although not particularly successful in its own terms, Yugoslav Marxism does represent an experiment that is being imitated in other parts of the Communist world.*

Democracy and Socialism

THE socio-economic pattern and the stability of the production relations in every society determine the character and form of its political system and the degree of its social and political stability. It is the quality of this relationship which gives real content to a Socialist society, and for this reason, in Yugoslavia, the Constitution, the Law on Associated Labour, and a series of other structural laws have given the workers in associated labour a direct economic claim on their own socialized past labour. What I mean by 'past labour' in this context is that portion of the value created by a worker's labour which is not paid out to him as personal income but is 'socialized'—that is, remains for the benefit of the society or social group in which he lives and works, serving either as accumulation, in the form of social capital, or as a source for collective social consumption of various kinds. Because of the social character of labour, the past labour of an individual worker cannot be expressed in individual terms, but only generally, as contributing to the value created by the joint labour of all workers. This is why the socialization of past labour is an absolute necessity, and is in the interest of the entire society as well as of the individual worker in associated labour. If the socialization of past labour does not alienate from the worker the product of his own efforts, but provides for its joint management by the workers themselves, it will become a material base for the equal personal rights of all workers in deciding on the use of the value created by their own joint labour.

It is in fact the character of this relationship between the worker and his past labour that most clearly and consistently expresses the class character of a society's social relations. The class character of production relations among any people depends on who controls and manages the funds of past labour—whether social capital or funds for collective or public consumption—and in what manner. No one denies that the appropriation of capital by private entrepreneurs is what determines the class character of capitalism. Yet its antithesis, namely the socialization of the workers' surplus labour, or surplus value as a part of their socialized past labour, does not necessarily produce the same results in the elimination of all forms of appropriation based on the monopolistic control over capital. In this case the alienation from the worker of his own socialized past labour, which arises from the techno-bureaucratic monopoly of government and the economic apparatus, except when it is under the efficient, democratic control of the working people, in fact retains very strong elements of wage-labour relationships. The consistent abolition of any form of private or monopolistic appropriation of capital or of the management of capital is, therefore, possible when socialized past labour is under the joint control of all workers, under their joint management, within a democratic system of equitable personal rights and obligations of workers. This is a means of extending the material base of their labour and increasing their productivity and thereby also their remuneration.

It would, of course, be absurd to claim that the form of Socialist self-management developed in Yugoslavia is the only possible form of democratic and direct management by workers of their joint past labour. In general terms, it might be said that the degree of democracy in this kind of Socialist society, one which provides for a direct influence by workers and working people on decision-making in production and society, is the true gauge of the workers' situation. Without any exaggeration we can claim that Yugoslav Socialist self-management is consistently and unambiguously opening up prospects for precisely this kind of development.

The relations of Socialist self-management do not sever the economic and political link between the worker in associated labour and his socialized past labour. Together with other workers, he supervises and even controls this portion of the fruit of his labour, while the results of management of past labour in the process of social reproduction belong jointly to all workers, so that through these economic channels they can return as the earnings of the basic organization of associated labour and thence as the personal income of the workers. This economic dependence of workers on the total process of social reproduction is at the same time the source of their economic and political control over the aggregate social capital representing the socialized past labour of individual and associated workers, which is no longer private property nor a monopoly right belonging to the State apparatus but remains in their own hands. This is why, under our system, all manner of earnings must flow into the hands of the workers of the basic organization of associated labour and no one other than the workers of that organization can control their earnings without their consent. Income thus serves as a means of enlarging and enriching the material base of labour of the basic organization of associated labour, while through various forms of the pooling of labour and income, subject to agreement by the basic organization of associated labour, it may also be concentrated in associated labour in general for the purpose of developing the material base, production techniques, and technology as a whole. For the same reasons, the banks, any more than the large companies, cannot earn social capital which they may dispose of without restriction; they can only operate with the resources invested by the basic organizations of associated labour and must use them subject to consent from the investors, following a jointly established plan. Neither can the State acquire any social capital except, in the cases provided for by law, for the purpose of carrying out large-scale projects of general public interest. And, even in such exceptional cases, the State does not acquire property by managing these facilities; they remain under the control of the working people who work with the socially owned means of production from which they derive.

In this manner, workers and working people are in principle guaranteed—although in practice it is not always so—a direct control not only over their current labour but also over their socialized past labour, i.e., over the resources for social reproduction, and, consequently, over the means for expanding and improving the material prerequisites of their labour; or, to put it succinctly, the workers control social capital in their basic organizations of associated labour, and through them have a say on all other forms of the pooling and movement of this capital. In this way, the worker has a real opportunity to exercise an effective economic and political control over the means, terms, and fruits of his labour, and, thanks to the delegate system, also over the resources for collective and public consumption, even in those areas where objective circumstances require government intervention.

This is why it became imperative for the entire socio-economic system to insure that the aggregate income of associated labour is earned in the basic organization of associated labour, to be subsequently pooled—under the direct control of the workers involved, on the basis of their decisive democratic vote—with the income of other basic organizations of associated labour within the process of social reproduction. This, of course, subject to corresponding socially-determined guidelines and to suitable moral and material incentives which also benefit the personal income of the workers. This is how control and the management of labour and social capital are concentrated in the hands of the working people; in the early phases of the development of Socialism this is the most radical way of abolishing the appropriation of workers' surplus labour by private entrepreneurs or bureaucratic-monopolistic forces and thereby also of doing away with the remnants of capitalism in all its forms. This is what is basically new and represents the true essence of the socio-economic system of Socialist self-management. This is how our working people and their self-managing communities become, or at least will become, the decisive framers of social policy in all areas of social life.

By the term workers in the above sense, I mean all people engaged in either physical or mental work—either in material production or in public services, in unskilled jobs or at the most responsible posts—who work with socially-owned means of production.

The second fundamental innovation in our socio-economic system is the introduction of a democratic self-management mechanism which enables the workers and their basic self-managing organizations and communities to bring their particular interests into conjunction with the shared, higher interests of associated labour, and with the common interests of society. This economic mechanism is what we call the pooling of labour and resources under workers' control, i.e., the system of integrated income in social reproduction. These relationships make it possible for the basic and other organizations of associated labour, without any mediation by the State—that is, freely and voluntarily—to join in all forms of mutual relationship of co-operation and association which promote their production, economic, social, and other interests. This is how, always provided the worker has full freedom of decision-making, the necessary concentration of social capital is obtained for the requirements of the modern technology of production. Finally, economic planning ceases to be an instrument of the centralist monopoly of the State and becomes a matter of a free conclusion of social compacts and self-management agreements between organizations of associated labour, in accordance with their actual interests.

As a result, such things as the management of current labour, income earning, its allocation, and its reinvestment in the process of reproduction on an extended scale are no longer dealt with by the State apparatus or by an arbitrarily appointed technocratic monopoly, but are dependent on the conditions which the workers themselves, guided by their own interests, can secure in the market place, in their free exchange of labour through social compacts and self-management agreements through the pooling of labour and resources, through the collective earning and allocation of income, etc. Although this is happening in the system of democratic worker self-management, these interests cannot, in fact, be pursued with absolute freedom, or through uncontrolled competition in the market place or mutual relationships, or in any of the forms of classical economic liberalism.

Self-management relationships have definite economic, political and democratic forms in which workers can freely and independently pursue their interests. Yet this pursuance of their interests, and their co-ordination with the general social interests, must be subordinated to a mutual democratic responsibility of self-management communities or of the working people themselves. This mutual democratic responsibility, which relies on the concept of State power, is an essential element of the system.

Such relationships are forced by the obligation on the part of workers in self-management to conclude self-management agreements on basic plans in each organization of associated labour, and also between basic organizations of associated labour and higher organizations and self-management communities. Self-managers are also steered in this direction by the right of the State to control market trends and prices, and thereby also the allocation of income, and by the right and duty of organizations of associated labour to negotiate whatever concerns market relationships and particularly prices. Our workers are also guided in this direction by the growth of large manufacturing-sales-financial communities in which organizations of associated labour are linked through labour, production, science, and resources for expanded reproduction and where, in keeping with their common interests, organizations of associated labour co-ordinate their internal and mutual relationships. We are, of course, only at the beginning of this process, but it is obvious that such self-management communities will assume an increasingly important place in our social life. And our working people are once again directed along this course by the self-managing interest unions where in a free discussion on an equal footing various self-management communities decide on the allocation of income, on the tenor of their joint plans, on prices, etc.

Consequently, self-management is not some kind of economic liberalism nor does pluralism imply an unrestrained competition among various partial interests. It is an economic, political, and democratic system which enables the workers to freely pursue their authentic interests, but which at the same time organizes them to be able to co-ordinate these interests, to deal with conflicts and to moderate public interests. This process may take place through social compacts, or through self-management agreements, or through majority decisions in delegate assemblies, or through self-management courts, or through the mediation of trade unions and other socio-political organizations, or through the intervention of the State as the organ of coercion. An essential part of the system of self-management is its built-in ability to shed the vestiges of the old non-Socialist relationships and forms which deny equality to workers, and of various economic and social deformations which, despite the existence of the system of self-management, may arise through the social practice of a multitude of social subjects.

There is no denying that in our society, too, such contradictory interests are possible, and they cannot be resolved either through self-management channels or through State coercion. Such contradictions include, for instance, differences in interests between the developed and underdeveloped parts of the country, different conditions of income earning resulting from the different organic composition of the factors of production, social differences due to remuneration according to work done, i.e., on the basis of a difference in the quantity and quality of labour expended, difference between the more productive and less productive organizations of associated labour, etc. Such contradictions are the source of certain forms of inequality still persisting in some areas of social life, particularly as regards the economic and social position of the working people.

These contradictions cannot be simply abolished, they must be transcended through history, through the further development of productive forces, through a further emancipation of man from material need, and, of course, through a further development of the system which will become increasingly capable of dealing with problems of this kind. This will be achieved through self-management practice by workers themselves, subject to the continuous influence of Socialist progressive, ideological, and political forces. What we would venture to say about this process is exactly what Marx himself said about Socialism in general—namely, that Communists must not presume to establish an ideal society, but must fight for the liberation of the workers so that they themselves can fight for better days. Our entire political and self-management system is actually geared to achieve this end in so far as existing social contradictions are concerned.

It is true that in some points the system does not provide sufficient incentives, but it is up to our self-managing and social practice to remove such weaknesses.

The role of the State in the system of social planning has certainly not disappeared. What is more, it is still highly significant and even pretty sizeable. The role of the State is nevertheless restricted to those areas and those cases which concern the most essential common social interests and where the system of a free and voluntary negotiation of agreements between organizations of associated labour cannot be a substitute for the still indispensable State coercion or arbitration. Even a 'State' plan cannot be passed without the consent of the assemblies of the socio-political community (commune, province, republic, or the federation) which are composed of associated labour delegates. In this manner, Socialist self-management has become an integral social system, and even the State itself with all its political institutions is increasingly becoming a direct instrument of self-management.

As a result, the new socio-economic status of the basic organization of associated labour and the workers associated within it, as well as other changes which have been effected pursuant to the principles of the Constitution and the Law on Associated Labour, have altered the direction of flow of the surplus labour of workers in extended and general social reproduction. Today the aggregate income, in all its forms, which is realized in the market place or at any given point in the pooling of labour and income and social reproduction always flows into the hands of the workers in the basic organization of associated labour. Surplus labour as a socio-economic category, i.e. as an element of class distinction, in fact ceases to exist because the worker can himself control it as part of the total value which he has produced by his labour, and the material results of the business management of this value are realized exclusively in the income of the basic organization of associated labour, which in turn is managed by the workers themselves.

It is true that surplus labour continues to exist as a purely economic, or material, category, and is manifested as capital accumulation, in other words, as the proceeds of the total social reproduction. However, within his basic organization of associated labour or under the terms agreed upon in self-management agreements and social compacts, the worker himself determines the amount of these proceeds, with the exception of those funds which are earmarked, by government decision, either for planned development or for general public expenditure. But even these decisions are taken in delegate assemblies, in which the delegates of associated labour have the final say on the allocation of income, for a decision may only be adopted by a majority vote in the chamber of delegates of associated labour.

Conversely, the worker's status in the basic organization of associated labour thus becomes dependent, in an economic and self-management sense, not only on the productivity of his current labour, but also on his management of social capital, i.e., the socialized past labour which is controlled and managed by the workers in the basic organization of associated labour. This dependence is becoming the material inducement for a voluntary democratic pooling of labour and resources within social reproduction on the basis of self-management. In this way, the personal interests of the worker and the specific interests of the basic organization of associated labour, as well as the shared interests of associated labour and society in general, are reconciled within the mechanism of economic relations, which in practice is kept running by none other than these very associated workers as the managers of socially owned resources, except, of course, in cases involving general social conditions which require government intervention. These economic relations have been established and are guaranteed under the Constitution. Given such economic relations, all basic organizations of associated labour have a material and creative interest that the joint revenues, which are earned by the pooling of labour and resources, should be as large as possible, so that the income of each and every one of them would be proportionately higher. And a higher income presupposes a higher productivity of associated labour, both in general social terms and individually. Such economic relations also provide the material and political prerequisites for the genuine personal creative freedom and initiative of the worker and the working man in general, and of the creative social forces in the entire system of self-management. These relations are not based merely on some limited democratic rights of the citizen, as is characteristic of the democracy of bourgeois society, but rather on real economic relations, i.e., on material resources, on the income which the worker earns and which he controls in the basic organization of associated labour. In this way, the worker has a direct voice in decision-making on the fate of the aggregate social income and wealth.

This evolution in socio-economic relations in the self-management of associated labour has produced major changes in the character of remuneration according to work performed. The idea that workers should receive approximately equal personal incomes for equal work can no longer be interpreted simply as meaning that the actual output of current labour of a worker in his job reflects his total labour. The criteria for the distribution of resources for the remuneration of a worker according to his performance should reflect the quantity and quality of labour in a broader sense, i.e., not only the immediate result of current labour but also the results obtained from the utilization of the means of production, from the management of past labour, the amount of capital accumulation and the returns accruing from it, the results of the pooling of labour and income, not to mention business policy, technological innovation, and so forth. In other words, the new socio-economic status of workers in the production relations of self-management is also exemplified in the criteria used in the allocation of personal incomes. In this way, in order to increase his personal income the worker is stimulated to become involved not only in his job but also in the management of past labour, as well as in the entire democratic system of managing labour and income, both in his basic organization of associated labour and at all other levels of management on which the results of labour of his own basic organization of associated labour depend.

All these changes in socio-economic relations have substantially strengthened the self-management and democratic prerogatives of the worker, the working man, and the citizen, and have extended his democratic rights in our society. As a result, Yugoslav workers and working people in general are progressively shedding the mentality of wage-labour subservience to monopolistic control over social capital.

The distinction between physical and mental work is also rapidly beginning to disappear. The worker is no longer just a manual labourer. As soon as he starts to take an equal part in decision-making on the business of his basic organization, and through it on the business of the larger communities and organizations on which his basic organization in some way or other depends, as soon as he can freely decide on the character of relationships and co-operation with other organizations and communities of associated labour, as soon as he begins to decide on the affairs concerning expanded and social reproduction, on the conditions and objectives of a free exchange of labour between various areas of associated labour, and as soon as he is in a position—through his delegations and delegates—to participate in decision-making on common interests at different levels of associated labour and society, his labour is no longer just physical but is also intellectual. Indeed, the economic relations of the entire system of social reproduction are designed to make his life and work depend less and less on the current physical labour expended on his job in production, and more and more on the results of overall associated labour, i.e., on his informed and responsible management of socialized past labour within overall social reproduction in the development of productive forces and, in general, on his creative initiative. The worker can enjoy such a position thanks to the fusion of labour and social capital in the income which he manages on behalf of his own interests, but also on behalf of the interests of all workers and society. It is this fusion, in fact, that provides the key to the gradual transition from remuneration according to work performed to remuneration according to need. Although this is a long-term process, the economic position of the worker and the basic organization of associated labour in the system of extended and social reproduction has already given him a head start in this direction.

Income sharing, as formulated and institutionalized by the Constitution and Law on Associated Labour, clearly and consistently postulates social ownership as a system of relations among people and not as a relationship between a person and things. Social ownership does not imply the monopoly of any single entity in society, nor of the State, work collective, or individual workers. Socially owned property is the common property of all people who work, i.e., at first it will be their class property, but in due course—as remuneration according to need comes into being—it will increasingly become the property of all members of society. To a very large extent, this is what it is already today. As labour with socially owned means of production is at the same time the source of personal property as expressed in personal income, no one can alienate from the worker those rights which belong to him *on the basis of the right to work using socially owned resources*, including the right of owning resources for personal consumption, on equal terms with other workers.

The necessary economic and social conditions have therefore been created for the intensive, rational, and efficient construction of a social and cultural superstructure of society that should further the interests of the working class and all working people, of all citizens. The material base of this superstructure is the free exchange of labour between workers in direct production and workers in the public services. By introducing and fostering relations based on the free exchange of labour, the system of self-management has begun to remove the age-old political monopoly of the State as regards this superstructure, i.e., the public services. This monopoly placed the State in a position where—in addition to privately owned capital—it was not only an arbitrator in the relations between the producing sector and the social superstructure, but also the authority which, by its decisions, fixed the volume and method of allocating and transferring resources from production in the sphere of public expenditure and public services.

I certainly have no intention of denying the important progressive role which the State has performed in the development of public services, e.g. in public health, education, culture, science, etc., particularly in removing the deformations which were introduced into those activities by propertied interests. Socialist States in particular have made great progress as compared with their earlier position, because the socialization of these services had made them available to the entire nation. However, such a system of political monopoly by the State would today not only hamper the working man in the pursuance of his interests within the system of self-management, but would also greatly fetter his creative freedom and his role as a framer of public policy. Such a system does not suit us, because to be viable, the self-managing Socialist society must allow the working man to control all income. For this reason, our self-managing society can no longer tolerate the old worker–State–public services formula, but must establish a direct link between the workers in direct production and the workers in the public services.

This principle has given rise to the concept of self-managing interest unions as free self-managing communities of workers in production and workers in the public services, who draw up self-management agreements to regulate their mutual relations. The State appears in this instance, or should appear, as an arbiter only in cases when self-management agreements cannot be reached, yet it is vital to the public interest that a decision should be taken. Such democratic self-management relations are one more source of man's freedom in society, for they mean that the working man can either have a direct say in decisions in the sphere of public services or can influence decision-making on matters affecting his interests and needs. The free exchange of labour helps erase the line between physical and mental work. Intellectual work in such relations is no longer superior to manual labour, but is merely one of the parts of freely associated labour and the free exchange of different forms of the results of labour.

Finally, I should like to take up the third sphere of socio-economic relations: the relationship between self-managing associated labour and the self-managing interest unions in the public services on the one hand, and the socio-economic and political functions of the State on the other. Self-management in associated labour is now becoming the true material base both for self-management in society, i.e., in the socio-political communities which exercise the authority of the State, from the commune to the federation, and for the exercise of their democratic rights by working people and citizens as regards the administration of the country. Self-management is also the material framework within which a worker can develop his creative faculties in the utilization of all social resources, both on behalf of his own interests connected with his life and work, and in the interests of the advancement of society as a whole.

The prime factor in the practical realization and co-ordination of these relations is the delegate system, based on the democratic right of the individual to elect delegations and delegates in all the self-managing communities—from basic organizations of associated labour and local communities on up—through whom he can directly express his authentic interests, and to influence and decide on the satisfaction of these interests in all spheres of social life. The structure of the delegate assemblies and the system of decision-making in them have been so designed as to secure, in principle, the leading role of associated labour in the overall system of governmental decision-making. In this manner, the citizens exercising their right of self-management have a real voice in the decisions of government bodies as a democratically organized force, and not a voice only at second remove, through alienated political parties. When delegate assemblies exercise power, they at the same time act as direct representatives of associated labour and other public interests and interest unions organized on the principle of self-management.

Having opted for this system of socio-economic and political relations, the Yugoslav society is evolving—on its own impetus—in the direction of a comprehensive system of a self-managing and democratic Socialist society, in which the state itself becomes only an instrument of the self-managing democracy. A worker's labour and self-management, i.e., his activities in self-management, are no longer confined to his current work on the job or to the fields of technology or economics, but also imply intellectual endeavour and political decision-making in all the spheres of public policy with which his personal interests are directly concerned. Politics, which in the system of bourgeois parliamentary democracy is practically alienated from the citizen or inaccessible to him, because it has been taken over by the top leaderships of the political parties, becomes in the self-managing socialist society a component part of self-management, of various forms of the democratic activity of self-managing subjects in pursuing their interests in various domains of social life. Politics in this manner ceases to be the preserve of professional politicians and behind-the-scenes political cartels and becomes a realm for action and direct decision-making by self-managers and their democratic delegate bodies. Of course, if workers are to enjoy such a status, not only must they be able to exercise all the democratic rights of self-management but there must be an economic acknowledgement of all the aspects of such work in the criteria for the distribution of income and the personal income of the workers, which means that these criteria must be correlated with the principle of workers' solidarity. However, these principles must never be enforced in such a way as to discourage the workers from making a rational allocation of their labour and income.

Obviously, such developments cannot be left to chance. In day-to-day practice, serious distortions are liable to occur in self-management relations, to the detriment of the acknowledged economic, social, and democratic rights of the workers, working people, and citizens. Pressure by techno-bureaucratic tendencies through various cracks in the system is persistent and unceasing. This is why in many areas the new relations have still not taken root or are not functioning in a satisfactory manner. In some places we are still at the very beginning. In promoting these relations, we come up against the most diverse objective and subjective obstacles. For this reason, in practice not everything will go according to the new system. There will be deviations and pressures from the technocracy, the extreme left, and other conservative and reactionary tendencies and forces. However, this will not disconcert us. The process of which we are speaking is one that is expected to go on for a long time. What is important is that this process follows a definite direction and that possibilities exist for the laws governing the socio-economic and production relations endorsed under the constitutional system to operate freely within the bounds of realistic social guidance. Our society has before it a clear concept of the system and line of development of democratic socio-economic and political relations in the system of self-management, and there is no reason whatever to fear that our practice will not eventually make such relations a reality. Obviously, all this requires the organized leading forces of our society, led by the League of Communists, to apply this directive in their daily practice, so that there should be no discrepancies in this respect.

We must also bear in mind those objective difficulties arising from the fact that in our society there is still a certain discrepancy between the social position of the worker as defined under the constitutional system and the ability of society to provide all the necessary material, social cultural, educational, and other prerequisites for the worker to enjoy such a position in full measure. For example, in principle the worker in a basic organization of associated labour is in a position to use, for the purpose of raising his labour productivity and asserting his creative abilities, both social capital and technical expertise, science and culture, which determine whether or not his basic organization of associated labour will be economically successful in the process of social reproduction. However, for many objective and sbujective reasons—including, of course, the still relatively low level of education and culture and application of science—the worker is still not able, with the greatest possible degree of social consciousness and creativity, to master, guide, and control all the processes which vitally affect his socio-economic position. This fact is particularly true of his mastery of the technical, scientific, and cultural components of labour, i.e., of his appreciation of scientific, cultural, and political endeavour and his own participation in it. Situations, therefore, arise in practice where the better-educated and professionally trained section of the working people in associated labour have a greater influence in the taking of decisions for which a certain degree of expertise is required than the work collective as a whole. I primarily have in mind here the management in organizations of associated labour and the intelligentsia as a whole, who have knowledge and a cultural background, and by virtue of this fact alone often have the deciding voice when decisions are being made. Because this section of the population has such a position, which is decreed by objective circumstances, there are recurrent tendencies for them to make a bid to acquire a certain monopoly by the 'élite' in the system of management.

Under the constraint of all these contingencies, in practice the self-management and democratic rights of the working man may become more or less fictitious in certain types and phases of decision-making, while the real decisions are made by a relatively small group of people. The elements of such a monopoly of political decision-making are still—deliberately or unconsciously—with us. This monopoly has simply been transplanted from the sphere of political power to the sphere of self-management, and it is manifested in the predominant role of management in decision-making, the existence of a technocracy, a neglect of democratic forms of decision-making in the collective, the passivity of workers in the exercise of their right of self-management, and so forth. Nevertheless, self-management is becoming strong enough to overcome these negative tendencies and pitfalls. In the system of self-management, these processes are coming under the ever greater control of the workers and progressive Socialist forces. As self-management strengthens, as self-managing collectives gain more technical and professional expertise, and as physical and mental works are more strongly linked in self-management, this control will become even more effective. The political system of society must likewise become a strong support in the establishment and strengthening of such a position for the worker.

If the negative processes we have mentioned were allowed to continue unchecked, they might lead to serious deformations in the Socialist self-managing democracy. That is why we are concerned not just with the general need to bring the political system into conformity with the altered socio-economic base of our society, but with making the political system capable of fostering conditions within which the workers can truly exercise the self-management and democratic rights that have been guaranteed them in principle by the social system. Obviously, in connection with such processes the question arises of the role of organized social forces in the system of Socialist self-managing democracy, and particularly in the delegate system. The workers, or citizens, must not be abandoned to the blind working of the processes set in motion by the present economic, social, and political structure of our society. In other words, democratic decision-making on the principle of self-management within self-managing organizations and communities cannot be a process which takes place exclusively within the confines of these communities. It must be open to the influence and co-operation of all those social forces which can introduce more general social considerations into local decision-making and in this way assist the self-management decision-making of the worker when he is not able by himself to appreciate the wider implications of his decisions for society.

Furthermore, it must be remembered that self-management is not a system that can operate through its own built-in automatism. Socialist self-management is a socio-economic and political relationship among people. At the present level of social development and in view of its class and social feature, such socio-economic and political relations can exist and function properly only if they rely upon a suitable system of political power, in this case upon the system in which the working class has the leading role in conjunction with all other working people. What this means is that even though some of the State's functions are gradually being cast off in favour of Socialist self-management, State power still is an indispensable instrument of maintaining the system of Socialist self-management. For workers could not exercise their leading role in the system of Socialist self-management unless they have a leading role in the system of government on which Socialist self-management itself must rely. Therefore, when we refer to the political system of Socialist self-management, we do not speak only of the social role of Socialist forces in general, but also of the position and social role of the Socialist forces in the system of government, including the specific and responsible role of the League of Communists in sustaining the development of the State power by the working class and working people.

Source: E. Kardelj, *Democracy and Socialism*, (London, 1978), pp. 21 ff.

16

ERNESTO GUEVARA:
LATIN AMERICAN MARXISM

To build Communism, a new man must be created simultaneously with the material base. That is why it is important to choose correctly the instrument of mass mobilization. That instrument must be fundamentally of a moral character.

'Socialism and Man in Cuba', *Che: Selected Works of Ernesto Guevara*, ed. R. E. Bonachea and N. P. Valdes (Cambridge, Mass., 1970), p. 159.

ERNESTO CHE GUEVARA (1928–67) *was born in Argentina and studied medicine in Buenos Aires. After travelling widely in Latin America, Guevara joined Fidel Castro in Mexico in 1955 and participated in the guerrilla war in Cuba which culminated in the overthrow of the Batista regime in 1959. After serving as Minister for Industry in Castro's government, Guevara departed in 1965 to join guerrilla bands in Bolivia where he was killed two years later. The first of the extracts below, from* Guerrilla Warfare (1961), *illustrates the idea that revolutionaries in Latin America should not rely too much on objective conditions but try rather to create these conditions by forming a guerrilla* foco *as the armed vanguard of the people. This attitude obviously elevates the military above the political in a manner quite uncharacteristic of mainstream Marxism. The second extract, from* Socialism and Man in Cuba (1968), *stresses both the role of personality and how the building of a Communist society is not necessarily tied to the development of the production forces since it does not follow separate stages.*

1. On Guerrilla Strategy

WE have already described the guerrilla fighter as one who shares the longing of the people for liberation and who, once peaceful means are exhausted, initiates the fight and converts himself into an armed vanguard of the fighting people. From the very beginning of the struggle he has the intention of destroying an unjust order and therefore an intention, more or less hidden, to replace the old with something new.

We have also already said that in the conditions that prevail, at least in America and in almost all countries with deficient economic development, it is the countryside that offers ideal conditions for the fight. Therefore the foundation of the social structure that the guerrilla fighter will build begins with changes in the ownership of agrarian property.

The banner of the fight throughout this period will be agrarian reform. At first this goal may or may not be completely delineated in its extent and limits; it may simply refer to the age-old hunger of the peasant for the land on which he works or wishes to work.

The conditions in which the agrarian reform will be realized depend upon the conditions which existed before the struggle began, and on the social depth of the struggle. But the guerrilla fighter, as a person conscious of a role in the vanguard of the people, must have a moral conduct that shows him to be a true priest of the reform to which he aspires. To the stoicism imposed by the difficult conditions of warfare should be added an austerity born of rigid self-control that will prevent a single excess, a single slip, whatever the circumstances. The guerrilla soldier should be an ascetic.

As for social relations, these will vary with the development of the war. At the beginning it will not be possible to attempt any changes in the social order.

Merchandise that cannot be paid for in cash will be paid for with bonds; and these should be redeemed at the first opportunity.

The peasant must always be helped technically, economically, morally, and culturally. The guerrilla fighter will be a sort of guiding angel who has fallen into the zone, helping the poor always and bothering the rich as little as possible in the first phases of the war. But this war will continue on its course; contradictions will continuously become sharper; the moment will arrive when many of those who regarded the revolution with a certain sympathy at the outset will place themselves in a position diametrically opposed; and they will take the first step into battle against the popular forces. At that moment the guerrilla fighter should act to make himself into the standard bearer of the cause of the people, punishing every betrayal with justice. Private property should acquire in the war zones its social function. For example, excess land and livestock not essential for the maintenance of a wealthy family should pass into the hands of the people and be distributed equitably and justly.

The right of the owners to receive payment for possessions used for the social good ought always to be respected; but this payment will be made in bonds ('bonds of hope', as they were called by our teacher, General Bayo, referring to the common interest that is thereby established between debtor and creditor).

The land and property of notorious and active enemies of the revolution should pass immediately into the hands of the revolutionary forces. Furthermore, taking advantage of the heat of the war—those moments in which human fraternity reaches its highest intensity—all kinds of co-operative work, as much as the mentality of the inhabitants will permit, ought to be stimulated.

The guerrilla fighter as a social reformer should not only provide an example in his own life but he ought also constantly to give orientation in ideological problems, explaining what he knows and what he wishes to do at the right time. He will also make use of what he learns as the months or years of the war strengthen his revolutionary convictions, making him more radical as the potency of arms is demonstrated, as the outlook of the inhabitants becomes a part of his spirit and of his own life, and as he understands the justice and the vital necessity of a series of changes, of which the theoretical importance appeared to him before, but devoid of practical urgency.

This development occurs very often, because the initiators of guerrilla warfare, or rather the directors of guerrilla warfare, are not men who have bent their backs day after day over the furrow. They are men who understand the necessity for changes in the social treatment accorded peasants, without having suffered in the usual case this bitter treatment in their own persons. It happens then (I am drawing on the Cuban experience and enlarging) that a genuine interaction is produced between these leaders, who with their acts teach the people the fundamental importance of the armed fight, and the people themselves, who rise in rebellion and teach the leaders these practical necessities of which we speak. Thus, as a product of this interaction between the guerrilla fighter and his people, a progressive radicalization appears which further accentuates the revolutionary characteristics of the movement and gives it a national scope. . . .

We have now abundantly defined the nature of guerrilla warfare. Let us next describe the ideal development of such a war from its beginning as a rising by a single nucleus on favourable ground.

In other words, we are going to theorize once more on the basis of the Cuban experience. At the outset there is a more or less homogeneous group, with some arms, that devotes itself almost exclusively to hiding in the wildest and most inaccessible places, making little contact with the peasants, dispossessed of their land or engaged in a struggle to conserve it, and young idealists of other classes join the nucleus; it acquires greater audacity and starts to operate in inhabited places, making more contact with the people of the zone; it repeats attacks, always fleeing after making them; suddenly it engages in combat with some column or other and destroys its vanguard. Men continue to join it; it has increased in number, but its organization remains exactly the same; its caution diminishes, and it ventures into more populous zones.

Later it sets up temporary camps for several days; it abandons these upon receiving news of the approach of the enemy army, or upon suffering bombardments, or simply upon becoming suspicious that such risks have arisen. The numbers in the guerrilla band increase as work among the masses operates to make of each peasant an enthusiast for the war of liberation. Finally, an inaccessible place is chosen, a settled life is initiated, and the first small industries begin to be established: a shoe factory, a cigar and cigarette factory, a clothing factory, an arms factory, bakery, hospitals, possibly a radio transmitter, a printing press, etc.

The guerrilla band now has an organization, a new structure. It is the head of a large movement with all the characteristics of a small government. A court is established for the administration of justice, possibly laws are promulgated, and the work of indoctrination of the peasant masses continues, extended also to workers if there are any near, to draw them to the cause. An enemy action is launched and defeated; the number of rifles increases; with these the number of men fighting with the guerrilla band increases. A moment arrives when its radius of action will not have increased in the same proportion as its personnel; at that moment a force of appropriate size is separated, a column or a platoon, perhaps, and this goes to another place of combat.

The work of this second group will begin with somewhat different characteristics because of the experience that it brings and because of the influence of the troops of liberation on the war zone. The original nucleus also continues to grow; it has now received substantial support in food, sometimes in guns, from various places; men continue to arrive; the administration of government, with the promulgation of laws, continues; schools are established, permitting the indoctrination and training of recruits. The leaders learn steadily as the war develops, and their capacity of command grows under the added responsibilities of the qualitative and quantitative increases in their forces.

If there are distant territories, a group departs for them at a certain moment, in order to confirm the advances that have been made and to continue the cycle.

But there will also exist an enemy territory, unfavourable for guerrilla warfare. There small groups begin to penetrate, assaulting the roads, destroying bridges, planting mines, sowing disquiet. With the ups and downs characteristic of warfare the movement continues to grow; by this time the extensive work among the masses makes easy movement of the forces possible in unfavourable territory and so opens the final stage, which is suburban guerrilla warfare.

Sabotage increases considerably in the whole zone. Life is paralysed; the zone is conquered. The guerrillas then go into other zones, where they fight with the enemy army along defined fronts; by now heavy arms have been captured, perhaps even some tanks; the fight is more equal. The enemy falls when the process of partial victories becomes transformed into final victories, that is to say, when the enemy is brought to accept battle in conditions imposed by the guerrilla band; there he is annihilated and his surrender compelled.

This is a sketch that describes what occurred in the different stages of the Cuban war of liberation; but it has a content approximating the universal. Nevertheless, it will not always by possible to count on the degree of intimacy with the people, the conditions, and the leadership that existed in our war. It is unnecessary to say that Fidel Castro possesses the high qualities of a fighter and statesman: our path, our struggle, and our triumph we owed to his vision. We cannot say that without him the victory of the people would not have been achieved; but that victory would certainly have cost much more and would have been less complete.

2.   Socialist Humanism and Revolution

I am finishing these notes during my trip through Africa, stimulated by the desire to fulfil, though tardily, my promise to you. I shall deal with the theme of the title as I believe it will be interesting to your Uruguayan readers.

It is common to hear from capitalistic spokesmen, as an argument in the ideological struggle against Socialism, the statement that this social system or the period of Socialist construction which Cuba has entered is characterized by the abolition of the individual for the sake of the state. I shall not try to refute this assertion on a merely theoretical basis, but shall attempt to establish the facts as they are experienced in Cuba and to add some general comments of my own. First, I shall outline a brief history of our revolutionary struggle before and after taking power.

As is known, 26 July 1953 is the precise date when the revolutionary actions were initiated which would eventually lead to triumph in January 1959. A group of men led by Fidel Castro in the dawn hours of that day attacked the Moncada Barracks in Oriente Province. The attack was a failure, and the failure became a disaster. The survivors wound up in prison and after a subsequent amnesty again undertook the revolutionary struggle.

During this process when there existed only the seeds of Socialism, man was the fundamental factor. On him we relied, individualized, specific, with first and last name, and on his capacity for action depended the success or failure of a given mission.

Then the phase of guerrilla struggle began. It developed in two distinct environments: the people, a dormant mass yet in need of mobilization, and the vanguard, the guerrilla, the generator of revolutionary consciousness and enthusiasm. This vanguard was the catalyst which created the subjective conditions for victory. Throughout the proletarianization of our thought and the revolution that was taking place in our habits and our minds, the individual was the fundamental factor. Every fighter of the Sierra Maestra who achieved a high rank in the revolutionary forces had a history of notable achievements to his credit. Based on these, he attained his rank.

This was the first heroic stage in which men vied to achieve a place of greater responsibility, of greater danger, and without any other satisfaction than that of fulfilling their duty. In our work in revolutionary education we return often to this instructive theme. In the attitude of our fighters, we could glimpse the man of the future.

Total commitment to the revolutionary cause has been repeated at other times during our history. Throughout the October Crisis and the days of hurricane Flora, we saw acts of exceptional sacrifice and courage performed by all of our people. From an ideological viewpoint, our fundamental task is to find the formula which will perpetuate in daily life these heroic attitudes.

In January 1959, the revolutionary government was established with the participation of the various members of the submissive bourgeoisie. The presence of the Rebel Army as the fundamental factor of force constituted the guarantee of power. Serious contradictions arose which were resolved in the first instance in February 1959, when Fidel Castro assumed the leadership of the government in the post of prime minister. This process culminated in July of the same year with the resignation of President Urrutia due to pressure from the masses.

Into the history of the Cuban came a personage with clearly defined features which would systematically reassert itself: the masses. This multidimensional entity is not, as is sometimes pretended, the sum of the elements of one category (reduced to the same uniformity by the system imposed) acting as a docile herd. It is true that it follows its leaders, primarily Fidel Castro, without vacillating. But the degree to which he has earned this trust corresponds precisely with his ability to interpret the desires and aspirations of the people and with his sincere endeavour to fulfil the promises made.

The masses participated in the agrarian reform and in the difficult task of managing State enterprises, they went through the heroic experience of Playa Girón, forged themselves in combat against the different groups of CIA armed bandits, lived through one of the most important self-definitions in modern times during the October Crisis, and today work on toward the construction of Socialism.

Looking at things from a superficial viewpoint, it would seem that those who speak about the subordination of the individual to the State are correct; the masses perform with unequalled enthusiasm and discipline the tasks assigned by the government, be they of an economic, cultural, defensive, or athletic nature. Generally the initiative comes from Fidel or those in the revolutionary high command and is explained to the people, who make it their own. At other times local experience which is thought to be valuable is picked up by the party and the government and generalized following the same procedure.

However, the State makes mistakes at times. When this occurs, the collective enthusiasm diminishes due to a quantitative diminishing that takes place in each of the elements that make up the collective and work becomes paralysed until it is reduced to an insignificant magnitude. This is the time to rectify the error.

That is what happened in March 1962, due to the sectarian line imposed on the party by Aníbal Escalante.

It is evident that the mechanism is not sufficient to assure a sequence of sensible measures. What is missing is a more structured relationship with the masses. We must improve this in years to come but for now, in the case of the initiatives arising on the top levels of government, we are using an almost intuitive method of listening to the general reactions in the face of the problems posed.

Fidel is a master at this, and his particular mode of integration with the people can only be appreciated by seeing him in action. At the great mass meetings, one can observe something like the dialogue of two tuning forks whose vibrations summon forth new vibrations each in the other. Fidel and the masses begin to vibrate in a dialogue of increasing intensity until it reaches an abrupt climax crowned by cries of struggle and of victory.

For one who has not lived the revolutionary experience, it is difficult to understand the close dialectical unity that exists between the individual and the mass, in which both are interrelated, and the mass, as a whole composed of individuals, is in turn interrelated with the leader.

Under capitalism, phenomena of this sort are observed when politicians appear who are capable of popular mobilization; but if it is not an authentic social movement, in which case it is not completely accurate to speak of capitalism, the movement will have the same lifespan as its promoter or until the popular illusions imposed by the rigours of capitalism are ended. In this type of society, man is directed by a cold mechanism which habitually escapes his comprehension. The alienated man has an invisible umbilical cord which ties him to the whole society: the law of value. It acts on all facets of his life, shaping his road and destiny.

The laws of capitalism, invisible and blind for most people, act on the individual even though he is not aware of them. He sees only a horizon that appears infinite. This is how capitalistic propaganda presents it, pretending to extract from the 'Rockefeller' story—whether it is true or not—a lesson of the possibility of success. Yet the misery which necessarily accumulates in order that an example of this sort arise and the sum total of vileness resulting from a fortune of this magnitude do not appear in the picture. It is not always possible to clarify these concepts for the popular forces. (It would be fitting at this point to study how the workers of the imperialist countries gradually lost their international class spirit under the influence of a certain complicity in the exploitation of the dependent countries and how this fact at the same time wears away the masses' spirit of struggle within their own country, but this is a topic which is not within the intention of these notes.)

In any case the path is very difficult and apparently an individual with the proper qualities can overcome it to achieve the final goal. The prize is glimpsed in the distance; the road is solitary. Moreover, this is a race of wolves: One can only arrive by means of the failure of others.

I shall now attempt to define the individual, this actor in the strange and passionate drama that is the building of Socialism, in his twofold existence as a unique being and as a member of the community.

I believe that the simplest way to begin is to recognize his unmade quality: man as an unfinished product. The prejudices of the past are carried into the present in the individual's consciousness and a continual effort has to be made in order to eradicate them. It is a twofold process. On the one hand, society acts with its direct and indirect education; and on the other, the individual submits himself to a conscious process of self-education.

The newly forming society has a hard competition with the past. This is so not only on the level of individual consciousness, with the residue of a systematic education orientated toward the isolation of the individual, but also on the economic level where, because of the very nature of the transitional period, mercantile relationships persist. *Mercancía* is the economic cell of capitalistic society; as long as it exists, its effects will be felt in the organization of production and, hence, in the individual's consciousness.

In Marx's scheme, the period of transition was conceived as the result of the explosive transformation of the capitalist system destroyed by its own contradictions; subsequent reality has shown how some countries which constitute the weak branches detach themselves from the capitalist tree, a phenomenon foreseen by Lenin. In those countries, capitalism has developed sufficiently for its effects to be felt in one way or another on the people, but it is not its own inner contradictions that explode the system after having exhausted all of its possibilities. The struggle for liberation against a foreign oppressor, misery provoked by strange accidents such as war, whose consequences make the privileged classes fall upon the exploited, and liberation movements to overthrow neo-colonial regimes are the habitual unchaining factors. Conscious action does the rest.

In these countries there has not been a complete education, for social work and wealth through the simple process of appropriation is far away from the reach of the masses. Underdevelopment on the one hand and the usual flight of capital to the 'civilized' countries on the other make a rapid change impossible without sacrifice. There is still a long stretch to be covered in the construction of an economic base, and the temptation to take the beaten path of material interest as the lever of accelerated development is very great.

There is the danger of not seeing the forest because of the trees. Pursuing the wild idea of trying to realize Socialism with the aid of the worn-out weapons left by capitalism (the market-place as the basic economic cell, profit making, individual material incentives, and so forth), one can arrive at a dead end. And one arrives there after having travelled a long distance with many forked roads where it is difficult to perceive the moment when the wrong path was taken. Meanwhile, the adapted economic base has undermined the development of consciousness. To construct Communism simultaneously with the material base of our society, we must create a new man.

This is why it is so important to choose correctly the instrument for the mobilization of the masses. That instrument must be of a fundamentally moral nature, without forgetting the correct utilization of material incentives, especially those of a social nature.

As I have stated before, it is easy to activate moral incentives in times of extreme danger. To maintain their permanence, it is necessary to develop a consciousness in which values acquire new categories. Society as a whole must become a gigantic school.

In general the phenomenon is similar to the process of the formation of a capitalist consciousness in the system's first stage. Capitalism resorts to force but also educates the people in the system. Direct propaganda is carried out by those who explain to the people the inevitability of the class system, whether it be of divine origin or due to the imposition of nature as a mechanical entity. This appeases the masses, who find themselves oppressed by an evil impossible to fight.

This is followed by hope, which differentiates capitalism from the previous caste regimes that offered no way out.

For some, the caste system continues in force: the obedient will be rewarded in the afterlife by the arrival in other wonderful worlds where the good are requited, and thus the old tradition is continued. For others, innovation: the division of classes is a matter of fate, but individuals can leave the class to which they belong through work, initiative, and so on. This process, and that of self-education for success, must be deeply hypocritical; it is the interested demonstration that a lie is true.

In our case, direct education acquires much greater importance. Explanations are convincing because they are true; there is no need for subterfuge. It is carried out through the state's educational apparatus in the form of general, technical, and ideological culture, by means of bodies such as the Ministry of Education and the party's information apparatus. Education takes among the masses, and the new attitude that is patronized tends to become a habit; the masses incorporate the attitude as their own and exert pressure on those who still have not become educated. This is the indirect way of educating the masses, as powerful as the other one.

But the process is a conscious one; the individual receives a continuous impact from the new social power and perceives that he is not completely adequate to it. Under the influence of the pressure implied in indirect education, the individual tries to accommodate to a situation which he feels is just while recognizing that his lack of development has impeded him in doing so until now. He educates himself.

In this period of the construction of Socialism, we can see the new man being born. His image is as yet unfinished; in fact, it will never be finished, for the process advances parallel to the development of new economic forms. Discounting those whose lack of education makes them tend toward the solitary road, toward the satisfaction of their ambitions, there are others who, even within this new panorama of overall advances, tend to march in isolation from the accompanying mass. What is important is that men acquire more awareness every day of the need to incorporate themselves into society, and, at the same time, of their importance as motors of that society.

They no longer march in complete solitude along lost paths toward distant longings. They follow their vanguard constituted of the party, of the most advanced workers, of the advanced men who move along bound in close communion to the masses. The vanguard has their sight on the future and its rewards, but these are not envisioned as something individual;

the reward is the new society where men will have different characteristics—
the society of Communist man.

The road is long and full of difficulties. Sometimes it is necessary to
retreat, having lost the way; at times, because of a rapid pace, we separate
ourselves from the masses; and on occasion, because of our slow pace, we
feel the close breath of those who follow on our heels. In our ambition as
revolutionaries, we try to move as quickly as possible, clearing the path,
understanding that we receive our nourishment from the masses and that
they will advance more rapidly if we encourage them by our example.

In spite of the importance given to moral incentives, the fact that there
exist two principal groups (excluding, of course, the minority fraction of
those who do not participate for one reason or another in the construction of
Socialism) indicates the relative lack of development of social consciousness.
The vanguard groups are ideologically more advanced than the mass. The
latter is acquainted with the new values, but insufficiently. Whereas in the
former a qualitative change occurs which permits them to make sacrifices
as a function of their vanguard character, the latter see only by halves and
must be subjected to incentives and pressures of some intensity; it is the
dictatorship of the proletariat operating not only over the defeated class
but also individually over the victorious class.

All of this entails, for its total success, a series of revolutionary institutions. The image of the multitudes marching toward the future fits the
concept of institutionalization as a harmonic unit of canals, steps, dams,
well-oiled apparatus which make the march possible, which will permit the
natural selection of those who are destined to march in the vanguard and
who will dispense rewards and punishments to those who fulfil their duty
or act against the society under construction.

The institutionalization of the revolution has still not been achieved. We
are searching for something new which will allow perfect identification
between the government and the community as a whole, adjusted to the
peculiar conditions of the building of Socialism and avoiding to the utmost
the commonplaces of bourgeois democracy transplanted to the society in
formation (such as legislative houses, for example). Some experiments have
been carried out with the aim of gradually creating the institutionalization
of the revolution, but without too much hurry. Our greatest restraint has
been the fear that any formal aspect might separate us from the masses and
the individual, making us lose sight of the ultimate and most important
revolutionary ambition: to see man liberated from his alienation.

Notwithstanding the lack of institutions (which must be overcome gradually), the masses now make history as a conscious aggregate of individuals
who struggle for the same cause. The individual's possibilities for expressing

himself and making himself heard in the social apparatus are infinitely greater, in spite of the lack of a perfect mechanism to do so.

It is still necessary to accentuate his conscious, individual, and collective participation in all the mechanisms of direction and production and tie them in with the idea of the need for technical and ideological education, so that the individual will grasp how these processes are closely inter-dependent and their advances parallel. He will thus achieve total awareness of his social being, which is equivalent to his full realization as a human creature, having broken the chains of alienation.

This will be translated concretely into the reappropriation of his nature through freed work and the expression of his own human condition through culture and art.

In order for it to attain the characteristic of being freed, work must acquire a new condition; man as a commodity ceases to exist and a system is established which grants a quota for the fulfilment of social duty. The means of production belong to society and the machine is only the front line where duty is performed. Man begins to liberate his thought from the bothersome fact that presupposes the need to satisfy his animal needs through work. He begins to see himself portrayed in his work and to understand its human magnitude through the created object, through the work carried out. This no longer entails leaving a part of his being in the form of labour, power, soul, which no longer belong to him, but rather it signifies an emanation from himself, a contribution to the life of society in which he is reflected, the fulfilment of his social duty.

We are doing everything possible to give work this new category of social duty and to unite it to the development of technology, on the one hand, which will provide the conditions for greater freedom, and to voluntary labour on the other, based on the Marxist concept that man truly achieves his full human condition when he produces without being compelled by the physical necessity of selling himself as a commodity.

It is clear that work still has coercive aspects, even when it is voluntary; man as yet has not transformed all the coercion surrounding him into conditioned reflexes of a social nature, and in many cases he still produces under the pressure of the environment (Fidel calls this moral compulsion). He still has to achieve complete spiritual re-creation in the presence of his own work, without the direct pressure of the social environment but bound to it by new habits. That will be Communism.

The change in consciousness is not produced automatically, just as it is not produced in the economy. The variations are slow and not rhythmic; there are periods of acceleration, others are measured, and some even involve a retreat.

We must also consider, as we have already pointed out, that we are not before a pure transition period such as that envisioned by Marx in the *Critique of the Gotha Program*, but rather a new phase not foreseen by him—the first period in the transition to Communism or in the construction of Socialism.

This process takes place in the midst of a violent class struggle; and elements of capitalism are present within it, which obscure the complete understanding of the essence of the process.

If to this be added the scholasticism that has delayed the development of Marxist philosophy and impeded the systematic treatment of the period in which the political economy has as yet not been developed, we must agree that we are still in diapers and it is urgent to investigate all the primordial characteristics of the period before elaborating a far-reaching economic and political theory.

The resulting theory will necessarily give pre-eminence to the two pillars of Socialist construction: the formation of the new man and the development of technology. In both aspects we have a great deal to accomplish still, but the delay is less justifiable regarding the conception of technology as the basis: here it is not a matter of advancing blindly, but rather of following for a considerable stretch the road opened up by the most advanced countries of the world. This is why Fidel harps with so much insistency on the necessity of the technological and scientific formation of all of our people and especially of the vanguard.

In the field of ideas that lead to non-productive activities, it is easier to see the division between material and spiritual needs. For a long time man has been trying to free himself from alienation through culture and art. He dies daily in the eight and more hours during which he performs as a commodity in order to be resuscitated in his spiritual creation. But this remedy bears the germs of the same disease: He is a solitary being who seeks communion with nature. He defends his oppressed individuality from the environment and reacts to esthetic ideas as a unique being whose aspiration is to remain immaculate.

It is only an attempt at flight. The law of value is no longer a mere reflection of production relations; the monopoly capitalists have surrounded it with a complicated scaffolding which makes of it a docile servant, even when the methods employed are purely empirical. The superstructure imposes a type of art in which the artist must be educated. The rebels are dominated by the apparatus, and only exceptional talents are able to create their own work. The remaining ones become shamefaced wage-workers, or they are crushed.

Artistic experimentation is invented and is taken as the definition of freedom. But this 'experimentation' has limits which are imperceptible until they are clashed with, that is to say, when the real problems of man and his alienation are dealt with. Senseless anguish or vulgar pastimes are comfortable safety valves for human uneasiness; the idea of making art a weapon of denunciation and accusation is combated.

If the rules of the game are respected, all honours are obtained—the honours that might be granted to a pirouette-creating monkey. The condition is to not attempt to escape from the invisible cage.

When the revolution took power, the exodus of the totally domesticated took place; the others, revolutionaries or not, saw a new road. Artistic experimentation gained new impulse. However, the roots were more or less traced, and the concept of flight was the hidden meaning behind the word freedom. This attitude, which was a reflection in consciousness of bourgeois idealism, was frequently maintained in the revolutionaries themselves.

In countries which have gone through a similar process, there was an attempt made to combat these tendencies with an exaggerated dogmatism. General culture became something like a taboo, and a formally exact representation of nature was proclaimed as the height of cultural aspiration. This later became a mechanical representation of social reality created by wishful thinking, the ideal society, almost without conflict or contradiction, that man was seeking to create.

Socialism is young and makes mistakes. We revolutionaries many times lack the knowledge and the necessary intellectual audacity to face the task of the development of the new human being by methods distinct from the conventional ones, and the conventional methods suffer from the influence of the society that created them. (Once again the topic of the relation between form and content appears.) Disorientation is great, and the problems of material construction absorb us. There are no artists of great authority who also have great revolutionary authority.

The men of the party must take this task on themselves and seek the achievements of the critical objective: to educate the people.

What is then sought is simplification, what everyone understands, what the functionaries understand. True artistic experimentation is annulled and the problem of general culture is reduced to the assimilation of the Socialist present and the dead (and therefore not dangerous) past. As such, Socialist realism is born on the foundation of the art of the last century.

But the realistic art of the nineteenth century is also class art, perhaps more purely capitalist than the decadent art of the twentieth century, where the anguish of alienated man shows through. In culture, capitalism has given all that it has to give and all that remains of it is the announcement of a bad-smelling corpse—in art, its present decadence. But why endeavour to seek in the frozen forms of Socialist realism the only valid recipe? 'Freedom' cannot be set against Socialist realism because the former does not yet exist and it will not come into existence until the complete development of the new society. But at all costs let us not attempt to condemn all post-mid-nineteenth-century art forms from the pontifical throne of realism. That would mean committing the Proudhonian error of the return to the past, and strait-jacketing the artistic expression of the man who is being born and constructed today.

An ideological and cultural mechanism which will permit experimentation and clear out the weeds that shoot up so easily in the fertilized soil of State subsidization is lacking.

In our country, the error of mechanical realism has not appeared, but rather the contrary. This has been because of the lack of understanding of the need to create the new man who will represent neither nineteenth-century ideas nor those of our decadent and morbid century. It is the twenty-first-century man whom we must create, although this is still a subjective and unsystematic aspiration. This is precisely one of the fundamental points of our studies and our work; to the extent that we make concrete achievements on a theoretical base or vice versa, that we come to theoretical conclusions of a broad character on the basis of our concrete studies, we will have made a valuable contribution to Marxism-Leninism, to the cause of mankind.

The reaction against nineteenth-century man has brought a recurrence of twentieth-century decadence. It is not a very grave error, but we must overcome it so as not to leave the doors wide open to revisionism.

The large multitudes of people are developing themselves, the new ideas are acquiring an adequate impetus within society, the material possibilities of the integral development of each and every one of its members make the task ever more fruitful. The present is one of struggle; the future is ours.

To summarize, the culpability of many of our intellectuals and artists lies in their original sin; they are not authentic revolutionaries. We can attempt to graft elm trees so they bear pears, but simultaneously we must plant pear trees. The new generations will arrive free of original sin. The possibility that exceptional artists will arise will be that much greater because of the enlargement of the cultural field and the possibilities for expression. Our task is to keep the present generation, maladjusted by its conflicts, from becoming perverted and perverting the new generations. We do not want to create salaried workers docile to official thinking or 'scholars' who live under the wing of the budget, exercising a freedom in quotation marks. Revolutionaries will come to sing the song of the new man with the authentic voice of the people. It is a process that requires time.

In our society the youth and the party play a large role. The former is particularly important because it is malleable clay with which the new man, without any of the previous defects, can be constructed.

They receive treatment which is in consonance with our ambitions. Education is increasingly more complete, and we do not forget the incorporation of the students into work from the very first. Our scholarship students do physical work during vacation or simultaneously with their studies. In some cases work is a prize, in others it is an educational tool; it is never a punishment. A new generation is being born.

The party is a vanguard organization. The best workers are proposed by their comrades for membership. The party is a minority, but the quality of its cadres gives it great authority. Our aspiration is that the party become a mass one, but only when the masses have attained the level of development of the vanguard, that is, when they are educated for Communism. Our work is aimed at providing that education. The party is the living example; its cadres must be lecturers of assiduity and sacrifice; with their acts they must lead the masses to end the revolutionary task, which entails years of struggle against the difficulties of construction, class enemies, the defects of the past, imperialism.

I would like to explain now the role played by the personality, man as the individual who leads the masses that make history. This is our experience, and not a recipe.

Fidel gave impulse to the revolution in its first years, he has given leadership to it always and has set the tone; but there is a good group of revolutionaries developing in the same direction as the maximum leader and a great mass that follows its leaders because it has faith in them. It has faith in them because these leaders have known how to interpret the longings of the masses.

It is not a question of how many kilograms of meat are eaten or how many times a year someone may go on holiday to the seashore or how many pretty imported things can be bought with present wages. It is rather that the individual feels greater fulfilment, that he has greater inner wealth and many more responsibilities. In our country the individual knows that the glorious period in which it has fallen to him to live is one of sacrifice; he is familiar with sacrifice.

The first ones came to know it in the Sierra Maestra and wherever there was fighting; later we have known it in all Cuba. Cuba is the vanguard of America and must make sacrifices because it occupies the advance position, because it points out to the masses of Latin America the road to full freedom.

Within the country the leaders must fulfil their vanguard role; and it must be said with all sincerity that in a true revolution, to which one gives oneself completely, from which one expects no material compensation, the task of the vanguard revolutionary is both magnificent and anguishing.

Let me say, with the risk of appearing ridiculous, that the true revolutionary is guided by strong feelings of love. It is impossible to think of an authentic revolutionary without this quality. This is perhaps one of the greatest dramas of a leader; he must combine an impassioned spirit with a cold mind and make painful decisions without flinching one muscle. Our vanguard revolutionaries must idealize their love for the people, for the most sacred causes, and make it one and indivisible. They cannot descend, with small doses of daily affection, to the places where ordinary men put their love into practice.

The leaders of the revolution have children who do not learn to call their father with their first faltering words; they have wives who must be part of the general sacrifice of their lives to carry the revolution to its destiny; their friends are strictly limited to their comrades in revolution. There is no life outside it.

In these conditions, one must have a large dose of humanity, a large dose of a sense of justice and truth, to avoid falling in dogmatic extremes, into cold scholasticism, into isolation from the masses. Every day we must struggle so that this love of living humanity is transformed into concrete facts, into acts that will serve as an example, as a mobilizing factor.

The revolutionary, ideological motor of the revolution within his party, is consumed by this uninterrupted activity that has no other end but death, unless construction be achieved on a world-wide scale. If his revolutionary eagerness becomes dulled when the most urgent tasks are realized on a local scale, and if he forgets about proletarian internationalism, the revolution that he leads ceases to be a driving force and it becomes a comfortable drowsiness which is taken advantage of by our irreconcilable enemy, by imperialism, which gains ground. Proletarian internationalism is a duty, but it is also a revolutionary need. This is how we educate our people.

It is clear that there are dangers in the present circumstances. Not only that of dogmatism, not only that of the freezing up of relations with the masses in the midst of the great task, but there also exists the danger of weaknesses in which it is possible to fall. If a man thinks that, in order to dedicate his entire life to the revolution, he cannot be distracted by the worry that one of his children lacks a certain product, that the children's shoes are in poor condition, that his family lacks some very necessary item, beneath this reasoning the germs of future corruption are allowed to filter through.

In our case we have maintained that our children must have, or lack, what the children of the ordinary citizen have or lack; our family must understand this and struggle for it. The revolution is made through man, but man must forge day by day his revolutionary spirit.

Thus we go forward. At the head of the immense column—we are neither ashamed nor afraid to say so—is Fidel, followed by the best party cadres and, immediately after, so close that their great strength is felt, come the people as a whole, a solid conglomeration of individualities moving toward a common objective: individuals who have achieved the awareness of what must be done, men who struggle to leave the domain of necessity and enter that of freedom.

That immense multitude is ordering itself; its order responds to an awareness of the need for order; it is no more a dispersed force, divisible in thousands of fractions shot into space like the fragments of a grenade, trying through any means, in a fierce struggle with their equals, to attain a position that would give them support in the face of an uncertain future.

We know that we have sacrifices ahead of us and that we must pay a price for the heroic act of constituting a vanguard as a nation. We, the leaders, know that we must pay a price for having the right to say that we are at the head of the people who are at the head of America.

Each and every one of us punctually pays his quota of sacrifice, aware of receiving our reward in the satisfaction of fulfilling our duty, conscious of advancing with everyone toward the new man who is glimpsed on the horizon.

Allow me to attempt to come to some conclusions:

We Socialists are more free because we are more fulfilled; we are more fulfilled because we are more free.

The skeleton of our freedom is formed, but it lacks the protein substance and the draperies; we shall create them.

Our freedom and its daily sustenance are the colour of blood and are swollen with sacrifice.

Our sacrifice is a conscious one; it is the payment for the freedom we are constructing.

The road is long and unknown in part; we are aware of our limitations. We shall make the twenty-first-century man, we ourselves.

We shall be forged in daily action, creating a new man with a new technology.

The personality plays the role of mobilization and leadership in so far as it incarnates the highest virtues and aspirations of the people and is not detoured.

The road is opened up by the vanguard group, the best among the good, the party.

The fundamental clay of our work is the youth; in it we have deposited our hopes and we are preparing it to take the banner from our hands.

If this faltering letter has made some things clear, it will have fulfilled my purpose in sending it.

Accept our ritual greetings, as a handshake or an 'Ave Maria Purisima.'

Patria o muerte

Sources: C. Guevara, *Guerrilla Warfare* (London, 1969), pp. 45 ff., 81 ff. *Che: Selected Works of Ernesto Guevara*, ed. R. E. Bonachea and N. P. Valdes (Cambridge, 1970), pp. 155 ff.

FURTHER NOTES

WORKS

GUEVARA, CHE, *Guevara on Revolution: A Documentary Overview*, ed. J. Mallin (Miami, 1969).

—— *Che: Selected Works of Ernesto Guevara*, ed. R. E. Bonachea and N. P. Valdes (Cambridge, Mass., 1970).

COMMENTARIES

BARKIN, DAVID P., and NITA R. MANITZAS (eds.), *Cuba: The Logic of the Revolution* (New York, 1973).

DEBRAY, REGIS, *A Critique of Arms*, 2 vols. (London, 1977–).

KAROL, K. S., *Guerrillas in Power: The Course of the Cuban Revolution* (New York, 1970).

LOWY, M., *The Marxism of Che Guevara* (New York, 1973).

O'CONNOR, J., *The Origins of Socialism in Cuba* (Ithaca, NY, 1970).

RAMM, H., *The Marxism of Regis Debray: Between Lenin and Guevara* (Lawrence, 1978).

17

AMILCAR CABRAL: AFRICAN MARXISM

In colonial conditions it is the petty bourgeousie which is the inheritor
of State power (though I wish I could be wrong). The moment national
liberation comes and the petty bourgeousie takes power we enter or
rather return to history, and thus the internal contradictions break out
again.

Revolution in Guinea: An African People's Struggle (London, 1969),

p. 57.

AMILCAR CABRAL (1924–73) *is thought by many to be the most original
of Africa's Marxist thinkers. Brought up in the Cape Verde Islands, then
under Portuguese control, he trained as an agricultural engineer and, in
1956, launched his revolutionary party—the African Party for the Independence of Guinea and Cape Verde, commonly abbreviated from its
Portuguese form to PAIGC. Guerilla war against the colonial power was
launched in 1963 and achieved success in 1974 following the Portuguese
officers' overthrow of the Salazar dictatorship in Lisbon. Cabral himself
had been assassinated at Portuguese instigation a year earlier. The excerpt
below is from an address to a Third World Conference in Cuba in 1966
and did much to establish Cabral's reputation as a political theorist. Its
rather cumbersome title, 'Presuppositions and Objectives of National Liberation in Relation to Social Structure', sufficiently indicates its scope,
which included a precise analysis of the local economy, of the balance of
class forces, and, in particular, of the crucial and ambivalent role of the
petty bourgeoisie in the national liberation movement. It is this realistic
and self-critical aspect to his writings which explains why Cabral is such a
powerful influence in Third World political thought.*

Class and Revolution in Africa

W E are not going to use this platform to rail against imperialism. An African
saying very common in our country says: 'When your house is burning, it's
no use beating the tom-toms.' On a Tricontinental level, this means that
we are not going to eliminate imperialism by shouting insults against it.
For us, the best or worst shout against imperialism, whatever its form, is
to take up arms and fight. This is what we are doing, and this is what we

will go on doing until all foreign domination of our African homelands has been totally eliminated.

Our agenda includes subjects whose meaning and importance are beyond question and which show a fundamental preoccupation with *struggle*. We note, however, that one form of struggle which we consider to be fundamental has not been explicitly mentioned in this programme, although we are certain that it was present in the minds of those who drew up the programme. We refer here to the *struggle against our own weaknesses*. Obviously, other cases differ from that of Guinea; but our experience has shown us that in the general framework of daily struggle this battle against ourselves—no matter what difficulties the enemy may create—is the most difficult of all, whether for the present or the future of our peoples. This battle is the expression of the internal contradictions in the economic, social, cultural (and therefore historical) reality of each of our countries. We are convinced that any national or social revolution which is not based on knowledge of this fundamental reality runs grave risk of being condemned to failure.

When the African peoples say in their simple language that 'no matter how hot the water from your well, it will not cook your rice', they express with singular simplicity a fundamental principle, not only of physics, but also of political science. We know that the development of a phenomenon in movement, whatever its external appearance, depends mainly on its internal characteristics. We also know that on the political level our own reality—however fine and attractive the reality of others may be—can only be transformed by detailed knowledge of it, by our own efforts, by our own sacrifices. It is useful to recall in this Tricontinental gathering, so rich in experience and example, that however great the similarity between our various cases and however identical our enemies, national liberation and social revolution are not exportable commodities; they are, and increasingly so every day, the outcome of local and national elaboration, more or less influenced by external factors (be they favourable or unfavourable) but essentially determined and formed by the historical reality of each people, and carried to success by the overcoming or correct solution of the internal contradictions between the various categories characterizing this reality. The success of the Cuban revolution, taking place only 90 miles from the greatest imperialist and anti-Socialist power of all time, seems to us, in its context and its way of evolution, to be a practical and conclusive illustration of the validity of this principle.

However we must recognize that we ourselves and the other liberation movements in general (referring here above all to the African experience) have not managed to pay sufficient attention to this important problem of our common struggle.

The ideological deficiency, not to say the total lack of ideology, within the national liberation movements—which is basically due to ignorance of the historical reality which these movements claim to transform—constitutes one of the greatest weaknesses of our struggle against imperialism, if not the greatest weakness of all. We believe, however, that a sufficient number of different experiences has already been accumulated to enable us to define a general line of thought and action with the aim of eliminating this deficiency. A full discussion of this subject could be useful and would enable this conference to make a valuable contribution towards strengthening the present and future actions of the national liberation movements. This would be a concrete way of helping these movements, and in our opinion no less important than political support or financial assistance for arms and suchlike.

It is with the intention of making a contribution, however modest, to this debate that we present here our opinion of *the foundations and objectives of national liberation in relation to the social structure.* This opinion is the result of our own experiences of the struggle and of a critical appreciation of the experiences of others. To those who see in it a theoretical character, we would recall that every practice produces a theory, and that if it is true that a revolution can fail even though it be based on perfectly conceived theories, nobody has yet made a successful revolution without a revolutionary theory.

Those who affirm—in our case correctly—that the motive force of history is the class struggle would certainly agree to a revision of this affirmation to make it more precise and give it an even wider field of application if they had a better knowledge of the essential characteristics of certain colonized peoples, that is to say peoples dominated by imperialism. In fact in the general evolution of humanity and of each of the peoples of which it is composed, classes appear neither as a generalized and simultaneous phenomenon throughout the totality of these groups, nor as a finished, perfect, uniform, and spontaneous whole. The definition of classes within one or several human groups is a fundamental consequence of the progressive development of the productive forces and of the characteristics of the distribution of the wealth produced by the group or usurped from others. That is to say that the socio-economic phenomenon 'class' is created and develops as a function of at least two essential and interdependent variables—the level of productive forces and the pattern of ownership of the means of production. This development takes place slowly, gradually, and unevenly, by quantitative and generally imperceptible variations in the fundamental components; once a certain degree of accumulation is reached, this process then leads to a *qualitative jump* characterized by the appearance of classes and of conflict between them.

Factors external to the socio-economic whole can influence, more or less significantly, the process of development of classes, accelerating it, slowing it down, and even causing regressions. When, for whatever reason, the influence of these factors ceases, the process reassumes its independence and its rhythm is then determined not only by the specific internal characteristics of the whole, but also by the resultant of the effect produced in it by the temporary action of the external factors. On a strictly internal level the rhythm of the process may vary, but it remains continuous and progressive. Sudden progress is only possible as a function of violent alterations—mutations—in the level of productive forces or in the pattern of ownership. These violent transformations carried out within the process of development of classes, as a result of mutations in the level of productive forces or in the pattern of ownership, are generally called, in economic and political language, *revolutions*.

Clearly, however, the possibilities of this process are noticeably influenced by external factors, and particularly by the interaction of human groups. This interaction is considerably increased by the development of means of transport and communication which has created the modern world, eliminating the isolation of human groups within one area, of areas within one continent, and between continents. This development, characteristic of a long historical period which began with the invention of the first means of transport, was already more evident at the time of the Punic voyages and in the Greek colonization, and was accentuated by maritime discoveries, the invention of the steam engine, and the discovery of electricity. And in our own times, with the progressive domesticization of atomic energy, it is possible to promise, if not to take men to the stars, at least to humanize the universe.

This leads us to pose the following question: does history begin only with the development of the phenomenon of 'class', and consequently of class struggle? To reply in the affirmative would be to place outside history the whole period of life of human groups from the discovery of hunting, and later of nomadic and sedentary agriculture, to the organization of herds and the private appropriation of land. It would also be to consider—and this we refuse to accept—that various human groups in African, Asia, and Latin America were living without history, or outside history, at the time when they were subjected to the yoke of imperialism. It would be to consider that the peoples of our countries, such as the Balacantes of Guinea, the Coaniamas of Angola, and the Macondes of Mozambique, are still living today—if we abstract the slight influence of colonialism to which they have been subjected—outside history, or that they have no history.

Our refusal, based as it is on concrete knowledge of the socio-economic reality of our countries and on the analysis of the process of development of the phenomenon 'class', as we have seen earlier, leads us to conclude that if class struggle is the motive force of history, it is so only in a specific historical period. This means that *before* the class struggle—and necessarily *after* it, since in this world there is no before without an after—one or several factors was and will be the motive force of history. It is not difficult to see that this factor in the history of each human group is the *mode of production*—the level of productive forces and the pattern of ownership—characteristic of that group. Furthermore, as we have seen, classes themselves, class struggle, and their subsequent definition, are the result of the development of the productive forces in conjunction with the pattern of ownership of the means of production. It therefore seems correct to conclude that the level of productive forces, the essential determining element in the content and form of class struggle, is the true and permanent motive force of history.

If we accept this conclusion, then the doubts in our minds are cleared away. Because if on the one hand we can see that the existence of history before the class struggle is guaranteed, and thus avoid for some human groups in our countries—and perhaps in our continent—the sad position of being peoples without any history, then on the other hand we can see that history has continuity, even after the disappearance of class struggle or of classes themselves. And as it was not we who postulated—on a scientific basis—the fact of the disappearance of classes as a historical inevitability, we can feel satisfied at having reached this conclusion which, to a certain extent, re-establishes coherence and at the same time gives to those peoples who, like the people of Cuba, are building Socialism, the agreeable certainty that they will not cease to have a history when they complete the process of elimination of the phenomenon of 'class' and class struggle within their socio-economic whole. Eternity is not of this world, but man will outlive classes and will continue to produce and make history, since he can never free himself from the burden of his needs, both of mind and of body, which are the basis of the development of the forces of production.

The foregoing, and the reality of our times, allow us to state that the history of one human group or of humanity goes through at least three stages. The first is characterized by a low level of productive forces—of man's domination over nature; the mode of production is of a rudimentary character, private appropriation of the means of production does not yet exist, there are no classes, nor, consequently, is there any class struggle. In the second stage, the increased level of productive forces leads to private appropriation of the means of production, progressively complicates the mode of production, provokes conflicts of interests within the socio-economic whole in movement, and makes possible the appearance of the phenomenon 'class' and hence of class struggle, the social expression of the contradiction in the economic field between the mode of production and private appropriation of the means of production. In the third stage, once a certain level of productive forces is reached, the elimination of private appropriation of the means of production is made possible, and is carried out, together with the elimination of the phenomenon 'class', and hence of class struggle; new and hitherto unknown forces in the historical process of the socio-economic whole are then unleashed.

In politico-economic language, the first stage would correspond to the communal agricultural and cattle-raising society, in which the social struc-ture is horizontal, without any state; the second to feudal or assimilated agricultural or agro-industrial bourgeois societies, with a vertical social structure and a State; the third to Socialist or Communist societies, in which the economy is mainly, if not exclusively, industrial (since agriculture itself becomes a form of industry) and in which the State tends to progressively disappear, or actually disappears, and where the social structure returns to horizontality, at a higher level of productive forces, social relations, and appreciation of human values.

At the level of humanity or of part of humanity (human groups within one area, of one or several continents) these three stages (or two of them) can be simultaneous, as is shown as much by the present as by the past. This is a result of the uneven development of human societies, whether caused by internal reasons or by one or more external factors exerting an accelerating or slowing-down influence on their evolution. On the other hand, in the historical process of a given socio-economic whole, each of the above mentioned stages contains, once a certain level of transformation is reached, the seeds of the following stage.

We should also note that in the present phase of the life of humanity, and for a given socio-economic whole, the time sequence of the three characteristic stages is not indispensable. Whatever its level of productive forces and present social structure, a society can pass rapidly through the defined stages appropriate to the concrete local realities (both historical and human) and reach a higher state of existence. This progress depends on the concrete possibilities of development of the society's productive forces and is governed mainly by the nature of the political power ruling the society, that is to say, by the type of State or, if one likes, by the character of the dominant class or classes within the society.

A more detailed analysis would show that the possibility of such a jump in the historical process arises mainly, in the economic field, from the power of the means available to man at the time for dominating nature, and in the political field, from the new event which has radically changed the face of the world and the development of history, *the creation of Socialist States.*

Thus we see that our peoples have their own history regardless of the stage of their economic development. When they were subjected to imperialist domination, the historical process of each of our peoples (or of the human groups of which they are composed) was subject to the violent action of an external factor. This action—the impact of imperialism on our societies—could not fail to influence the process of development of the productive forces in our countries and the social structures of our countries, as well as the content and form of our national liberation struggles.

But we also see that in the historical context of the development of these struggles, our peoples have the concrete possibility of going from their present situation of exploitation and underdevelopment to a new stage of their historical process which can lead them to a higher form of economic, social, and cultural existence.

The political statement drawn up by the international preparatory committee of this conference, for which we reaffirm our complete support, placed imperialism, by clear and succinct analysis, in its economic context and historical co-ordinates. We will not repeat here what has already been said in the assembly. We will simply state that imperialism can be defined as a world-wide expression of the search for profits and the ever-increasing accumulation of *surplus value* by monopoly financial capital, centred in two parts of the world: first in Europe, and then in North America. And if we wish to place the fact of imperialism within the general trajectory of the evolution of the transcendental factor which has changed the face of the world, namely capital and the process of is accumulation, we can say that imperialism is piracy transplanted from the seas to dry land, piracy reorganized, consolidated, and adapted to the aim of exploiting the natural and human resources of our peoples. But if we can calmly analyse the imperialist phenomenon, we will not shock anybody by admitting that imperialism—and everything goes to prove that it is in fact the last phase in the evolution of capitalism—has been a historical necessity, a consequence of the impetus given by the productive forces and of the transformations of the means of production in the general context of humanity, considered as one movement, that is to say a necessity like those today of the national liberation of peoples, the destruction of capital, and the advent of Socialism.

The important thing for our peoples is to know whether imperialism, in its role as capital in action, has fulfilled in our countries its historical mission: the acceleration of the process of development of the productive forces and their transformation in the sense of increasing complexity in the means of production; increasing the differentiation between the classes with the development of the bourgeoisie, and intensifying the class struggle; and appreciably increasing the level of economic, social, and cultural life of the peoples. It is also worth examining the influences and effects of imperialist action on the social structure and historical processes of our peoples.

We will not condemn nor justify imperialism here: we will simply state that, as much on the economic level as on the social and cultural level, imperialist capital has not remotely fulfilled the historical mission carried out by capital in the countries of accumulation. This means that if, on the one hand, imperialist capital has had, in the great majority of the dominated countries, the simple function of multiplying surplus value, it can be seen on the other hand that the historical capacity of capital (as indestructible accelerator of the process of development of productive forces) depends strictly on its freedom that is to say on the degree of independence with which it is utilized. We must however recognize that in certain cases imperialist capital or moribund capitalism has had sufficient self-interest, strength, and time to increase the level of productive forces (as well as building towns) and to allow a minority of the local population to attain a higher and even privileged standard of living, thus contributing to a process which some would call dialectical, by widening the contradictions within the societies in question. In other, even rarer cases, there has existed the possibility of accumulation of capital, creating the conditions for the development of a local bourgeoisie.

On the question of the effects of imperialist domination on the social structure and historical process of our peoples, we should first of all examine the general forms of imperialist domination. There are at least two forms: the first is direct domination, by means of a political power made up of people foreign to the dominated people (armed forces, police, administrative agents, and settlers); this is generally called *classical colonialism or colonialism*. The second form is indirect domination, by a political power made up mainly or completely of native agents; this is called *neo-colonialism*.

In the first case, the social structure of the dominated people, whatever its stage of development, can suffer the following consequence: (*a*) total destruction, generally accompanied by immediate or gradual elimination of the native population and, consequently, by the substitution of a population from outside; (*b*) partial destruction, generally accompanied by a greater or lesser influx of population from outside; (*c*) apparent conservation, conditioned by confining the native society to zones or reserves generally offering no possibilities of living, accompanied by massive implantation of population from outside.

The two latter cases are those which we must consider in the framework of the problematic national liberation, and they are extensively present in Africa. One can say that in either case the influence of imperialism on the historical process of the dominated people produces paralysis, stagnation, and even in some cases regression in this process. However this paralysis is not complete. In one section or another of the socio-economic whole in question, noticeable transformations can be expected, caused by the permanent action of some internal (local) factors or by the action of new factors introduced by the colonial domination, such as the introduction of money and the development of urban centres. Among these transformations we should particularly note, in certain cases, the progressive loss of prestige of the ruling native classes or sectors, the forced or voluntary exodus of part of the peasant population to the urban centres, with the consequent development of new social strata; salaried workers, clerks, employees in commerce and the liberal professions, and an unstable stratum of unemployed. In the countryside there develops, with very varied intensity and always linked to the urban milieu, a stratum made up of small landowners. In the case of neo-colonialism, whether the majority of the colonized population is of native or foreign origin, the imperialist action takes the form of creating a local bourgeoisie or pseudo-bourgeoisie, controlled by the ruling class of the dominating country.

The transformations in the social structure are not so marked in the lower strata, above all in the countryside, which retains the characteristics of the colonial phase; but the creation of a native pseudo-bourgeoisie which generally develops out of a petty bourgeoisie of bureaucrats and accentuates the differentiation between the social strata and intermediaries in the commercial system (compradores), by strengthening the economic activity of local elements, opens up new perspectives in the social dynamic, mainly by the development of an urban working class, the introduction of private agricultural property, and the progressive appearance of an agricultural proletariat. These more or less noticeable transformations of the social structure, produced by a significant increase in the level of productive forces, have a direct influence on the historical process of the socio-economic whole in question. While in classical colonialism this process is paralysed, neo-colonialist domination, by allowing the social dynamic to awaken (conflicts of interests between native social strata or class struggles), creates the illusion that the historical process is returning to its normal evolution. This illusion will be reinforced by the existence of a political power (national

State) composed of native elements. In reality it is scarcely even an illusion, since the submission of the local 'ruling' class to the ruling class of the dominating country limits or prevents the development of the national productive forces. But in the concrete conditions of the present-day world economy this dependence is fatal and thus the local pseudo-bourgeoisie, however strongly nationalist it may be, cannot effectively fulfil its historical function; it cannot *freely* direct the development of the productive forces; in brief, it cannot be a national bourgeoisie. For as we have seen, the productive forces are the motive force of history, and total freedom of the process of their development is a indispensable condition for their proper functioning.

We therefore see that both in colonialism and in neo-colonialism the essential characteristic of imperialistic domination remains the same: the negation of the historical process of the dominated people by means of violent usurpation of the freedom of development of the national productive forces. This observation, which identifies the essence of the two apparent forms of imperialist domination, seems to us to be of major importance for the thought and action of liberation movements, both in the course of struggle and after the winning of independence.

On the basis of this, we can state that national liberation is the phenomenon in which a given socio-economic whole rejects the negation of its historical process. In other words, the national liberation of a people is the regaining of the historical personality of that people, its return to history through the destruction of the imperialist domination to which it was subjected.

We have seen that violent usurpation of the freedom of the process of development of the productive forces of the dominated socio-economic whole constitutes the principal and permanent characteristic of imperialist domination, whatever its form. We have also seen that this freedom alone can guarantee the normal development of the historical process of a people. We can therefore conclude that national liberation exists only when the national productive forces have been completely freed from every kind of foreign domination.

It is often said that national liberation is based on the right of every people freely to control its own destiny and that the objective of this liberation is national independence. Although we do not disagree with this vague and subjective way of expressing a complex reality, we prefer to be objective, since for us the basis of national liberation, whatever the formulas adopted on the level of international law, is the inalienable right of every people to have its own history, and the objective of national liberation is to regain this right usurped by imperialism, that is to say, to free the process of development of the national productive forces.

For this reason, in our opinion, any national liberation movement which does not take into consideration this basis and this objective may certainly struggle against imperialism, but will surely not be struggling for national liberation.

This means that, bearing in mind the essential characteristics of the present world economy, as well as experiences already gained in the field of anti-imperialist struggle, the principal aspect of national liberation struggle is the struggle against neo-colonialism. Furthermore, if we accept that national liberation demands a profound mutation in the process of development of the productive forces, we see that this phenomenon of *national liberation* necessarily corresponds to a *revolution*. The important thing is to be conscious of the objective and subjective conditions in which this revolution can be made and to know the type or types of struggle most appropriate for its realization.

We are not going to repeat here that these conditions are favourable in the present phase of the history of humanity: it is sufficient to recall that unfavourable conditions also exist, just as much on the international level as on the internal level of each nation struggling for liberation.

On the international level, it seems to us that the following factors, at least, are unfavourable to national liberation movements: the neo-colonial situation of a great number of states which, having won political independence, are now tending to join up with others already in that situation; the progress made by neo-capitalism, particularly in Europe, where imperialism is adopting preferential investments, encouraging the development of a privileged proletariat and thus lowering the revolutionary level of the working classes, the open or concealed neo-colonial position of some European states which, like Portugal, still have colonies; the so-called policy of 'aid for undeveloped countries' adopted by imperialism with the aim of creating or reinforcing native pseudo-bourgeoisies which are necessarily dependent on the international bourgeoisie, and thus obstructing the path of revolution: the claustrophobia and revolutionary timidity which have led some recently independent states whose internal economic and political conditions are favourable to revolution to accept compromises with the enemy or its agents; the growing contradictions between anti-imperialist states; and, finally the threat to world peace posed by the prospect of atomic war on the part of imperialism. All these factors reinforce the action of imperialism against the national liberation movements.

If the repeated interventions and growing aggressiveness of imperialism against the peoples can be interpreted as a sign of desperation faced with the size of the national liberation movements, they can also be explained to a certain extent by the weaknesses produced by these unfavourable factors within the general front of the anti-imperialist struggle.

On the internal level, we believe that the most important weaknesses or unfavourable factors are inherent in the socio-economic structure and in the tendencies of its evolution under imperialist pressure, or to be more precise in the little or no attention paid to the characteristics of this structure and these tendencies by the national liberation movements in deciding on the strategy of their struggles.

By saying this we do not wish to diminish the importance of other internal factors which are unfavourable to national liberation, such as economic underdevelopment, the consequent social and cultural backwardness of the popular masses, tribalism, and other contradictions of lesser importance. It should however be pointed out that the existence of tribes only manifests itself as an important contradiction as a function of opportunistic attitudes, generally on the part of detribalized individuals or groups, within the national liberation movements. Contradictions between classes, even when only embryonic, are of far greater importance than contradictions between tribes.

Although the colonial and neo-colonial situations are identical in essence, and the main aspect of the struggle against imperialism is neo-colonialist, we feel it is vital to distinguish in practice these two situations. In fact the horizontal structure, however it may differ from the native society, and the absence of a political power composed of national elements in the colonial situation make possible the creation of a wide front of unity and struggle, which is vital to the success of the national liberation movement. But this possibility does not remove the need for a rigorous analysis of the native social structure, of the tendencies of its evolution, and for ensuring true national liberation. While recognizing that each movement knows best what to do in its own case, one of these measures seems to us indispensable, namely the creation of a firmly united vanguard, conscious of the true meaning and objective of the national liberation struggle which it must lead. This necessity is all the more urgent since we know that with rare exceptions the colonial situation neither permits nor needs the existence of significant vanguard classes (working class conscious of its existence and rural proletariat) which could ensure the vigilance of the popular masses over the evolution of the liberation movement. On the contrary, the generally embryonic character of the working classes and the economic, social, and cultural situation of the physical force of most importance in the national liberation struggle—the peasantry—do not allow these two main forces to distinguish true national independence from fictitious political independence. Only a revolutionary vanguard, generally an active minority, can be aware of this distinction from the start and make it known, through the struggle, to the popular masses. This explains the fundamentally political nature of the national liberation struggle and to a certain extent makes the form of struggle important in the final result of the phenomenon of national liberation.

In the neo-colonial situation the more or less vertical structure of the native society and the existence of a political power composed of native elements—national state—already worsen the contradictions within that society and make difficult if not impossible the creation of as wide a front as in the colonial situation. On the one hand the material effects (mainly the nationalization of cadres and the increased economic initiative of the native elements, particularly in the commercial field) and the psychological effects (pride in the belief of being ruled by one's own compatriots, exploitation of religious or tribal solidarity between some leaders and a fraction of the masses) together demobilize a considerable part of the nationalist forces. But on the other hand the necessarily repressive nature of the neo-colonial state against the national liberation forces, the sharpening of contradictions between classes, the objective permanence of signs and agents of foreign domination (settlers who retain their privileges, armed forces, racial discrimination), the growing poverty of the peasantry and the more or less notorious influence of external factors all contribute towards keeping the flame of nationalism alive, towards progressively raising the consciousness of wide popular sectors, and towards reuniting the majority of the population, on the very basis of awareness of neo-colonialist frustration, around the ideal of national liberation. In addition, while the native ruling class becomes progressively more bourgeois, the development of a working class composed of urban workers and agricultural proletarians, all exploited by the indirect domination of imperialism, opens up new perspectives for the evolution of national liberation. This working class, whatever the level of its political consciousness (given a certain minimum, namely *the awareness of its own needs*), seems to constitute the true popular vanguard of the national liberation struggle in the neo-colonial case. However it will not be able to completely fulfil its mission in this struggle (which does not end with the gaining of independence) unless it firmly unites with the other exploited strata, the peasants in general (hired men, sharecroppers, tenants, and small farmers) and the nationalist petty bourgeoisie. The creation of this alliance demands the mobilization and organization of the nationalist forces within the framework (or by the action) of a strong and well-structured political organization.

Another important distinction between the colonial and neo-colonial situations is in the prospects for the struggle. The colonial situation (in which the *national class* fights the repressive forces of the bourgeoisie of the colonizing country) can lead, apparently at least, to a nationalist solution (national revolution); the nation gains its independence and theoretically adopts the economic structure which best suits it. The neo-colonial situation (in which the working classes and their allies struggle simultaneously against the imperialist bourgeoisie and the native ruling class) is not resolved by a nationalist solution; it demands the destruction of the capitalist structure implanted in the national territory by imperialism, and correctly postulates a Socialist solution.

This distinction arises mainly from the different levels of the productive forces in the two cases and the consequent sharpening of the class struggle.

It would not be difficult to show that in time the distinction becomes scarcely apparent. It is sufficient to recall that in our present historical situation—elimination of imperialism which uses every means to perpetuate its domination over our peoples, and consolidation of Socialism throughout a large part of the world—there are only two possible paths for an independent nation: to return to imperialist domination (neo-colonialism, capitalism, state capitalism), or to take the way of Socialism. This operation, on which depends the compensation for the efforts and sacrifices of the popular masses during the struggle, is considerably influenced by the form of struggle and the degree of revolutionary consciousness of those who lead it. The facts make it unnecessary for us to prove that the essential instrument of imperialist domination is violence. It we accept the principle that the *liberation struggle is a revolution* and that it does not finish at the moment when the national flag is raised and the national anthem played, we will see that there is not, and cannot be, national liberation without the use of liberating violence by the nationalist forces, to answer the criminal violence of the agents of imperialism. Nobody can doubt that, whatever its local characteristics, imperialist domination implies a state of permanent violence against the nationalist forces. There is no people on earth which, having been subjected to the imperialist yoke (colonialist or neo-colonialist), has managed to gain its independence (nominal or effective) without victims. The important thing is to determine which forms of violence have to be used by the national liberation forces in order not only to answer the violence of imperialism but also to ensure through the struggle the final victory of their cause, true national independence. The past and present experiences of various peoples, the present situation of national liberation struggles in the word (especially in Vietnam, the Congo and Zimbabwe) as well as the situation of permanent violence, or at least of contradictions and upheavals, in certain countries which have gained their independence by the so-called peaceful way, show us not only that compromises with imperialism do not work, but also that the normal way of national liber-ation, imposed on peoples by imperialist repression, is *armed struggle*.

We do not think we will shock this assembly by stating that the only effective way of definitively fulfilling the aspirations of the peoples, that is to say of attaining national liberation, is by armed struggle. This is the great lesson which the contemporary history of liberation struggle teaches all those who are truly committed to the effort of liberating their peoples.

It is obvious that both the effectiveness of this way and the stability of the situation to which it leads after liberation depend not only on the characteristics of the organization of the struggle but also on the political and moral awareness of those who, for historical reasons, are capable of being the immediate heirs of the colonial or neo-colonial state. For events have shown that the only social sector capable of being aware of the reality of imperialist domination and of directing the state apparatus inherited from this domination is the native petty bourgeoisie. If we bear in mind the aleatory characteristics and the complexity of the tendencies naturally inherent in the economic situation of this social stratum or class, we will see that this specific inevitability in our situation constitutes one of the weaknesses of the national liberation movement.

The colonial situation, which does not permit the development of a native pseudo-bourgeoisie and in which the popular masses do not generally reach the necessary level of political consciousness before the advent of the phenomenon of national liberation, offers the petty bourgeoisie the his-torical opportunity of leading the struggle against foreign domination, since by nature of its objective and subjective position (higher standard of living than that of the masses, more frequent contact with the agents of colo-nialism, and hence more chances of being humiliated, higher level of edu-cation and political awareness, etc.) it is the stratum which most rapidly becomes aware of the need to free itself from foreign domination. This historical reponsibility is assumed by the sector of the petty bourgeoisie which, in the colonial context, can be called *revolutionary*, while other sectors retain the doubts characteristic of these classes or ally themselves to colonialism so as to defend, albeit illusorily, their social situation.

The neo-colonial situation, which demands the elimination of the native pseudo-bourgeoisie so that national liberation can be attained, also offers the petty bourgeoisie the chance of playing a role of major and even decisive importance in the struggle for the elimination of foreign domination. But in this case, by virtue of the progress made in the social structure, the function of leading the struggle, is shared (to a greater or lesser extent) with the more educated sectors of the working classes and even with some elements of the national pseudo-bourgeoisie who are inspired by patriotic sentiments. The role of the sector of the petty bourgeoisie which participates in leading the struggle is all the more important since it is a fact that in the neo-colonial situation it is the most suitable sector to assume these functions, both because of the economic and cultural limitations of the working masses, and because of the complexes and limitations of an ideological nature which characterize the section of the national pseudo-bourgeoise which supports the struggle. In this case it is important to note that the role with which it is entrusted demands from this sector of the petty bourgeoisie a greater revolutionary consciousness, and the capacity for faithfully interpreting the aspirations of the masses in each phase of the struggle and for identifying themselves more and more with the masses.

But however high the degree of revolutionary consciousness of the sector of the petty bourgeoisie called on to fulfil this historical function, it cannot free itself from one objective reality: the petty bourgeoisie as a service class (that is to say that a class not directly involved in the process of production), does not possess the economic base to guarantee the taking over of power. In fact history has shown that whatever the role—sometimes important—played by individuals coming from the petty bourgeoisie in the process of a revolution, this class has never possessed political control. And it could never possess it, since political control (the State) is based on the economic capacity of the ruling class, and in the conditions of colonial and neo-colonial society this capacity is retained by two entities: imperialist capital and the native working class.

To retain the power which national liberation puts in its hands, the petty bourgeoisie has only one path: to give free rein to its natural tendencies to become more bourgeois, to permit the development of a bureaucratic and intermediary bourgeoisie in the commercial cycle, in order to transform itself into a national pseudo-bourgeoisie, that is to say in order to negate the revolution and necessarily ally itself with imperialist capital. Now all this corresponds to the neo-colonial situation, that is, to the betrayal of the objectives of national liberation. In order not to betray these objectives, the petty bourgeoisie has only one choice: to strengthen its revolutionary consciousness, to reject the temptations of becoming more bourgeois and the natural concerns of its class mentality, to identify itself with the working classes, and not to oppose the normal development of the process of revolution. This means that in order to truly fulfil the role in the national liberation struggle, the revolutionary petty bourgeoisie must be capable of committing suicide as a class in order to be reborn as revolutionary workers, completely identified with the deepest aspirations of the people to which they belong.

This alternative—to betray the revolution or to commit suicide as a class—constitutes the dilemma of the petty bourgeoisie in the general framework of the national liberation struggle. The positive solution in favour of the revolution depends on what Fidel Castro recently correctly called *the development of revolutionary consciousness*. This dependence necessarily calls our attention to the capacity of the leader of the national liberation struggle to remain faithful to the principles and to the fundamental cause of this struggle. This shows us, to a certain extent, that if national liberation is essentially a political problem, the conditions for its development give it certain characteristics which belong to the sphere of morals.

We will not shout hurrahs or proclaim here our solidarity with this or that people in struggle. Our presence is in itself a cry of condemnation of imperialism and a proof of solidarity with all peoples who want to banish from their country the imperialist yoke, and in particular with the heroic people of Vietnam. But we firmly believe that the best proof we can give of our anti-imperialist position and of our active solidarity with our comrades in this common struggle is to return to our countries, to further develop this struggle, and to remain faithful to the principles and objectives of national liberation.

Our wish is that every national liberation movement represented here may be able to repeat in its own country, arms in hand, in unison with its people, the already legendary cry of Cuba:

PATRIA O MUERTE, VENCEREMOS!

DEATH TO THE FORCES OF IMPERIALISM!

FREE, PROSPEROUS AND HAPPY COUNTRY FOR EACH OF OUR PEOPLES!

VENCEREMOS!

Source: A. Cabral, *Revolution in Guinea: An African People's Struggle* (London, 1969), pp. 73 ff.

FURTHER NOTES

WORKS

CABRAL, A., *Revolution in Guinea: An African People's Struggle* (London, 1969).
—— *Unity and Struggle: Speeches and Writings* (London 1980).

COMMENTARIES

AABY, P., *The State of Guinea-Bissau: African Socialism or Socialism in Africa?* (Uppsala, 1978).

BIENEN, H., 'State and Revolution: The Work of Amilcar Cabral,' *Journal of Modern African Studies*, 5, no. 4 (1977).

BREWER, A., *Marxist Theories of Imperialism* (London, 1980).

CHABAL, P., *Amilcar Cabral: Revolutionary Leadership and People's War* (Cambridge, 1983).

DAVIDSON, B., Introduction to *Unity and Struggle*, above.

—— *No Fist is Big Enough to Hide the Sky: The Liberation of Guinea-Bissau and Cape Verde* (London, 1981).

McCOLLESTER, C., 'The Political Thought of Amilcar Cabral', *Monthly Review* (Mar. 1977).

McCULLOCH, J., *In the Twilight of Revolution: The Political Theory of Amilcar Cabral* (London, 1983).

ROSEBERG, C., AND T. CALLAGHY, *Socialism in Sub-Saharan Africa: A New Assessment* (Berkeley, 1979).

NAME INDEX

Adorno, Theodor W. (1903–69) 340
  German sociologist and music critic; member
  of Frankfurt School
Agnelli, Giovanni
  Head of Fiat car company 265
Aristotle (384–322 BC) 225, 265
  Greek philosopher and scientist; pupil of
  Plato
Augustus (63 BC–AD 14) 14
  First emperor of Rome

Babeuf, François Noël (1760–97) 44, 84, 203
  French revolutionary and communist
Bakunin, Mikhail (1814–76) 106
  Russian anarchist and writer
Barras, Paul-François (1755–1829) 203
  French revolutionary politican, member of
  Directorate
Barrot, Odilon (1791–1873) 73
  French liberal politician
Batista, Fulgencio (1901–73) 373
  Cuban military leader and president
  (1940–4, 1952–9); overthrown by Castro
Bauer, Bruno (1809–82) 11, 14
  German philosopher and Young Hegelian
Bauer, Otto (1881–1938) 252
  Austrian socialist of the Austro-Marxist
  tendency
Bayo, Alberto (1892–?) 374
  Cuban-born soldier, served in Morocco and
  the Spanish Civil War; taught guerrillas
  in Cuba
Bebel, August (1840–1913) 167
  German socialist leader; a founder of the
  Social Democratic Party
Benjamin, Walter (1892–1940) 356
  German essayist; member of Frankfurt
  School
Benni, Alfano 265
  Co-founder of Italian National Liberal Party
  (1925); member of Mussolini's
  government
Bernadotte, Jean Baptiste Jules (1764–1844) 136
  French marshal under Napoleon but later
  fought against him; king of Norway and
  Sweden (1818–44) as Charles XIV
Bernstein, Eduard (1850–1932) 3, 76–86, 87,
  108–15, 117, 118–19, 120, 248, 252
Blanchot, Maurice (b. 1907) 354
  French novelist and writer on literature

Blanqui, Louis Auguste (1805–81) 114, 119, 121
  French socialist and revolutionary
Bloch, Joseph (1871–1936) 62
  Journalist and publisher; editor of
  Sozialistische Monatshefte
Bömelburg, Theodor (1862–1912) 120
  German trade union leader; president of
  Construction Workers' Union; opposed
  mass strike at Cologne trades union
  congress
Brissot, Jacques Pierre (1754–93) 203
  French revolutionary Girondist politician
Bronstein, Lev, Davidovich *see* Trotsky
Buckle, Henry Thomas (1821–62) 142
  British historian and positivist sociologist
Bukharin, Nikolai (1888–1938) 201, 210, **226–
  245**, 285

Cabral, Amilcar (1924–73) **392–409**
Cadorna, Luigi (1850–1928) 270
  Italian military leader in First World War
Calvin, John (1509–64) 275, 276
  French theologian; leader of Protestant
  Reformation in France and Switzerland
Carlyle, Thomas (1795–1881) 137
  Scottish essayist and historian
Castro, Fidel (b. 1927) 373, 377, 378, 379, 384,
  385, 388, 390, 408
  President of Cuba; led overthrow of Batista
  in 1959
Chang Kuo-tao 317
  Former Chinese Communist Party member;
  opposed party line in 1930s and later
  joined the Kuomintang (1938)
Cherbuliez, Antoine Elisée (1797–1869) 16
  Swiss economist; follower of Sismondi
Chernov, Viktor Mikhailovich (1876–1952)
  173, 174
  A leader and theoretician of the Social
  Revolutionary Party in Russia; minister of
  agriculture in Provisional Government
  (May–August 1917)
Chiang Kai-shek (1897–1975) 205, 314, 320
  Chinese general and president (1928–31;
  1943–9); defeated in Civil War (1949) and
  withdrew to Taiwan where he was
  president until his death; succeeded Sun
  Yat-sen as leader of Kuomintang (1925)
Constantine I (AD ?280–337) 74
  First Christian Roman emperor (306–37)

Croce, Benedetto (1866–1952)—*cont*
   Italian philosopher and statesman; opposed
   fascism and helped re-establish liberalism
   in post-war Italy
Cunow, Heinrich Wilhelm Karl (1862–1936)
   131, 157
   German Marxist historian, sociologist, and
   ethnographer

Darwin, Charles Robert (1809–82) 87, 134
   English naturalist; formulated theory of
   evolution by natural selection
David, Eduard (1863–1930) 100
   Leader of German Social Democratic Party;
   a founder of the journal *Sozialitische
   Monatshefte*
De Ruggiero, Guido 277
   Italian philosopher and historian; co-founder
   of antifascist National Union (1924);
   minister of education after fall of
   Mussolini
Democritus (?460–?370 BC) 138
   Greek philosopher; developed the atomist
   theory of matter originated by his teacher
   Leucippus
Denikin, Anton Ivanovich (1872–1947) 182
   General of Russian Tsarist army; launched
   unsuccessful offensive on Moscow (1919–
   20)
Dickens, Charles (1812–70) 121
   English novelist; noted for his portrayals of
   life in Victorian England and criticism of
   social injustice
Dietzgen, Joseph (1828–88) 182
   German philosopher and Social Democrat;
   arrived independently at basic
   propositions of dialectical materialism
Dilthey, Wilhelm (1833–1911) 248
   German philosopher of the Neo-Kantian
   school
Diocletian (AD 245–313) 74
   Roman emperor (284–305); instigated last
   severe persecution of Christians (303)
Djilas, Milovan (b. 1911) 357
   Yugoslav politician and political theorist;
   developed theory of the Soviet
   bureaucracy as a 'new class'
Dühring, Eugen (1833–1921) 71, 139
   German economist and philosopher

Eisner, Kurt (1867–1919) 120
   German Social Democratic journalist and
   editor of *Vorwärts* (1898–1905); after
   imprisonment for anti-war activities and
   led short-lived Bavarian republic (1919);
   assassinated

Engels, Friedrich (1820–95) 3, **62–75**, 76, 77,
   78–9, 87, 94, 101, 118, 138, 139, 143, 167,
   172, 174, 214, 243, 244, 255, 261, 286, 302
Erasmus, Desiderius (?1466–1536) 277
   Dutch humanist and leading scholar of the
   Renaissance
Escalante, Anibal 379
   One-time leader of the Popular Socialist
   Party in Cuba

Feuerbach, Ludwig Andreas (1804–72) 13–15,
   71, 248, 252, 276
   German materialist philosopher; argued in
   The Essence of Christianity (1841) that
   God is a projection of the human's inner
   being
Fourier, Charles (1772–1837) 44, 46
   French social reformer; advocated
   cooperativist organization of society

Gentile, Giovanni (1875–1944) 265, 280
   Italian idealist philosopher and Fascist
   politician; minister of education (1922–4)
Ghe, Alexander (1879–1919) 174
   Russian anarchist; member of All-Russian
   Central Executive Committee of Soviets
   after October 1917
Goethe, Johann Wolfgang von (1749–1832)
   225, 280, 350
   German poet, novelist, dramatist, and
   thinker
Goldmann, Lucien (1913–70) 249
   Marxist philosopher and literary critic; pupil
   of Lukacs
Gramsci, Antonio (1891–1937) **264–82**
Grave, Jean (1854–1939) 174
   French socialist and anarchist
Guesde, Jules (1845–1922) 195
   French socialist leader
Guevara, Ernesto 'Che' (1928–67) **373–391**
Guizot, François Pierre Guillaume (1787–1874)
   19, 20
   French historian and statesman; chief
   minister (1840–8) whose policies helped
   provoke the revolution of 1848

Hegel, Georg Wilhelm Friedrich (1770–1831)
   13, 19, 139, 140, 144, 232, 234, 237–**238**,
   248, 249, 250, 252, 258, 276, 278
   German idealist philosopher with profound
   influence on the origins of Marxism
Heidegger, Martin (1889–1976) 249, 251
   German existentialist philosopher; most
   noted work *Being and Time* (1927)
Henderson, Arthur (1863–1935) 301
   British Labour politician, born Scotland;
   president of World Disarmament

Conference (1932–5); Nobel peace prize (1934)

Heraclitus (?540–?475 BC) 232
Greek philosopher whose theories of the universe in a state of perpetual flux influenced Plato, Aristotle, and Hegel

Hess, Moses (1812–75) 249
German writer and journalist and sometime friend of Marx; converted Engels and Bakunin to communism

Hitler, Adolf (1889–1945) 206, 311
German dictator, born Austria; leader of National Socialist (Nazi) party; chancellor of Germany from 1933 transforming it into the totalitarian Third Reich of which he was Führer from 1934

Hobson, John Atkinson (1858–1940) 156
British economist whose pioneering work on modern imperialism (Imperialism, 1902) was later taken up and developed by Lenin

Höglund, Carl Zeth Konstantin (1884–1956) 184
Swedish professor, author, and politician; leader of the Social Democratic Party and the socialist youth movement

Horkheimer, Max (1895–1973) 340
German philosopher and sociologist; member of Frankfurt School

Hutten, Ulrich von (1488–1523) 133
German humanist and poet; advocated reform of the empire by secularization of church property

Huysmans, Camille (1871–1968) 162
Veteran leader of Belgian labour movement; secretary of International Socialist Bureau of the Second International (1904–19)

Jansen, Cornelius (1585–1638) 275
Dutch Roman Catholic theologian; in Augustinus (1640) defended teachings of St Augustine on free will, grace, and predestination

Jaurès, Jean Leon (1859–1914) 125, 195
French politician and writer; founded socialist paper l'Humanité (1904) and united French socialist movement into single party (1905); dominated French left from 1890s until his assassination

Jourdan, J. B. 136
General under Napoleon; created marshal 1804; served under Louis XVIII 1814; recreated marshal 1815

Kamenev, Lev Borisovich (1883–1936) 209, 210
Bolshevik leader; supported Zinoviev and Stalin against Trotsky and later supported

Trotsky against Stalin (1926–7); arrested 1935, tried and shot

Kant, Immanuel (1724–1804) 248
German idealist philosopher; in Critique of Pure Reason (1781) sought to determine the limits of human knowledge

Kardelj, Edward (1910–79) 357–72

Kautsky, Karl (1854–1938) 87–106, 108, 130, 155–8, 162, 163, 248, 263, 288, 294, 302

Kerensky, Aleksander Fyodorovich (1881–1970) 199, 202
Russian liberal revolutionary leader; prime minister in the Provisional Government (July–October 1917); overthrown by the Bolsheviks

Khinchuk, Lev (b. 1868) 206
Menshevik then Bolshevik; leader of Moscow soviet after the Revolution and later ambassador to Britain and then Germany

Kierkegaard, Søren Aabye (1813–55) 248
Danish philosopher and theologian; originator of Christian existentialism

Kolchak, Alexander Vasilyevich (1873–1920) 182
Tsarist admiral; proclaimed himself supreme Ruler of Russia (1918) and headed counter-revolution in Siberia until defeat in 1920

Köller, Ernst Matthias (1841–1928) 74
German statesman; member of the Reichstag (1881–8) and Prussian Minister of the Interior (1894–5)

Krasnov, P. N. (1869–1947) 271
Russian general; led Cossack attack on Petrograd (1917) and imprisoned; after October aided the 'white' counter-revolutionary movement

Kroner, Richard 248
German historian of philosophy

Kropotkin, Peter (1842–1921) 174
Russian anarchist and writer

Krupskaya, Nadezhda K. (1869–1939) 207, 210
Russian Bolshevik; married Lenin during his exile in Siberia (1896–1900); backed Zinoviev and Kamenev against Stalin (1925); published memoirs Memories of Lenin

Kugelmann, Ludwig (1830–1902) 302
German physician and friend of Marx's family; participant in 1848–9 revolution and member of the International

Lafargue, Paul (1842–1911) 3
Disciple and associate of Marx and Engels; active member of the General Council of

Lafargue, Paul (1842–1911) 3
  Disciple and associate of Marx and Engels;
  active member of the General Council of
  the International and a founder of French
  Workers' Party
Lassalle, Ferdinand (1825–64) 169, 170
  German lawyer, writer and socialist; a
  founder of the General German Workers'
  Union (1863) forerunner of the Social
  Democratic Party
Lavoisier, Antoine Laurent (1743–94) 279
  French scientist and a founder of modern
  chemistry; named oxygen
Lenin, Vladimir Ilich (Ulyanov) (1870–1924)
  87, 100, 108, 124, 127–32, 134, 145–87,
  188, 198, 202, 207, 208, 209, 210, 211–12,
  214, 219, 226, 246, 248, 249, 264, 271, 279,
  280, 285–300, 302, 303–4, 305–6, 308, 311,
  318, 321, 324, 332, 381
Lensch, Paul (1873–1926) 131
  German Social Democrat politician
Li Hsüeh-feng (?1906–?) 338
  Sometime First Secretary of North China
  Bureau of Communist party; disappeared
  from political scene (1970) and later
  identified by Mao as a conspirator with
  Lin Biao against him
Liebknecht, Karl (1871–1919) 178, 184
  German politician; radical member of Social
  Democratic Party and later its Spartacist
  off-shoot; murdered with Luxemburg
  during an abortive revolt
Liu Shao-ch'i (1898–1974) 338
  Chinese Communist statesman; chairman of
  People's Republic of China (1959–68),
  deposed during Cultural Revolution
Louis Phillippe (1773–1850) 137
  King of France (1830–48); forced to abdicate
  by revolution of 1848
Löwith, Karl, (b. 1897) 248
  German philosopher of history
Lukacs, Georg (1885–1971) 246–263
Luther, Martin (1483–1546) 275
  German leader of the Protestant
  Reformation
Luxemburg, Rosa (1871–1919) 87, 108–133,
  178, 271

MacDonald, Etienne 136
  General under Napoleon; served in Italy,
  Hungary, Russia, and Germany
MacDonald, James Ramsay (1855–1937) 301
  British statesman; led first Labour
  government (1924; 1929–31), and a
  coalition government (1931–5) which
  majority of Labour Party refused to
  support

Maisky, Ivan Mikhailovich (1884–1975) 206
  Soviet diplomat, historian, and academician;
  formerly a Menshevik, joined Communist
  Party in 1921; ambassador to Britain
  (1932–43)
Mann, Thomas (1875–1955) 350
  German novelist whose works deal mainly
  with the problem of the artist in bourgeois
  society; Nobel literature prize (1929)
Mao Zedong (1893–1979) 314–39
Marcuse, Herbert (1898–1979) 246, 340–56
Marshall, George Catlett (1880–1959) 310
  US general and statesman; army chief-of-staff
  (1939–45) and secretary of state (1947–9);
  proposed Marshall Plan for European
  recovery (1947); Nobel peace prize
  (1953)
Martov, I. O., (1873–1923) 162
  Menshevik leader and member of Petrograd
  Soviet Executive Committee after
  February 1917
Marx, Karl (1818–83) 3–61, 62, 63, 69, 70, 71,
  72, 76, 77, 78, 87, 94, 99, 100, 101, 102,
  103, 118, 122, 134, 138, 139, 140, 142, 143,
  144, 163, 164, 165, 167, 169, 172, 174, 175,
  195, 202, 225, 228, 240, 247, 248, 249, 251,
  252, 256, 280–1, 288, 295, 296, 300, 302–
  3, 321, 363, 380, 385
Metternich, Klemens (1773–1859) 20
  Austrian statesman; foreign minister and
  chancellor of Austria, noted for his
  defence of the aristocracy
Michels, Robert (1876–1936) 243, 244
  German sociologist and economist, noted
  particularly for his work on the theory of
  élites
Mirabeau, Comte de (Honoré-Gabriel Riqueti)
  (1749–91) 135, 203
  Prominent leader of French Revolution
Missiroli, Mario 275
  Italian liberal journalist
Molotov, Vyacheslav Mikhailovich (1890–
  1986) 211, 213, 214, 215
  Bolshevik activist and Soviet statesman,
  serving under Stalin as Premier (1930–41)
  and Commissar for Foreign Affairs (1939–
  49); 1953–6); fell politically (1957) ·
  becoming ambassador to Mongolia;
  retired on a pension
Moreau, J. V. (d. 1814) 136
  General under Napoleon; banished
  (1804) to America; took up arms against
  Napoleon (1813), killed at battle of
  Dresden
Morgan, Lewis Henry (1818–81) 139
  American social scientist and historian of
  primitive society

Napoleon I (Napoleon Bonaparte) (1769–1821) 135, 136, 137, 203, 210
French general and emperor (1804–15) establishing extensive empire in Europe; undefeated in battle until Leipzig (1813) and finally Waterloo (1815); the *Code Napoléon* remains the basis of French law

Napoleon III (Louis Napoleon Bonaparte) (1808–73) 70, 222
Nephew of Napoleon I and emperor of French 1852–70, being deposed after Franco-Prussian War; president of Second Republic 1848–51

Nicholas of Cusa (1401–64) 258
German cardinal, mathematician, and philosopher; asserted, before Copernicus, that the earth revolved around the sun

Nieh Yüan-tzu 335
Lecturer at Peking University; prepared a big-character poster which (June 1966) Mao ordered to be widely circulated, heralding beginning of Cultural Revolution

Noske, Gustav (1868–1946) 301
Right-wing German Social Democrat; organized assassination (1919) of Luxemburg and Liebknecht

Owen, Robert (1771–1858) 44, 46
Welsh industrialist and social reformer; formed model industrial community in Scotland and pioneered co-operative societies

Pareto, Vilfredo (1848–1923) 243
Italian sociologist and economist; noted for pioneering work on the theory of élites

Parvus (Alexander Lazarevitch Helphand) (1869–1924) 131
Originally on left-wing of German Social Democratic Party; later a Menshevik and contributor to Menshevik publications

Pecqueur, Constantin (1801–87) 59
French economist and utopian socialist

Pilsudski, Jozef (1867–1935) 205
Polish nationalist leader and statesman; president (1918–21), premier (1926–8; 1930)

Plato (?427–?347 BC) 258, 265
Greek thinker, regarded as a founder of western philosophy, pupil of Socrates

Plekhanov, Georgy (1856–1918) 85, **134–44**, 145, 174, 196, 248, 285, 287

Pomyalovsky, Nikolai Gerasimovich (1835–63) 173
Russian writer; in *Sketches of Seminary Life* depicted adverse conditions under which children of the clergy and the urban poor studied

Postyshev, Pavel (1888–1940) 214–15
Russian Bolshevik; Politbureau member (1926) and Communist Party secretary in Ukraine; arrested (1938) and executed

Potemkin, Vladmimir (1878–1946) 206
Soviet diplomat; ambassador to Greece, Italy, and France; sometime deputy minister of foreign affairs and commissar of public education

Proudhon, Pierre Joseph (1809–65) 43, 387
French sociologist, economist, and anarchist; declared in *What is Property* (1840) that property is theft

Puttkammer, Robert von (1828–1900) 122
Prussian statesman, holding number of high government posts; appointed Minister of Interior (1881) and opposed trade-union activity

Rakovsky, Christian (1873–1941) 203, 210–11
Premier of the Ukraine and opponent of Stalin after the Revolution; deported to Siberia (1928), tried (1938) and sentenced to 20 years' imprisonment

Rankovic, Aleksandar 357
Yugoslav statesman, sometime Minister of the Interior and chief of police; a Communist Party leader, ousted (1966)

Ratzel, Friedrich (1844–1904) 140, 141, 142
German geographer and ethnographer; noted for pioneering research into culture areas and human influence on the environment

Renan, Ernest (1823–92) 275
French philosopher, historian, and essayist; wrote much on religious topics including a *Life of Jesus* (1863)

Robespierre, Maximilien François Marie Isidore de (1758–94) 135, 136, 203
French revolutionary and Jacobin leader; established Reign of Terror (1793–4) and was executed in the coup d'état of Thermidor

Rockefellers 302, 380
Family of US industrialists and philanthropists founded by John Davison Rockefeller (1839–1937)

Rosmer, Alfred 271
Founder of French Communist Party, later sympathetic to Trotsky

Rothschilds 302
Family of European Jewish bankers founded by Meyer Amschel Rothschild (1743–1812)

Rousseau, Jean Jacques (1712–78) 84

Rousseau, Jean Jacques (1712–78)—*cont*
    French philosopher and writer whose work
    influenced the French Revolution
Rykov, Alexei Ivanovich (1881–1938) 210
    Bolshevik activist who played leading role in
    economic affairs after Revolution;
    Premier (1924–9); dropped from
    Politbureau (1929), later arrested, tried,
    and shot

Saint-Simon, Comte de (Claude Henri de
    Rouvroy) (1760–1825) 44
    French utopian socialist thinker
Sainte-Beuve, Pierre-Henri (1819–55) 134, 135
    French manufacturer and landowner;
    member of the Constituent and the
    Legislative Assembly during Second
    Republic; representative of Party of Order
Salazar, Antonio de Oliveira (1889–1970) 392
    Prime minister and virtual dictator of
    Portugal (1932–68); drafted 1933
    constitution based on that of fascist Italy
Sartre, Jean-Paul (1905–80) 249
    French existentialist philosopher, novelist,
    and dramatist
Scheidemann, Philipp (1865–1939) 301
    Right-wing leader of German Social
    Democratic Party; head of German
    government (February–June 1919)
Schelling, Friedrich Wilhelm Joseph von (1775–
    1854) 250
    German philosopher; maintained the
    existence of one reality, and regarded
    nature as an absolute working towards
    self-consciousness
Schmidt, Konrad (1863–1932) 110
    German economist and philosopher whose
    work influenced later Marxist
    'revisionists'
Schopenhauer, Arthur (1788–1860) 280
    German philosopher; stressed the
    importance of the will, as creative primary
    factor, over the idea, as secondary
    receptive factor
Schulze-Gävernitz, Gerhardt von (1864–1943)
    85
    German economist; professor of political
    economy at Freiburg University
Schweinfurth, Dr 139, 140
    Late nineteenth-century German naturalist,
    noted for his work in Africa
Siéyès (Abbé) Emmanuel Joseph (1748–1836)
    136
    French statesman, political theorist, and
    churchman; instrumental in bringing
    Napoleon I to power (1799)

Sismondi, Jean Charles Léonard Simonde de
    (1773–1842) 16, 40
    Swiss historian and economist; his work
    contributed to the movement for Italian
    unification
Smith, Adam (1723–90) 56
    Scottish economist and philosopher;
    advocated in *The Wealth of Nations* free
    trade and private enterprise as opposed to
    state regulation of the economy
Sorel, Georges (1847–1922) 275
    French syndicalist philosopher; developed
    theory of direct action for political change
Stakhanov, A. G. (1906–77) 212
    Soviet coal-miner; in 1935 was first worker
    to benefit from system (later known as
    Stakhanovism) designed to raise
    production by offering incentives for
    higher output
Stalin (Joseph Vissarionovich Djugashvili)
    (1879–1953) 188, 201, 202, 206, 208, 209,
    210, 212, 213, 222, 223, 226, 285–313, 321,
    335
Sun Yat-sen (1866–1925) 323
    Chinese revolutionary and early leader of
    Kuomintang
Suritz, Jacob (1881–1952) 206
    Soviet ambassador to Germany and France,
    and one of few diplomats to survive the
    purges; former Menshevik
Szabó, Erwin (d. 1918) 255
    Hungarian Marxist writer and librarian

Taylor, Frederick Winslow (1856–1915) 266
    US engineer; coined term 'scientific
    management' (1911) to describe his
    techniques for increasing output of
    workers
Teng Hsiao-p'ing (b. 1904) 338
    Chinese elder statesman; appointed
    Communist Party general secretary (1956)
    but fell from power during Cultural
    Revolution (1966–9); vice-premier 1973–6
    and 1977–80; chairman of Central Military
    Commission from 1982
Thiers, Louis Adolphe (1797–1877) 56
    French statesman and historian; first
    president of Third Republic (1871–3);
    responsible for suppressing Paris
    Commune
Tito, Josip Broz (1892–1980) 357
    Yugoslav statesman, prime minister (1945–
    53), president (1953–80); led guerrilla
    resistance to German occupation in World
    War II
Tolstoi, Leo (1828–1910) 239
    Russian novelist and philosopher; author of

War and Peace (1865–9) and *Anna Karenina* (1875–7)

Tomsky, Mikhail Pavlovich (1880–1936) 210
  Bolshevik activist and trade union leader; supported Bukharin against Stalin (1928–9) and was dismissed from union posts; committed suicide to avoid arrest

Trotsky, Leon (1879–1940) 87, 108, 128, 130, 131, 132, **188–225**, 263, 271, 272, 285

Troyanovsky, Alexander (1882–1955) 206
  Soviet ambassador to US (1934–9); former Menshevik

Tsereteli, Irakly Georgievich (1882–1959) 173, 174, 202
  Menshevik leader; after February 1917 member of Central Executive Committee of Soviets and held senior posts in Provisional Government; left the country 1921

Tugan-Baranovsky, Mikhail Ivanovich (1865–1919) 171, 252
  Russian Marxist intellectual, follower of Bernstein, and one of the 'legal Marxists'; author of standard work on Russian factories

Urrutia, Manuel 378
  Senior Cuban judge who accepted the legality of the revolution and was nominated president by the 26 July Movement in succession to Batista; resigned shortly afterwards (July 1959)

Uspensky, Gleb Ivanovich (1840–1902) 115
  Russian novelist of peasant life

Voltaire (Françoise Marie Arovet) (1694–1778) 275

French writer of the Enlightenment period; noted for his belief in religious, political, and social liberty

Waitz, Georg (1813–86) 139
  German historian, particularly of the medieval period

Weitling, Wilhelm (1808–71) 95
  German utopian socialist; a tailor and activist in workers' movements in Germany, France, and Switzerland

Wu Han (b. 1909) 335, 336
  Deputy mayor of Peking in early 1960s; wrote a play (1961) considered to be a thinly-veiled attack on Mao, a public attack on which (in November 1965) signalled the beginning of the Cultural Revolution

Yagoda, Genrikh Grigoryevich (1891–1938) 212
  Soviet police chief; deputy head of Cheka from 1920 and appointed Commissar and Head of Police in 1934; prepared first purge trials but later was himself purged, tried, and executed

Zinoviev, Grigori Evseyevich (1883–1936) 209, 210, 285
  Bolshevik leader, worked closely with Lenin before Revolution; chairman of Comintern until removed (1926); sided with Stalin against Trotsky but later joined Trotsky; expelled from Party, arrested (1935) and shot

SUBJECT INDEX

alienation 9, 14, 249 ff., 259, 347, 350 ff., 358, 380, 384, 386
anarchism 121 f., 174, 236, 242 f.
art 142 f., 224, 349 ff., 384 ff.
Asiatic mode of production 20

Bonapartism 222 f.
bourgeoisie 22 ff., 150 f., 400
bureaucracy 84 f., 88, 130, 146, 173, 184 ff., 188, 202, 204 ff., 303

capitalism 16 ff., 22 ff., 49 ff., 68 f., 117, 150, 153 ff., 257
    state 217, 221
caste system 13, 381 f.
Catholicism 275 ff., 281 f.
centralism 124 ff., 149, 328 f., 361
Christianity 36, 39 f., 74
class 5, 65 f., 89, 217, 243, 341, 358, 394 ff.
    consciousness 147, 179, 194, 246, 248, 253 ff., 273
    polarization of 27, 57, 77
    ruling 7, 15, 19, 29, 45, 66, 91, 217 f., 241, 301, 401, 407
    struggle 21, 29, 45, 92, 99, 114, 123, 152, 197, 231, 236, 239 ff., 256, 326, 328, 385, 394 ff., 404 f.
classless society 19, 37 f., 64 f., 69, 175 f., 214, 232, 250, 262, 306, 348, 396
combined and uneven development 185
Communism 3, 10 ff., 20 f., 96
Communist Manifesto 20 ff., 76 ff., 85, 116, 165, 262
Communist Party 31
Communist society 67, 169 ff., 174 ff., 213, 223 ff., 229 f., 232, 243 f., 306, 373, 381, 384, 397
contradiction 110 f., 160 f., 232 ff., 259, 278 f., 310, 314, 318 ff., 325 ff., 392, 399, 403
credit 38, 53, 112 f.
crises 269
Cultural Revolution 314, 335 ff.
culture 185, 348 ff., 386

Darwinism 134
democracy 37, 77, 80 f., 84 f., 87, 90 ff., 102, 108, 113, 118, 130, 151 ff., 166 ff., 175 f., 196, 207 f., 302, 328 f., 364, 368
democratic centralism 207 ff.,
determinism 64, 69 f., 87, 138 ff.

dialectic 138 f., 174, 232 ff., 248, 258 f., 274, 318 ff., 335, 354, 399
dictatorship of proletariat 37 f., 82, 87, 94 ff., 108, 118 f., 128 ff., 151 ff., 164 ff., 191, 193, 199 ff., 227 ff., 244 f., 285, 293, 297 ff., 327, 331

economism 146 ff.
élitism 243, 368
equality 131, 170 f., 175, 215 f., 279, 302
ethics 11, 252, 274, 373 f., 381
exploitation 23, 27, 58 ff., 109

false consciousness, 253, 341 f.
family 5, 23, 34 f., 45, 88, 220, 389
Fascism 222 f.
feudalism 6, 20 ff., 25, 38 f., 63 f., 67 f.
finance capitalism 153 ff.
freedom 128, 230, 261, 279, 302, 328 f., 340, 343 ff., 364, 386 f., 389
French Revolution 134 ff.

German Social Democracy 73 f., 78 ff., 90, 106, 125, 166
guerrilla warfare 314, 317 f., 373 ff.

hegemony 255, 264 ff., 268, 272 f., 276
humanism 348 f., 377 ff.

idealism 11 f., 144, 205, 318 f.
ideas 7 f., 15, 36 f.
ideology 8, 20, 143, 146, 248 f., 279 f., 342, 394
imperialism 102, 145, 153 ff., 226, 288 ff., 311 f., 320 f., 392 ff., 399 ff.
individuals 4, 7 f., 11, 16 f., 70, 84, 134, 203, 247, 379 ff., 388 f.
industrial reserve army 52 ff., 68, 192
inheritance 38
intellectuals 264 ff., 273 f.,

labour 16 ff., 33, 172
labour, division of 4 ff., 12 f., 16 f., 201, 340, 358 ff.
law 29, 78, 114, 141 f., 193
liberalism 42, 83, 276, 361
lumpenproletariat 29, 54

Marxism, nature of 3
materialism 15, 138, 143 f., 276 f., 280, 286, 318 ff., 324

materialist conception of history 3 ff., 62 ff., 69 ff., 122, 138 ff., 280 f, 308 f., 324
Messianism 246 f.
Middle Ages 6
mode of production 4, 11, 19 f., 23, 59, 63 f.,
monopoly capitalism 153 ff., 303
morality 7 f., 29, 84

nationalism 31, 36, 155, 197, 271
nature 12, 14, 42, 385
 human 15, 278 ff., 384
needs 345 ff., 352, 385
New Economic Policy 204

opportunism 161 f., 165, 287
overproduction 25

Paris Commune 48, 73, 78, 93, 119, 263
parliamentary activity 78, 86 ff., 93, 108 ff., 118, 145, 177 ff., 193, 328, 383
party 28, 90 f., 101, 108, 125 f., 145, 148, 165, 178, 207 ff., 228, 243 f., 286 ff., 330, 388, 403 f.
peasantry 6, 49, 100 ff., 151, 184 ff., 189 f., 198 f., 230 f., 294 ff., 314 ff.
permanent revolution 188, 193 ff., 205 f., 271, 294 ff.
petty bourgeoisie 27, 29, 40, 42, 47, 151 f., 164, 199, 299, 392, 404, 406 ff.
philosophy 8 f., 11, 63, 276, 278, 332, 385
populism 157
practice 11
praxis 247 f., 252, 273
productive forces 9 f., 16 f., 19 f., 25, 63 f., 96, 140 ff., 172 f., 211, 227 f., 240 f., 244, 324, 333, 373, 394 ff., 401
proletariat 6, 10, 17 f., 21 ff., 27 ff., 43 ff., 89 f., 163 ff., 188 f., 228 ff., 252 f., 265 f., 341, 402
property, private 5 f., 15 ff., 58 ff., 374
 abolition of 10 f., 30, 32, 37 f.

reformism 76 ff., 110 ff., 157
reification 246, 256 f., 260, 355
religion 4, 7 f., 11, 20, 29, 36 f., 52, 277, 329, 349
revolution 9 ff., 17 f., 20, 37, 47, 49, 64, 69, 71, 100 f., 114 ff., 123, 145, 150 ff., 163 ff., 181 f., 203 f., 239, 285 ff., 308 f., 315 ff., 340, 355, 394
rights 96, 170 ff., 174 f., 401
Roman Empire 6, 21, 74

science 8, 248, 251 f.,
 natural 14, 20, 340, 385
Second International 155, 288, 292
self-management 357 ff.
separation of powers, 15
slavery 5, 141
Socialism 38 ff., 64, 80, 84, 87, 96 ff., 109 ff., 129, 171, 174 f., 196, 360, 363, 380 ff.
 in one country 200 ff., 223 ff., 243, 285, 297, 307 ff., 325 ff., 333 f., 380 ff.
Soviets 169, 174, 181 f., 210 ff., 304 ff.
Stalinism 222 f.
State 7, 11, 18, 22 f., 48, 87 f., 127, 145, 163 ff., 226 ff., 241 ff., 267 f., 271, 301, 305, 317, 361, 365 f., 379, 407
 abolition of 3, 65 ff., 165 ff., 168 f., 172 f., 177, 211 f., 217, 262, 306, 363, 397
strikes 93, 108, 120, 181, 192, 271 f.

totalitarianism 213, 344, 352, 354 f.
totality 11, 251 ff., 258 ff.
trade unions 28, 77, 80, 86, 109 ff., 121 f., 146 ff., 228, 230, 287, 367 f.

United States 48 f.
Utopia 40, 44 ff., 112 f., 121, 164, 169, 171, 173, 190, 200, 214, 262, 265, 279, 340, 345

violence 71 ff., 81, 100, 102 f., 303, 311 f., 317, 405